AF872533

New and Established Titles

for the Underground Technologist, Mining Specialist and Geologist

ROCK CHARACTERIZATION, TESTING AND MONITORING*

ISRM Suggested Methods

Editor: **E T BROWN**, *Imperial College of Science and Technology, London*

A unique and authoritative collection of suggested methods of carrying out rock mechanics tests prepared by a large number of international experts. Provides valuable guidelines to the practising engineer and geologist on methods of characterizing rock masses for engineering purposes, determining design parameters for rock by laboratory and field testing and monitoring the performance of engineering rock structures.

200pp May 1981
0 08 027308 4 (H) **£16.50 US$40.00**
0 08 027309 2 (F) **£ 8.00 US$20.00**

A CONCISE HISTORY OF MINING

C E GREGORY, *Launceston*

An easy-to-read text that in simple language embraces the broad canvas of mining history from the Stone Age to the present day, and a wide range of mineral production across the globe.

Contents:
What is mining? The eight ages of Man. Chronological development of particular aspects of mining. Traditions, customs and folklore of miners. Epilogue. Appendices. Acknowledgements. Bibliography. Glossary of Mining Terms. Index.

280pp 43 illus 56 lit refs February 1981
0 08 023882 3 (H) **£15.00 US$30.00**

ROCK BOLTING

R SCHACH, K GARSHOL and **A M HELTZEN**, *Norwegian Institute of Rock Blasting Techniques*

This is a practical handbook on rock bolting for mines and construction sites. It provides rules and guidelines for the treatment of bolt elements, bolt steel etc. The text is divided into bolt groups covering such areas as immediate and permanent support, and shuttering bolts and yielding bolts. Based on empirical evidence, this book is noteworthy for being concentrated in scope and practical in outlook. The information given is up-to-date and includes discussion of new advances in the field which have not been treated previously.

96pp 1979
0 08 022503 9 (H) **£6.00 US$13.75**

TUNNELLING UNDER DIFFICULT CONDITIONS

I KITAMURA, *Japan Tunnelling Association, Tokyo*

This proceedings volume of the International Tunnel Symposium, Tokyo, Japan, contains 53 papers and 150 literature references.

Independent Opinion
" . . . a study of this volume might serve as good preventive medicine for tunnelling and civil engineers."
Mine and Quarry

420pp 1979
0 08 024237 5 (H) **£33.00 US$75.00**

*** Textbook Inspection Copies available on request.**
H = Hardcover F = Flexicover
Prices include postage and insurance.
Prices subject to change without notice.
Sterling prices applicable to UK and Eire customers.

Pergamon Press Headington Hill Hall, Oxford OX3 0BW, UK
Fairview Park, Elmsford, New York 10523, USA

ISRM

LIST OF MEMBERS

1980

Société Internationale de Mécanique des Roches

International Society for Rock Mechanics

Internationale Gesellschaft für Felsmechanik

LIST OF MEMBERS
LISTE DES MEMBRES
LISTE VON MITGLIEDERN
1980

PUBLISHED ON BEHALF OF
THE INTERNATIONAL SOCIETY FOR ROCK MECHANICS

by

PERGAMON PRESS

OXFORD · NEW YORK · TORONTO · SYDNEY · PARIS · FRANKFURT

U.K.	Pergamon Press Ltd., Headington Hill Hall, Oxford OX3 0BW, England
U.S.A.	Pergamon Press Inc., Maxwell House, Fairview Park, Elmsford, New York 10523, U.S.A.
CANADA	Pergamon Press Canada Ltd., Suite 104, 150 Consumers Rd., Willowdale, Ontario M2J 1P9, Canada
AUSTRALIA	Pergamon Press (Aust.) Pty. Ltd., P.O. Box 544, Potts Point, N.S.W. 2011, Australia
FRANCE	Pergamon Press SARL, 24 rue des Ecoles, 75240 Paris, Cedex 05, France
FEDERAL REPUBLIC OF GERMANY	Pergamon Press GmbH, 6242 Kronberg-Taunus, Hammerweg 6, Federal Republic of Germany

First edition 1981

British Library Cataloguing in Publication Data

International Society for Rock-Mechanics
List of members. - 1980
1. Rock mechanics - Directories
I. List of members (*International Society for Rock Mechanics*)
624.1'5132'05 TA706

ISBN 0-08-027587-7

In order to make this volume available as economically and as rapidly as possible the typescript has been reproduced in its original form. This method unfortunately has its typographical limitations but it is hoped that they in no way distract the reader.

Printed in Great Britain by A. Wheaton & Co. Ltd., Exeter

ARGENTINA

GRUPO ARGENTINO DE MECANICA DE ROCAS

Sociedad Argentina de Mecanica de Suelos, Lavalleja 847, 1414 BUENOS AIRES

Secretary Dr J. C. Medici
Total number of Ordinary Members: 21

7866 BOLOGNESI, A.I.L., Prof.-Civil Engr
* U. Prof.
+ Luis Saenz Pena 250, 6º
(1110) BUENOS AIRES

7567 CAMPAGNUCCI, D.J., Prof.-Civil Engr
* U. Prof.
+ Neuquen, 480
(8000) BAHIA BLANCA

10831 CERUTTI, N.J., Mr - Civil Engr
* Consultant
Jose E. Uriburu 1426, 6º "A"
(1114) BUENOS AIRES

7868 DEL PECHO, J.E., Dr-Geologist
* Consultant
+ Calle 64, 820, 3º "8"
(1900) LA PLATA

7869 GOLDEMBERG, J.J., Mr - Civil Engr
* Consultant
Lavalleja 847
(1414) BUENOS AIRES

10832 GURI, R.J.L., Mr - Civil Engr
* Consultant
Marcelo Gailot 808, 8º Piso Jardin
(5000) CORDOBA

7870 LOPEZ, A.O., Mr - Civil Engr
* Consultant
Canale 1114
(1846) Adrogue - BUENOS AIRES

7871 MAURINO, V.E., Mr-Geologist
* Consultant
+ Calle 7, 65
(1896) City Bell - BUENOS AIRES

7872 MEDICI, J.C., Dr - Geologist
* Consultant
Lavalleja 847
(1414) BUENOS AIRES

10833 MENDIGUREN, E., Mr - Civil Engr
* Consultant
Alberti 352
(1714) Ituzaingo - BUENOS AIRES

7873 MICUCCI, C.A., Mr - Civil Engr
* Consultant
Luis Saenz Pena 250, 6º
(1110) BUENOS AIRES

10834 MOLL, L.L., Prof.-Civil Engr
* U.Prof
Buenos Aires 124, 3º
(5000) CORDOBA

7874 MORETTO, O., Prof.-Civil Engr
* U.Prof.
+ Luis Saenz Pena 250, 6º
(1110) BUENOS AIRES

7875 NUNEZ, E., Prof.-Civil Engr
* U. Prof
+ Maipu 42, 2º Of.224, 3º C
(1084) BUENOS AIRES

10835 RIMOLDI, H.V., Dr-Geologist
* Consultant
+ Lavalle 1625, 6º Of.61
(1048) BUENOS AIRES

10836 SUAREZ, J.A., Mr - Civil Engr
* Consultant
+ Calle 3, 687
(1900) LA PLATA

7879 TORRES, F.L., Prof.-Civil Engr
* U. Prof.
Maipu 548
(2000) ROSARIO

7878 TREVISAN, S.J., Prof.-Civil Engr
* U. Prof.
Diagonal 74, 2831
(1900) LA PLATA

7880 VARDE, O.A., Mr - Civil Engr
* Consultant
+ Belgrano 535, 3º, Of.A
(1092) BUENOS AIRES

7881 VERCELLI, H., Mr - Civil Engr
* Consultant
Maipu, 548
(2000) ROSARIO

7882 WIELAND, R., Mr - Civil Engr
* Tarragona 46
(5000) CORDOBA

AUSTRALIA

AUSTRALIAN GEOMECHANICS SOCIETY

The Institution of Engineers, Australia, 11 National Circuit, BARTON, A.C.T. 26C
Chairman
Chairman: Mr W. M. Bamford
Secretary: Mr J. Harding

Total number of Ordinary Members - 187

Total number of Corresponding Members - 1

Total number of Supporting Members - 8

ORDINARY MEMBERS

4236 * + ALLEN, F.J., Mr
P.O. Box 375
COOMA NORTH N.S.W. 2629

4239 * + APPLEYARD, L.D., Mr
P.O. Box 140
GORDON S.A. 2072

7719 ASKEW, J. E., Mr
66 Lucerne Crescent
ALPHINGTON N.S.W. 2880

10870 + BACZYNSKI, N.R.P., Mr
Department of Mining
University of Melbourne
PARKVILLE VIC. 3052

10871 * + BAKER, J. C., Mr
31 Cocupara Avenue
LINDFIELD N.S.W. 2070

4248 + BAMFORD, W. E., Mr
Mining Department
University of Melbourne
PARKVILLE VIC. 3052

6062 BARING, J.G., Mr
16 Imperial Avenue
MT WAVERLEY VIC. 3149

7708 BARNETT, R.H.W., Mr
263 Nelson Road
MT NELSON TAS. 7007

6063 BARRET, J.R., Mr
16 High Street
MONT ALBERT VIC. 3127

7709 * BEAVIS, F.C., Prof.
School of Applied Geology
University of N.S.W.
P.O. Box 1
KENSINGTON N.S.W. 2033

4256 BELL, G., Mr
62 Russell Street
SURREY HILLS VIC. 3127

4258 + BENNET, A.G., Dr
34 Stanton Crescent
ROSANNA VIC. 3084

7729 + BERGLIN, C.O., Mr
Main Road Department
P.O. Box 1089
TOWNSVILLE OLD. 4810

7730 + BEST, E.J., Mr
Canberra College of Advanced Education
P.O. Box 1
BELCONNEN ACT. 2616

10872 * + BEYNON, W. J., Mr
Box 69
Kincumber N.S.W. 2250

4262 BLACKWOOD, R.L., Mr
School of Mining
University of N.S.W.
P.O. Box 1
KENSINGTON N.S.W. 2033

10873 BOCK, H., Dr
James Cook University
TOWNSVILLE OLD. 4811

7734 * BOWLING, A.J., Mr
32 Parremore Street
ROSETTA TAS. 7010

6654 * + BOYCE, B.T., Mr
Queensland Institution of Technology
P.O. Box 2434
BRISBANE OLD 4001

7735 BRIDGES, M.C., Mr
19 Paterson Crescent
MT ISA OLD. 4825

10874 BROUG, N.W.A., Mr
29 Marilyn Street
DONCASTER VIC. 3108

7721 BURGESS, P.J., Mr
* McMahon, Burgess & Yeates
+ 4 Help Street
CHATSWOOD N.S.W. 2067

10875 BUTCHER, M.J., Mr
* 6 Kildare Avenue
ATHELSTONE S.A. 5076

10876 BYRNE, G.M., Mr
+ Geological Survey of P.N.G.
P.O. Box 778
PORT MORESBY, PAPUA NEW GUINEA

4285 CARTER, E.K., Dr
+ 10 Norfolk Street
RED HILL N.S.W. 2603

10877 CHAPMAN, G.A., Mr
3 Brine Street
HUGHESDALE VIC. 3166

7728 CHAU, E.P., Mr
* 35 Wongneichong Road
+ Floor 4, Flat 4
HAPPY VALLEY (HONG KONG)

7710 CHOWDHURY, R.N., Dr
* Wollongong University
+ P.O. Box 1144
WOLLONGONG N.S.W. 2500

4292 CLARKE, G.H., Mr
* 163 Swann Road
+ TARINGA QLD. 4068

4293 CLEGG, B., Dr
* Department of Civil Engineering
+ University of W.A.
NEDLANDS W.A. 6009

10878 COLE, B.R., Mr
* 26 Colchester Court
+ KEYSBOROUGH VIC. 3175

7738 CONNOR, G.B., Mr
The Zinc Corp. Ltd
BROKEN HILL N.S.W. 2880

10879 COONEY, A.M., Mr
+ 25 Adrian Street
BURWOOD VIC. 3125

10880 COUSINS, B.F., Mr
University of Tasmania
G.P.O. Box 252C
HOBART TAS. 7001

4307 COX, F., Mr
+ 28 Allister Street
MT WAVERLEY VIC. 3149

10881 CRICHTON, A.J., Mr
* 62 Orana Avenue
+ COOMA NORTH N.S.W. 2630

10882 CURRAN, G.A., Mr
* 6 Warringa Road
+ FRANKSTON VIC. 3199

6660 CURREY, D.T., Mr
* 67 Roslyn Street
+ BRIGHTON VIC. 3186

6471 DICKSON, R.J., Mr
* S.M.E.C.
P.O. Box 356
COOMA NORTH N.S.W. 2630

4325 DONALD, I.B., Mr
* Department of Civil Engineering
+ Monash University
CLAYTON VIC. 3168

4328 DOVER, J.D., Mr
* 33 Jeffery Street
BLACKBURN VIC. 3130

10883 DROUGAS, P., Mr
16 Loftus Road
PENNANT HILLS N.S.W. 2120

644 DUNKIN, H.H., Prof.
30 Serpentine Street
MONT. ALBERT VIC. 3127

647 DUTTON, A.H., Mr
S.A. Institute of Technology
North Terrace
ADELAIDE S.A. 5000

7746 EADES, G.W., Mr
+ Main Roads Department
G.P.O. Box 1412
BRISBANE QLD. 4001

6152 ELVISH, K.J., Mr
P.O. Box 51
NORSEMAN W.A. 6443

4345 ENEVER, J.R., Mr
59 Campbell Street
GLEN WAVERLEY VIC. 3150

6535 ERVIN, M.C., Mr
Coffey & Partners
P.O. Box 93
GLEN IRIS VIC. 3146

10883 EVANS, R.S., Mr
+ 16 Myrtle Avenue
KEW VIC. 3101

4348 FAIRWEATHER, D.F., Mr
Conzinc Riotinto of Aust.
G.P.O. Box 384 D
MELBOURNE VIC. 3001

10884 FEODOROFF, N., Mr
18 Centre Way
BELAIR S.A. 5052

10885 FITZGERALD, D.J., Mr
24 Halstead Street
CAULFIELD VIC. 3162

4351 FITZHARDINGE, C.F., Mr
* Coffey & Partners Pty. Ltd
+ 14A Henley Beach Road
MILE END S.A. 5031

7752 FLANN, R.C., Mr
27 Hopetown Street
ELSTERNWICK VIC. 3185

7753 FLUTTER, J., Mr
* 72 Eucumbene Drive
DUFFY ACT. 2611

10886 FORREST, G.R., Mr
Unit 2/19 Neilson Street
BAYSWATER VIC. 3153

10887 FRENDA, G.A., Mr
* 35 Albion Avenue
\+ PADDINGTON N.S.W. 2021

848 FRIDAY, R.G., Mr
c/- Golder Associates
466 Malvern Road
PRAHRAN VIC. 3181

10888 FULTON, M.J., Mr
* 21 Sauter Street
Heatley
TOWNSVILLE QLD 4814

7756 FURSTNER, J.M., Mr
\+ P.O. Box 150 ARAWA
Bougainville Island
PAPUA NEW GUINEA

4363 GERRARD, G.M., Mr
"Iluka"
108 Bluff Road
BLACK ROC. VIC. 3193

4380 GIUDICI, S., Dr
109 Lansdowne Crescent
WEST HOBART TAS. 7000

10930 GOAD, D.L., Mr
* 21 Lovelace Street
PRESTON VIC. 3072

6091 GORDON, F.R., Mr.
\+ 43 Williams Road
NEDLANDS W.A. 6009

4375 GRAY, N.M., Mr
\+ 1 Centenary Avenue
HUNTERS' HILL N.S.W. 2110

4378 GRIFFITHS, S.J., Mr
Becks Road
FREEMANS WATERHOLE N.S.W. 2323

4381 GUNN, I.L., Mr
2 Mardee Street
NEWMAN W.A. 6753

6539 HALE, G.E., Mr
\+ 15 Ferry Road
AUSTINS FERRY TAS. 7011

4382 HAMMER, J., Mr
8 Dumas Street
MOUNT BARKER S.A. 5251

4385 HARDCASTLE, R.T.A., Mr
* c/- Hardcastle & Richards P/L
Box 138 P.O.
PARKVILLE VIC. 3052

4388 HARGRAVES, A.J., Dr
* P.O. Box 1239
WOLLONGONG N.S.W. 2500

6478 HAUSMANN, M.R., Mr
* 37 Maroa Crescent
NORTH MANLY N.S.W. 2100

10889 HEBBLEWHITE, B.K., Mr
17 Keston Avenue
MOSMAN N.S.W. 2088

7763 HENDERSON, A.D., Mr
12 Redan Street
MOSMAN N.S.W. 2088

6541 HILL, G.B., Dr
\+ 62 Colin Street
WEST PERTH W.A. 6005

4402 HODGSON, J.D., Mr
* 9 Philip Road
\+ MONA VALE N.S.W. 2103

6095 HOLLINGSWORTH, P.C., Mr
* P.O. Box 251
\+ NORTH BRISBANE QLD. 4000

10890 HOLT, T., Mr
* 8 Mayfair Drive
\+ Newtown
GEELONG VIC. 3220

4407 HOSKING, A.D., Mr
* 3 Minawa Street
\+ COOMA NORTH N.S.W. 2630

646 HOULSBY, A.C., Mr
32 Apps Avenue
TURRAMURRA N.S.W. 2074

10891 HOWARTH, D.F., Mr
c/- School of Mining
University of N.S.W.
P.O. Box 1
KENSINGTON N.S.W. 2033

4411 IKIN, D.B., Mr
135 Derwent Avenue
LINDISFARNE TAS. 7015

4415 JAGGAR, F.E., Mr
15 Bulwarra Street
KEIRAVILLE N.S.W. 2500

4416 JAMES, J.S., Mr
* 2 Cranleigh Grove
\+ MT. WAVERLEY VIC. 3149

10892 JOHNSON, N.A., Mr
"Louden"
13 Ridge Road
STERLING S.A. 5152

6671 JONES, M.C., Mr
18 Lauramont Avenue
SANDY BAY TAS. 7005

4425 JORDAN, D.W., Mr
Engineering Geo Branch
Snowy Mountains Engineering
COOMA N.S.W. 2620

4426 JUST, G.D., Dr
Department of Mining and
Met. Engineering
University of Qld.
St. LUCIA QLD. 4067

10894 KAESEHAGEN, F.E., Mr
7 Marney Street
CHAPEL HILL QLD. 4069

10895 + KEMPLAY, E.G., Mr
44 Jells Road
CHELTENHAM N.S.W. 3192

4432 * + KIEK, S.N., Mr
Box 5053
Unitech. P.O.
PAPUA NEW GUINEA

4433 KING, R.W., Mr
8 Bass Place
ST. IVES N.S.W. 2075

10897 + KLENOWSKI, G., Mr
c/- S.M.E.C.
P.O. Box 356
COOMA NORTH N.S.W. 2630

4438 + KNIGHT, M.J., Dr
School of Applied Geology
University of N.S.W.
P.O. Box 1
KENSINGTON N.S.W. 2033

7772 KNOOP, B.P., Mr
102 Regent Street
SANDY BAY TAS. 7005

4445 LACKEY, S.F., Mr
25 Boorool Road
KEW EAST VIC. 3102

10898 * + LEGGO, T., Mr
c/- Dames & Moore
17 Myrtle Street
CROWS NEST N.S.W. 2065

10899 * + LOPES, D., Mr
Universal Soil Labs Pty. Ltd
37 Martin Place
GLEN WAVERLEY VIC. 3150

6163 LUKASZEWSKI, G.M., Dr
CSIRO Div. of Min. Chem.
P.O. Box 124
PT. MELBOURNE VIC. 3207

4471 * MADDOX, J.M., Mr
41 Fisher Avenue
SANDY BAY TAS. 7005

10904 + MARCOS, G.W.A., Mr
62 First Avenue
ROSSMOYNE W.A. 6155

4477 MARSHALL, C.W., Mr
C.W. Marshall Associates
1 Alfred Street
SYDNEY N.S.W. 2000

6109 MARSHALL, R., Mr
P.O. Box 13
SANDY BAY TAS. 7005

642 * + MATHER, R.P., Mr
63 Kingsway
NEDLANDS W.A. 6009

6542 MATTES, N.S., Dr
39 Hammond Avenue
NORMANHURST N.S.W. 2076

7778 MAVER, J.L., Mr
S.R. & W.S.C.
590 Orrong Road
ARMADALE VIC. 3143

10905 MAY, J.R., Mr
44 Victoria Avenue
MACLEOD VIC. 3085

10900 McCONNELL, A.J., Mr
P.O. Box 2
HAMILTON QLD. 4007

10901 * McDONALD, P., Mr
15 Beech Street
SURREY HILLS VIC. 3127

10902 * + McGregor, A.S., Mr
P.O. Box 256
LILYDALE VIC. 3140

4464 * McKenry, K.H., Mr
c/- Department of Civil Engng
P.O. Box 197
CAULFIELD EAST VIC. 3145

10903 * McKenzie, I.L., Mr
Coffey & Hollingsworth
1503 Malvern Road
GLEN IRIS VIC. 3146

4465 McKINNON, R.L., Mr
c/- B.H.P.
G.P.O. Box 86A
MELBOURNE VIC. 3001

3132 McMAHON, B.K., Dr
58 Bent Street
LINDFIELD N.S.W. 2070

7779 MERCER, J.K., Mr
I.C.I. Australia Ltd
G.P.O. Box 4311
MELBOURNE VIC. 3001

10906 * MILLS, K.G., Mr
46 Elimatta Drive
ASHGROVE QLD 4060

6112 MITCHELL, G.G., Mr
* 8 Awatea Street
ST. IVES N.S.W. 2075

10907 MOELLE, K.H.R., Dr
* Department of Geology
+ The University of Newcastle
Newcastle N.S.W. 2308

10908 MONYPENNY, R.G., Mr
+ 46 Hoon Street
EAST WAHROONGA N.S.W. 2076

10929 MORTIZ, P. K., Mr
c/- Frankipile Aust. Pty. Ltd.
3 Ord Street
WEST PERTH W.A. 6005

4496 MURRAY, K.J. Mr
550 Lane Street
BROKEN HILL
N.S.W. 2880

7784 NEYLAND, A.J., Mr
+ "POATINA"
Old Warrandyte Road
NORTH RINGWOOD VIC. 3134

10909 NORTON, R., Miss
6/399 Toorak Road
SOUTH YARRA VIC. 3141

10910 NYE, E.J., Mr
* c/- Mott, Hay & Anderson
+ Far East
18th Floor
1 Hysan Avenue
HONG KONG

7788 O'MALLEY, G.B., Mr
c/- The Aust. I.M.M.
Box 310
CARLTON SOUTH VIC. 3053

4510 OXENFORD, R.A., Mr
* 135 Victoria Street
+ GRAFTON N.S.W. 2460

10911 PAQUAY, R.D., Mr
P.O. Box 49
WITTENOOM W.A. 6752

850 POLAK, E.J., Mr
+ Bureau of Mineral Research
P.O. Box 378
CANBERRA CITY ACT. 2601

10896 POLLARD, G.E., Mr
* P.O. Box 392
+ WEIPA N.Q. 4874

4521 PRESS, M.J., Mr
* Department of Civil Engineering
+ W.A. Institute of Technology
Hayman Road
SOUTH BENTLEY W.A. 6102

6502 PULLEY, J.M., Mr
* Coffey & Partners
+ P.O. Box 118
EAST BRISBANE QLD. 4169

10912 PUMP, W.L., Mr
* 1159 Hoddle Street
EAST MELBOURNE VIC. 3002

10913 RAWLINGS, C.D., Mr
+ c/- Department of Geology
University of Newcastle
NEWCASTLE N.S.W. 2308

7792 REGAN, W.W., Mr
20 Sefton Street
PASCOE VALE VIC. 3044

10914 RENFREY, G.J. Mr
* Terrain Analysis Pty. Ltd
+ P.O. Box 271
KENMORE QLD 4069

7793 RICHARDS, A.B., Mr
12 Stradmore Avenue
TEMPLESTOWE VIC. 3106

4541 ROBERTS, B.C., Mr
23 Glen Road
ASHBURTON VIC. 3147

10928 ROBINSON, G.A., Mr
79 Stanley Street
NEDLANDS W.A. 6009

6125 ROBINSON, K.J., Mr
105B Kars Street
FRANKSTON VIC. 3199

4544 ROBSON, W.M., Mr
+ Gutteridge Haskins & Davey
GPO Box 668
BRISBANE QLD 4001

6508 RONAN, S.R., Mr
+ Engng Water Supply
GPO Box 641F
ADELAIDE S.A. 5001

4547 ROSENGREN, K.J., Mr
8 Domino Court
GLEN WAVERLEY VIC. 3150

10915 ROSS, A.H., Mr
* c/- Capricornia Institute of
+ Advanced Education
M 576
ROCKHAMPTON QLD. 4700

6128 ROWLANDS, D., Dr
Dept Mining and Met. Engineering
University of Qld
ST. LUCIA QLD. 4067

4550 RUSSO, R.J., Mr
86 Barlow Street
CLAYFIELD QLD 4001

10916 SANDERS, R.G., Mr
+ 161 Berringa Road
PARK ORCHARDS VIC. 3114

4551 SANDS, W.H., Mr
+ c/- H.E.C.
TULLAH TAS. 7321

7795 SCOTT, R.W., Mr
12 Forest Glen Crescent
BELROSE N.S.W. 2085

4555 SCOTT, W J., Mr
* P.O. Box 495
\+ PARRAMATTA N.S.W. 2150

4560 SHELDON B.E , Mr
* 87 Laver Road
\+ DAPTO N.S.W. 2530

7797 SHIRLEY, A.F., Mr
* 22 Gwydir Avenue
\+ NORTH TURRAMURRA N S.W. 2074

10917 SIMMONS J.V , Mr
Department of Civil Engineering
University of Alberta
EDMONTON ALBERTA T6G 2G7
CANADA

4569 SMITH, K.W.G., Mr
* Research House
\+ 28 Lauretta Street
NEWTOWN S.A. 5074

10918 SMITH, R.B., Mr
* 19 Wordsworth Avenue
\+ CAMPBELLTOWN N.S.W. 2560

10919 SMITH, R.L., Mr
* 44 Kuring-gai Chase Road
\+ MT. COLAH N.S.W. 2079

6686 STAPLEDON, D.H., Prof.
\+ South Australian Institute of Technology
P.O. Box 1, Ingle Farm
SOUTH AUSTRALIA 5098

10920 SULLIVAN, T.D., Mr
* P.O. Box 125
\+ NORTH RYDE N.S.W. 2113

4579 SVENSON, D., Mr
* P.O. Box 76
\+ NORTH QUAY BRIS 4000

4583 TAPP, B.A., Mr
\+ 6 Karima Court
NORTH RINGWOOD VIC. 3134

6516 TARVYDAS, R.K., Mr
* Hydro Electric Commission
\+ G.P.O. Box 355D
HOBART TAS 7001

4586 THOMAS, E.G., Dr
School of Mining Engng
University of N.S.W.
P.O. Box 1
KENSINGTON N.S.W. 2033

4588 THOMPSON, I.R., Mr
12 Jeffery Street
COBAR N.S.W. 2835

6689 THORNTON, P.N., Mr
\+ 20 Caringal Avenue
DONCASTER VIC. 3108

4593 TROLLOPE, D.H., Prof.
* James Cook University
TOWNSVILLE QLD. 4811

6548 TRUDINGER, J.P., Mr
\+ Dames & Moore
34 Richardson Street
WEST PERTH W.A. 6005

10921 TUCKER, R.C., Mr
* Divisional Office Mrd.
\+ P.O. Box 1089
TOWNSVILLE QLD. 4810

552 VANCE, W.E., Mr
19 Beard Street
ELTHAM VIC. 3095

7806 VUTUKURI, V.S., Mr
University of N.S.W.
Broken Hill Division
BROKEN HILL N.S.W. 2880

7808 WALKER, B.F., Mr
4 Alexander Parade
ROSEVILLE N.S.W. 2069

10922 WALLIS, P.F., Mr
\+ 39 Eaton Road
LINDFIELD N.S.W. 2070

6007 WARD, G.J., Mr
45 Bell Street
FITZROY VIC. 3065

10923 WASHUSEN, J.A., Mr
* 167 Danks Street
ALBERT PARK VIC. 3206

6135 WATKINS, J.R., Mr
\+ Department of Applied Geology
R.M.I.T.
124 Latrobe Street
MELBOURNE VIC. 3000

6522 WAWRYK, S., Mr
* 4 Ormonde Avenue
MILLSWOOD S.A. 5034

10924 WEBER, B., Mr
\+ P.O. Box 2330
ALICE SPRINGS N.T. 5470

6012 WICKHAM, A.E., Mr
* 1 Thorpe Street
BALMORAL QLD 4171

6013 WILLIAMS, D.C., Mr
MIM HOLDINGS Ltd
P.O. Box 1433
BRISBANE QLD 4001

6137 WILLOUGHBY, D.R., Mr
CSIRO
Division of Applied Geomechanics
P.O. Box 54
MT. WAVERLEY VIC. 3149

10925 WILSON, A.G., Mr
* 30 Greenways Road
GLEN WAVERLEY VIC. 3150

7712 WILSON, E.G., Mr
* P.O. Box 378
\+ CANBERRA CITY ACT 2601

6016 WILSON, R.G., Mr
\+ 10 Burradool Street
WEST WOLLONGONG N.S.W. 2500

6140 WOOD, C.C., Dr
* 25 Rokeby Terrace
\+ TARINGA QLD. 4068

10926 WORNER, N.M., Mr
P.O. Box 32
ARMADALE VIC. 3142

6527 WOROTNICKI, J., Mr
* 10 Compton Street
\+ GLEN WAVERLEY VIC. 3250

10927 YTTRUP, P.J., Mr
* 13 Cannes Court
HIGHTON VIC. 3216

CORRESPONDING MEMBERS

6686 STAPLEDON, D.H., Prof.
\+ Head of Applied Geology
11 Classic Court
WEST LAKES S.A. 5021

SUPPORTING MEMBERS

1277 THE BROKEN HILL PROPRIETARY CO. LTD
c/o Australian Iron & Steel Pty. Ltd
Collieries Central Office
P.O. Box 1239
WOLLONGONG N.S.W. 2500

844 COBAR MINES PTY. LTD
P.O. Box 31
COBAR N.S.W. 2835

13202 DEPARTMENT OF HOUSING & CONSTRUCTION
Attn: Librarian
P.O. Box 111
DICKSON ACT 2602

1274 DEPARTMENT OF WORKS
17 Yarra Street
Hawthorn
P.O. Box 2807AA
MELBOURNE VIC. 3001

1273 METROPOLITAN WATER SEWERAGE & DRAINAGE BOARD
Cnr. Pitt and Bathurst Streets
SYDNEY N.S.W. 2000

7181 MOUNT ISA MINES LIMITED
c/o Townsville Transportation & Services Pty. Ltd.
P.O. Box 533
TOWNSVILLE QLD. 4810

631 SNOWY MOUNTAINS ENGINEERING CORPORATION
Box 356, Post Office
COOMA NORTH N.S.W. 2630

841 THE ZINC CORPORATION LIMITED
P.O. Box 444
BROKEN HILL N.S.W. 2880

AUSTRIA

ÖSTERREICHISCHE GESELLSCHAFT FÜR GEOMECHANIK (OGG)

Paracelsus-Strasse 2, A-5020 SALZBURG

President - Prof. E. Pacher

Total Number of Ordinary Members - 220

Total Number of Supporting Members - 9

ORDINARY MEMBERS

11012 ANELLI-MONTI, O., Prok. Dipl.-Ing.
Pyhrn-Autobahn AG
Wilhelm-Raabe-gasse 24
8010 GRAZ

13205 ANGERER, H., Dr phil.
Forsttechn. Dienst f. Wildbach -u. Lawinenverbauung
Paracelsusstrasse 4
5020 SALZBURG

7183 ARTOPOULOS, S., Dr rer. nat.
Schildsheiderstrasse 63
4006 ERKRATH 2 (FRG)

9017 ARWANITAKIS, M., Dipl.-Ing. Dr. mont.
Nawarchu Wotsi 41, Kato Patissia
ATHEN 218 (GRIECHENLAND)

8134 AUFERBAUER, H., Dr Reg. BK
Amt d. Steiermark. Landesregierung
Fuchsenfeldweg 71-75
8042 GRAZ

10191 AYAYDIN, N., Dipl.-Ing.
Ingenieurbüro Dr Pacher
Franz-Josef-Strasse 3
5020 SALZBURG

4233 BAUER, F., Prof. Dr Ing.
Technische Universität Graz
Technikerstrasse 4
8010 GRAZ

13198 BAUMGARTNER, P., Dr
Büro für Bau-, Hydro-, und Therotische Geologie
im Winkl 7
4801 TRAUNKIRCHEN

7173 BECKER, L.P., Ing., Dr phil.
Institut f. Geologie u Paläontologie der Universität Graz
Heinrichstrasse 26
8010 GRAZ

6632 BERNOLD, J., techn. Ing.
8880 WALENSTADT (SWITZERLAND)

7100 BERTLE, H., DDr
Dorfstrasse 34
6780 SCHRUNS

61 BEYRER, C.H. Baurat h.c. Dipl.-Ing.
Zivilingenieur für Bauwesen
Institutsvorstand der Bautechnischen Versuchs-und Forschungsanstalt
Ulrichshöglweg 9
5020 SALZBURG

3406 BLINDOW, F.K., Dipl.-Ing.
Beton-und Monierbau Ges.m.b.H.
Zeuhausgasse 3
6020 INNSBRUCK

1830 BLÜMEL, O., Prof., Dr
Universtät Innsbruck
Innrain 32
6020 INNSBRUCK

10417 BÖHM, K., Ing.
Firma ALWAG
Wagram 49
4061 PASCHING

7957 BOLDT, H., Dr-Ing.
Bergwerksdirektor
Rheudter Strasse 64
4174 ISSUM 2 - SEVELEN (FRG)

6705 BONAPACE, B., Dipl.-Ing.
Tiroler Wasserkraftwerke AG
Landhausplatz 2
6010 INNSBRUCK

7230 BORBEREKY, H., Consulting Engr
Zivilingenieur
Dr.-Thomas-Strasse 11
4810 GMUNDEN

8143 BORM, G., Dipl. Geophys., Dr rer. nat.
Am Wingertsberg 10
6900 HEIDELBERG 1 (FRG)

7101 + BOROVICZENY, F., Dr, Geologe
Geologische Bundesanstalt
Rasumofskygasse 23
1031 WIEN

59 BRANDECKER, H., Dr, Geologe
Anton-Breitner-Strasse 1
5020 SALZBURG

8608 BRANDSTETTER, F.P., Dr
Geschäftsführer
Fa. Gebirgssicherung GmbH
Gnigler Strasse 69
5020 SALZBURG

10597 BREYMANN, H., Dipl.-Ing.
Zivilingenieur
5550 RADSTADT 454

361 BROILI, L., Dr, Geologe
1-33019 Tricesimo Via Aquileia 4
UDINE (ITALY)

7582 BRÜCKL, E., Dipl.-Ing. Dr phil.
Zivilingenieur f. Techn. Physik
Pfarrgasse 7
5400 HALLEIN

1831 BRUNNSTEINER, A., Dipl.-Ing.
Zivilingenieur für Bauwesen
Kirchweg
6600 REUTTE

8010 BRUNOLD, H., Dipl.-Ing.
Hubertusweg 7
8071 HAUSMANNSTÄTTEN

6223 CHOI, H.S., Dipl.-Ing.
Fa. Lahmeyer Intern. GmbH
Lyonerstr 22
6000 FRANKFURT/M.1 (FRG)

9005 + CZURDA, K., Dr phil., Geologe
Geol. Institut d. Universität Innsbruck
Universitätsstr. 4
6020 INNSBRUCK

10587 DALLER, J., Dipl.-Ing.
Geoconsult
Sterneckstrasse 55
5020 SALZBURG

1832 + DEMMER, W., Dr,
Konsulent f. Baugeologie
Hovengasse 6
2100 KORNEUBURG

7973 DIERA, N., Dipl-Ing.
Hochschulassistent
Puschmannstrasse 4
4400 STEYR

3822 DUDDECK, H., Prof. Dr Ing.
Institut f. Statik der TU Braunschweig
Pockelstrasse 4
D-3300 BRAUNSCHEIG (FRG)

1834 DWORSCHAK, G., Dipl.-Ing.
Zivilingenieur f. Bauwesen u. Kulturtechnik
Dürnvellach 53
9821 OBERVELLACH

6550 EGGER, P., Dr Ing.
ETH, Laboratoire de géotechnique
22 avenue de Provence
1007 LAUSANNE (SWITZERLAND)

6579 EPPENSTEINER, W., Dr-phil., Geologe
Geologisches Institut der TU Wien
Karlsplatz 13
1040 WIEN

4057 FECHER, P.,
Prokurist
Leiter des Unternehmensbereiches Berg- und Tunnelbau, F. Heitkamp GmbH
Langekampstrasse 36
4690 HERNE 2 (FRG)

10190 FECKER, E., Dipl.-Geol. Dr-Ing.
Institut f. Boden-u. Felsmechanik der Universität Karlsruhe
Kaiserstrasse 12
7500 KARLSRUHE 1 (FRG)

8600 FEDER, G., Prof. Dr techn
Institut f. Konstruktiven Tiefbau der Montanuniversität Leoben
Franz-Josef-Strasse 18
8700 LEOBEN

8606 + FEHLEISEN, F., Dr, Geologe
Erzgasse 16
8740 ZELTWEG

1835 FESSLER, W., Baurat h.c. Dipl.Ing. Dr
Max-Emanuel-Strasse 7-11
1180 WIEN

12 FETTWEIS, G.B., Prof. Dr-Ing.
Institut f. Bergbaukunde der Montanuniversität Leoben
Franz-Josef-Strasse 18
8700 LEOBEN

7456 FIEBIGER, G., Dipl.-Ing.
Hochschulassistent
Goldbacherstrasse 63
4400 STEYR

7 FINGER, W., Dipl.-Ing.
Wölfnitz, Schweitzerweg 3
9061 KLAGENFURT

187 FLÖGL, H., Baurat h.c. Dipl.-Ing. Dr techn
Stockhofstrasse 32
4020 LINZ

10598 FORNARO, M, Dipl., Ing.
Assistent
Instituto Arte Mineraria - Politecnico di Torino
Corso Duca degli Abruzzi 24
I-10129 TORINO (ITALY)

10569 FRICK, F., Dipl.-Geol. Dr rer.nat.
Pommernstrasse 36
3504 KAUFUNGEN (FRG)

1837 FRITSCH, V., Prof. Dr
Leiter d. Versuchsanstalt für Geoelektrik u. Blitzschutz
Arsenal, Objekt 3
1030 WIEN

8143 FRITZ, J., Dipl.-Ing.
Ingieurbüro für Wasserwirtschaft
Am Schönblick 1
7417 URACH 1 (FRG)

7239 FÜRLINGER, W., Dr, Geologe
Spitalskystrasse 10
4400 STEYR

7102 GATTINGER, T.E., Dr, Geologe
Geologische Bundesanstalt
Rasumofskygasse 23
1031 WIEN

1838 GEHRING, K., Dipl.Ing. Dr mont.
Vöest-Alpine
8740 ZELTWEG

10654 GLESSIG, M., Dipl.-Ing.
Josef-Messner-Strasse 8
5020 SALZBURG

7997 GLITZNER, N., Dipl.-Ing.
Faberstrasse 13
8000 MÜNCHEN 70 (FRG)

8133 GOBIET, W., Dr, Reg.Baurat
Amt d. Steiermärk. Landesregierung
Landesbaudirektion, Fachabt. IIc
Landhausgasse 7
8011 GRAZ

4056 GOLSER, J., Dipl.-Ing.
Ingenieurkonsulent f. Bauwesen
GEOCONSULT
Sterneckstrasse 55
5020 SALZBURG

7240 GÖTZ, H.-P., Dipl.-Ing. Dr
Fa. E. Heitkamp GmbH
Langekampstrasse 36
4690 HERNE 2

7182 GSCHAIDER, F., Dipl.-Ing.
Baudirektor
Rettenpacherstrasse 6/54
5020 SALZBURG

1833 GÜRTLER, M., Lawinenexpertin
Reisnerstrasse 3/1/11
1030 WIEN

8135 HAAS, H., Dir. Ing.
Techn. Geschäftsführer
Österr. Schacht-u. Tiefbauunternehmen Ges.m.b.H.
Heldengasse 12
8753 FOHNSDORF

2418 HABENICHT, H., Dipl.-Ing. Dr mont.
Drachenlochstrasse 18
5083 GARTENAU

4054 HACKL, E., Dipl. Ing.
Ingenieurkonsulent f. Bauwesen
Geoconsult
Sterneckstrasse 55
5020 SALZBURG

8184 HANKE, W.K., Dipl.-Ing.
Direktor
Rheinstrasse 2
7570 BADEN-BADEN (FRG)

1839 HASER, F.E.H., Dr phil., a.o. Prof.
Dept of Geology,
University of Malaya
Pantai Valley
KUALA LUMPUR (MALAYSIA)

3397 HAUSLER, H., Dr, Geologe
Tummelplatz 4
4020 LINZ

6142 HELLER, G., Dipl.-Ing.
Ingenieürburo Dr Pacher
Franz-Josef-Strasse 3
5020 SALZBURG

1840 HERZOG, U., Dr, Geologe
Amt d. Kärntner Landesregierung
Josef-Gruber-Str. 134
9020 KLAGENFURT

8610 HESSE, K.-H., Dr rer.nat.
Ingenieurgeologe
Lehrstuhl f. Ingenieur-geologie u.
Hydrogeologie
Kopernikusstrasse 6
5100 AACHEN (FRG)

1954 HÖFLER, G., Dipl.-Ing.
Montanuniversität Leoben
Franz-Josef-Strasse 18
8700 LEOBEN

8136 + HOMANN, O., Dr, OBR
Amt der Steiermärk. Landesregierung
LANDESBAUDIREKTION IIc

164 + HORNINGER, G. Prof. Dr
Institut f. Geologie der TU Wien
Karlsplatz 13
1040 WIEN

163 INNERHOFER, G., Dipl.-Ing.
Vorarlberger Illwerke AG
Silvrettastrasse 863
6780 SCHRUNS

1950 JABUREK, F., Prof. Dr techn.
Montanuniversität Leoben
Franz-Josef-Strasse 18
8700 LEOBEN

7453 JAGSCH, D., Dipl.-Ing.
Ingenieurbüro f. Tunnel-u. Felsbau GmbH
Rintheimer Strasse 48
7500 KARLSRUHE 1 (FRG)

10588 JAHN, R., Dipl-Ing.
Geoconsult
Sterneckstrasse 55
5020 SALZBURG

10589 JELITZKA, P., Ing.
8114 KLEINSTÜBING Nr. 59

460 JOHN, M., Dipl.-Ing. Dr techn.
Ingenieurgemeinschaft Lässer-Feizlmayr
Framsweg 16
6020 INNSBRUCK

7998 JÖSTL, J., Dipl.-Ing.
Weissenseestr. 77/1
D-8000 MÜNCHEN 90

5 KAHLER, F., Prof. Dr
W. Hofrat i.R.
Linsengasse 29
9020 KLAGENFURT

7174 + KAISER, J., Dr phil
6752 WALD AM ARLBERG 71a

301 KAPFHAMMER, F., Dipl.-Ing.
Waldburgergasse 13
5026 SALZBURG

10584 KAUTZ, H., Prok., Dipl.-Ing.
Fa. Austrobohr
Ditscheinergasse 3/111
1030 WIEN

75 * KIENZL, K., w. Hofrat i.R., Dipl.-Ing.
Hochheimgasse 2
1130 WIEN

151 KIESSLING, H., Dr techn.
Institut f. Mineralogie u. Petrographie
der Universität Salzburg
Akademiestrasse 26
5020 SALZBURG

7103 KIRSCHKE, D., Dipl.-Ing. Dr-Ing.
Beratender Ingenieur f. Felsmechanik und Tunnelbau
Hertzstrasse 12
D 7505 ETTLINGEN

7084 KIRCHNER, E., Dr phil.
Institut f. Mineralogie u. Petrographie
der Universität Salzburg
Akademiestrasse 26
5020 SALZBURG

6162 KIRSCHNER, E., Dipl.-Ing.
Direktor
STUAG Strassen-u. Tiefbau AG
Seilerstätte 18-20
1010 WIEN

7110 KOBEL, M., Dipl-Geol. Dr
Büro f. Technische Geologie E. Weber
7304 MAIENFELD (SWITZERLAND)

66 KOBLER, R., w. Hofrat, Dipl.-Ing.
Amt der Salzburger Landesregierung
Michael-Pacher-Strasse 36
5020 SALZBURG

3819 KOERNER, U., Dipl.-Geol. Dr
Oberlandesgeologe
Herlinstrasse 6
7830 EMMENDINGEN

7974 KRAUSE, H., Dipl.-Geol. Dr
Oberlandesgeologe
Geologisches Landesamt Baden-Württemberg
Urbanstrasse 53
7000 STUTTGART (FRG)

7457 KRONFELLNER-KRAUS, G., w. Hofrat, Dipl.-Ing. Dr
Forstl. Bundesversuchsanstalt,
Institut f. Wildbach-u. Lawinenverbauung
Tirolergarten
1131 WIEN-SCHÖNBRUNN

7241 KUHNHENN, K., Dipl.-Ing. Dr.-Ing.
Ingenieurbüro Müller-Hereth
Rintheimer Strasse 48
7500 KARLSRUHE 1 (FRG)

10593 KULOZIK, R., Dipl.-Geol.
Auf dem Knick 12
5810 WITTEN (FRG)

10201 KÜNZEL, H., Dipl.-Ing.
Mantlergasse 13
1130 WIEN

1952 KUTZBACH, H., Dipl.-Ing.
Teinfaltstrasse 3
1010 WIEN

4055 LAABMAYR, F., Dipl.-Ing.
Ingenieurkonsulent f. Bauwesen
Breitenfelderstrasse 47
5020 SALZBURG

10712 LANEVE, R., Dipl.-Ing.
Fa. Oberranzmeyer
Archenweg 10
A 6020 INNSBRUCK

10641 LANGE, R.
Grubenvorstandsmitglied der Gewerkschaft
Walter am Handelshof 1
4300 ESSEN (FRG)

7085 LÄSSER, P., Dipl.-Ing.
Ingenieurgemeinschaft Lässer-Feizlmayr
Framsweg 16
6020 INNSBRUCK

10556 LAUBBICHLER, K., Dipl.-Ing.
Geoconsult
Sterneckstrasse 55
5020 SALZBURG

63 LAUFFER, H., Dr techn.
Direktor
Tiroler Wasserkraftwerke AG
Landhausplatz 2
6010 INNSBRUCK

10601 LED, M., Dipl.-Ing.
St. Nikolai 8
9862 KREMSBRÜCKE

13203 LEINWEBER, D., Prok., Dipl.-Ing.
Fa. Austrobohr
Schmiedgasse 40
8010 GRAZ

9013 LIEGLER, K., Dr, Geologe
Zierleitengasse 50
1190 WIEN

10218 LITSCHAUER, H., Dipl.-Ing.
Unfallverhütungsdienst
Dr Franz-Rehrl-Platz 5
5020 SALZBURG

13201 LOACKER, H., Dr
Vorarlberger Illwerke AG
Postfach 117
6901 BREGENZ

7455 MAHRINGER W., Dr, Meteorologe
Wetterdienst Salzburg
Flughafen
5035 SALZBURG

8144 MAIDL, B., Prof. Dr -Ing.
Stiepel, Weidengrund 43
4630 BOCHUM (FRG)

10650 MARINI, C., Dr
Duca d'Aosta Str. 30
39100 BOZEN (ITALY)

10215 MARZOLA, A., Geologe
Via Risorgimento 34
VERONA (ITALY)

1848 MATL, F., Ministerialrat Dr techn.
Bundesministerium f. Bauten u. Technik
Stubenring 1
1010 WIEN

82 MIGNON, K., Dr, Geologe
Sonnenburgstrasse 3
6020 INNSBRUCK

10186 MIKURA, E., Dr phil., Ing., Geologe
Bautechniker
Hovengasse 5/2/13
2100 KORNEUBURG

8143 MOHN, J., cand. geol.
Ungererstrasse 19/708
8000 MÜNCHEN 40 (FRG)

1849 MOOSBRUGGER, M., Dipl.-Ing.
Zivilingenieur für Bauwesen
Eisengasse 18
6850 DORNBIRN

9000 MOSER, G., Dipl.-Ing.
6780 GANTSCHIER 447

8147 MÜHLHÄUSER, H.-P., Betriebswirt
Geschäftsführer d. Fa. K.-H. Mühlhäuser
Schillerstr. 4
6120 MICHELSTADT-STEINBACH (FRG)

8137 MÜLLER, K., Dipl.-Ing.
Kleingmainergasse 23 A
5020 SALZBURG

1 MÜLLER L., Prof., Baurat h.c.
Dr techn. Dr mont. h.c.
Paracelsusstrasse 2
5020 SALZBURG

9014 MÜLLER P.J., Dr Geol.
Geoconsult
Sterneckstrasse 55
5020 SALZBURG

7952 MÜLLER, S., Ing.
Reitsam 34
5452 PFARRWERFEN

7999 MUSSGER, K., Dipl.-Ing
Geoconsult
Sterneckstrasse 55
5020 SALZBURG

7196 NECAS, J., Dipl.-Ing.
Ingenieurbüro f. Tunnel-u. Stollenbau
Allmersbacherstr. 25
7000 STUTTGART 50 (FRG)

8148 NEUERBURG, R., Dipl.-Geol.
Schlesierstrasse 10
D-8190 WOLFRATHAUSEN

7453 NIEL, A., Ministerialrat, Dipl.-Ing.
Bundesministerium f. Land-und Forstwirtschaft
Stubenring 1
1010 WIEN

8612 NOWY, W., Geologe
Bergheim-Wöllersdorf 23
3053 LAABEN

1851 OBERLEITNER, P., Dipl.-Ing.
Zivilingenieur f. Bauwesen
Wegererstr. 51
4400 STEYR

26 OBERRANZMEYER, R., Dipl-Ing.
Zivilingenieur für Bauwesen
Archenweg 10
6020 INNSBRUCK

10583 OBERTHALER, B., Ing.
4061 PASCHING Nr. 104

6636 OISMÜLLER, H., Dipl.-Ing.
Klenzestrasse 58
8000 MÜNCHEN (FRG)

2 PACHER, F., Dipl.-Ing.
Dr-Ing. h.c.
Ingenieurkonsulent für Bauwesen
Franz-Josef-Strasse 3
5020 SALZBURG

10590 PAIL, F. J., Dipl.-Ing.
Haimlgasse 9/16
5020 SALZBURG

6634 PANTZARTZIS, P., Geologe
Papagou-Str. 43
NEA-LESVOS-AMARUSSION (GREECE)

7977 POISEL, R., Dipl.-Ing.
Institut f. Geologie der TU Wien
Karlsplatz 13
1040 WIEN

4058 PONTOW, K.-H., Prok., Dr-Ing.
Fa. E. Heitkamp GmbH, Niederl. München
Fraunhoferstrasse 11
8033 MARTINSRIED (FRG)

333 POSCH, H., Dipl.-Ing.
Direktor
Heiliggeiststrasse 21
6020 INNSBRUCK

10189 PRÄSENT, G., Dipl.-Ing.
Riefsiedlung, Waldweg 849
8101 GRATKORN

8142 PREIN, R., Dipl.-Ing.
Zivilingenieur für Bauwesen
Laintal 28a
8793 TROFAIACH

10632 PRINZ, H., Ing.
Fachverband der Bergwerke und Eisenerzeugenden Industrie
Goethegasse 3
1010 WIEN

1855 RADLER, S., Prof. Dr-Ing.
Universität für Bodenkultur
Gregor-Mendel-Strasse 33
1180 WIEN

8145 RAHN, W., Dipl-Geol.
Wohlfahrtstrasse 208
4630 BOCHUM (FRG)

12071 RAPP, R., Dr-Ing.
Steinbergstr. 19
D-8034 GERMERING

3229 RAUCH, W., Dipl.-Ing.
Zivilingenieur für Bauwesen
Schützenstrasse 41
6040 INNSBRUCK

7104 RAUTENSTRAUCH, R.W., Dipl-Geol.
Heintzmannscheide 41
4630 BOCHUM (FRG)

7997 REIK, G., Dipl-Geol. Dr
Institut f. Bodenmechanik u. Felsmechanik
Universität Karlsruhe
Kaiserstrasse 12
7500 KARLSRUHE 1 (FRG)

10629 REISCHL, S., Ing.
Geoconsult
Sterneckerstrasse 55
5020 SALZBURG

7958 REITZ, A., Dipl.-Ing. Dr-techn.
Zivilingenieur
Schablederweg 7
4020 LINZ

1856 RESCH, R., Dr techn.
Geschäftsführer
d. Fa. Austrobohr
Schmiedgasse 40
8010 GRAZ

6633 RESCHER, O.-J., Prof. Dr
Inst. f. konstruktiven Wasserbau
Techn. Universität Wien
Karlsplatz 13
1040 WIEN

8609 RIEDL, R., Ing. (grad.)
Bauleiter
Kraihammersiedlung 142
5301 EUGENDORF

6580 + RIEHL-HERWIRSCH, G., Dr
Institut f. Geologie der TU Wien
Karlsplatz 13
1040 WIEN

7953 RIENÖSSL, K., Ing.
Oberingenieur, Hbv.
Apothekerhofstrasse 11
5020 SALZBURG

7229 RODE, P., Dipl.-Ing.
Fa. Kapsreiter KG
Faberstrasse 2c
5020 SALZBURG

10627 RUCKENSTEINER, G., Dr-Ing.
c/o Fa. A. Tagger
5440 GOLLING/SALZACH

8011 RÜCKL, J., Dipl.-Ing.
Haydnstrasse 3
5020 SALZBURG

10592 SANIA, H.-P., Dipl.-Ing
Hustadtring 143
4630 BOCHUM (FRG)

6635 SAUER, G., Dipl.-Ing. Dr
Fels-u. Tunnelbauberatung und -planung
Habach 55
5020 SALZBURG-KOPPL

8607 + SCHÄFFER, G., Dr
Geologische Bundesanstalt
Rasumofskygasse 23
1031 WIEN

7113 SCHANTL, J., Dr phil.
Univ.-Ass.
Institut f. Mineralogie u. Petrographie
der Universität Salzburg
Akademiestrasse 26
5020 SALZBURG

31 SCHEIDEGGER, A.E. Prof. Dr
Institut f. Geophysik der TU Wien
Karlsplatz 13
1040 WIEN

1860 SCHIMMERL, J., Prof., Dipl.-Ing. Dr techn.
Techn. Universität Wien
Karlsplatz 13
1040 WIEN

4234 SCHMIDT, M., Prof. Dr-Ing.
Techn. Universität Berlin, Fachber.16
Hardenbergstr. 42 BH 10
1000 BERLIN 12 (FRG)

11003 SCHMIEDER, H., Dipl.-Ing.
Arge Pfandertunnel
Postfach 144
6901 BREGENZ

7532 SCHOBER, W., Prof. Dr techn.
Institut f. Bodenmechanik,
Felsmechanik und Grundbau
Technikerstrasse 13
6020 INNSBRUCK

10600 SCHRAMM, J.-M., Dr phil., Geologe
Christian-Doppler-Str. 5
5020 SALZBURG

13204 SCHUBERT, W., Dr, Dipl.-Ing.
GEOCONSULT
Sterneckstrasse 55
5020 SALZBURG

8145 SCHUERMANN, F., Dr Ing.
Am Kronenberg 8
4320 HATTINGEN 16 (FRG)

10565 SCHULTER, A., Dipl.-Ing.
Bauleiter
Waldgasse 1
8055 GRAZ

1861 SCHUR, F., Baurat h.c., Dipl.-Ing.
Zivilingenieur für Bauwesen
Gneisfeldstrasse 11
5020 SALZBURG

13100 SCHWINGENSCHLÖGL, R., Ing. Dr phil
Albertgasse 6/12a
1080 WIEN

1953 SEDLACEK, H., Baumeister
NEWAG
Johann-Steinböck-Strasse 1
2344 MARIA ENZERSDORF-SÜDSTADT

7123 SEEBER, G., Prof. Dr
Humboldtstrasse 14
6060 HALL IN TIROL

16 + SEELMEIER, H., Prof. Dr
Techn. Universität Graz
Rechbauerstrasse 12
8010 GRAZ

28 SELTENHAMMER, L., Dipl.-Ing.
Zivilingenieur für Bauwesen
Josef-Gangl-Gasse 32
1130 WIEN

1900 SIMMLER, H., Prof. Dr techn.
Institut f. Konstruktiven
Wasserbau der TU Graz
Stremayrgasse 10
8010 GRAZ

7228 SOMMER, A., Dipl.-Ing.
Vorsitzender des Grubenvorstandes
der Gewerkschaft Walter
Am Handelshof 1
4300 ESSEN (FRG)

1858 SPAUN, G., Dr, Geologe
Kühbergstrasse 48
5020 SALZBURG

1859 SPICKERNAGEL, H., Prof. Dr-Ing.
Institut f. Markscheidewesen der
Montanuniversität Leoben
Franz-Josef-Strasse 18
8700 LEOBEN

8171 STADLER, G., Dipl.-Ing.
Franz-Fischer-Strasse 24
A-5061 ELSBETHEN

3823 STAUB, A., Dipl.-Ing.
Fa. Swietelsky Bauges. m.b.H.
Museumstrasse 3
4010 LINZ

32 STEFKO, E., Dipl.-Ing.
Baudir. der Vorarlberger Illwerke AG
Josef-Huter-Strasse 35
6901 BREGENZ

9001 STEINHAUSER, P., Prof. Dr
Institut f. Meteorologie u. Geophysik
Universität Wien
Hohe Warte 38
1190 WIEN

10591 STOHL, K., Dipl.-Ing.
4481 ASTEN O.Ö.

10686 STROBL, B.
Ingenieurbüro Dr Pacher
Franz-Josef-Strasse 3
A-5020 SALZBURG

1864 TAGGER, G.K.
Direktor
Kalk-und Schotterwerke
5440 GOLLING/SALZACH

8166 TENTSCHERT, E., Dr, Geologe
Tiroler Wasserkraftwerke AG
Landhausplatz 2
6010 INNSBRUCK

7243 TOTH, L., Dr
Elsslergasse 17/8
1130 WIEN

423 TRAUZETTEL, G., Dipl.-Ing. Dr rer nat
Sachverständiger f. Baugrund u. Gründungen
Feuerbach, Fahrionstr. 13
7000 STUTTGART (FRG)

7171 TREMMEL, E., Prof., Dipl.-Ing. Dr techn.
Institut f. Baustatik der TU Wien
Karlsplatz 13
1040 WIEN

7959 TROPPER, W., Dipl.-Ing. Dr techn.
Höhenstrasse 13 a
6020 INNSBRUCK

1902 TSCHECH, E., Prof. Dr techn.
Techn. Universität Graz
Stremayrgasse 11
8010 GRAZ

1865 TSCHIERSKE, N.
Eckenhaid Vogelherd 6
8501 ECKENTAL 1

1867 ÜBLAGGER, A., Dipl.-Ing.
Reichenhallerstrasse 13
5020 SALZBURG

345 URBAN, G.
Baudirektor a.D.
Landolinsteige 1A
7300 ESSLINGEN/NECKAR (FRG)

9002 URBANSKI, H.-W., Ing. grad.
Prüfstellenleiter, Ingenieurbüro für Baustofftechnologie/Material-prüfstelle
Unckelstrasse 3
4400 MÜNSTER 48 (FRG)

10219 VARDAR, M., Dr-Ing.
Assistent
I.T.Ü. Maden Fakültesi
Tatbiki Jeoloji Kürsüsü, Tesvikiye
ISTANBUL/TURKEY

10687 VAVROVSKY, G.-M., Dipl.-Ing.
Ingenieurbüro Dr Pacher
Franz-Josef-Str. 3
5020 SALZBURG

42 * VEDER, C., Prof. Dr techn. Dr h.c.
Kohlergasse 15
1180 WIEN

6637 + VOGELTANZ, R., Dr phil.
Oberbaurat
Festungsgasse 7
5020 SALZBURG

7244 VOGLER, O.M., Dipl.-Ing.
Zivilingenieur für Bauwesen
Linke Wienzeile 10
1060 WIEN

10415 WAGNER, A., sen.
Fa. ALWAG
Wagram 49
4061 PASCHING

7112 WAGNER, H., Dipl.-Ing. Dr techn.
Beton-u. Monierbau GmbH
Abt. Untertagebau
Zeughausgasse 3
6020 INNSBRUCK

506 WAGNER, H., Dipl.-Ing. Dr mont.
Chamber of Mines, Mining Res. Lab.
P.O. Box 809
JOHANNESBURG (SOUTH AFRICA)

9003 WAHL, W., Dipl.-Ing.
Büro f. Baugenieurwesen Laabmayr
Rindermarkt 7
8000 MÜNCHEN (FRG)

19 WATZNAUER, A., Prof. Dr-Ing.
Sachsen, Postfach 67
92 FREIBERG (DDR)

7111 WEBER, E., Ing. Geologe
Büro fur Technische Geologie
Im Herrenfeld
7304 MAIENFELD (SWITZERLAND)

347 WEBER, H., Dipl.-Ing.
Zivilingenieur für Bergwesen
Schwarzstrasse 27
5020 SALZBURG

1870 + WEISS, E.H., Prof. Dr
Villacher Strasse 25
9020 KLAGENFURT

1906 WEISS, P., Obering.
Österr.-Amerikanische Magnesit AG
Magnesitbergbau Millstätter Alpe
Karnten
9545 RADENTHEIN

9004 WEISSBACH, G., Dipl.-Geol
Maximilianstrasse /
8034 GERMERING (FRG)

8611 WERANY, A., Bauing.
Ziegelhauser Landstrasse 13
6900 HEIDELBERG (FRG)

1907 WERNER, H.-U., Dr-Ing.
Mannhardtstrasse 7
D-8000 MÜNCHEN 22 (FRG)

7956 WIDERHOFER, R., Dipl.-Ing. Dr techn.
Tiroler Wasserkraftwerke AG
Landhausplatz 2
6010 INNSBRUCK

7445 WIDMANN, R., Dr techn.
Tauernkraftwerke A.G.
Rainerstrasse 29
5020 SALZBURG

10655 WIETEK, B., Dipl.-Ing.
Zivilingenieur für Bauwesen
Blasius-Hueber-Strasse 16
6020 INNSBRUCK

8196 WILDHAGEN, H., Dipl.-Ing.
Stockmannstrasse 24
8000 MÜNCHEN 71 (FRG)

10680 WOLLMANN, G., Dipl.-Ing.
Direktor
Fa. Hamberger Bauges.m.b.H.
Langäckergasse 10
A-2345 BRUNN a. Geb.

445 WYCITAL, H., Dipl.-Ing. Dr techn.
Argentinierstrasse 36
1040 WIEN

7985 ZACAS, M., Dipl.-Bauing.
Paisley College of Technology
High Street
Paisley
Renfrewshire PA1 2BE
SCOTLAND (UK)

SUPPORTING MEMBERS

261 ALLGEMEINE BAUGESELLSCHAFT A.
PORR AG
Engelsberggasse 4
1030 WIEN III

6830 ENNSKRAFTWERKE - AKTIENGESELLSCHAFT
Resthofstrasse 2
A 4400 STERY

6827 INTERFELS
Schwartzstrasse 27
5020 SALZBURG

88 ÖSTERREICHISCHE DRAUKRAFTWERKE AG
Baumbachplatz 2
9020 KLAGENFURT

126 ÖSTERREICHISCHES NATIONALKOMITEE
DER INTERNATIONALEN TALSPERREN-
KOMMISSION
an der Hülben 4
1010 WIEN

6825 REMBERT OBERRANZMEYER
A - 6020 Postfach 533
Archenweg 10
INNSBRUCK

95 SALZBURGER A.G. FÜR ELEKTRIZITÄTS-
WIRTSCHAFT
Postfach 170
Schwarzstrasse 44
5021 SALZBURG

403 TIROLER WASSERKRAFTWERKE AKTIEN-
GESELLSCHAFT
Landhausplatz 2
6020 INNSBRUCK

46 VORARLBERGER ILLWERKE AKTIEN-
GESELLSCHAFT
Josef Huter Strasse 35
A-6901 BREGENZ

BELGIUM

GROUPEMENT BELGE DE MECANIQUE DES ROCHES (GBMR)

Universite Catholique de Louvain, Place du Levant, Batiment Vinci, B-1348
LOUVAIN-LA-NEUVE, BELGIUM

President: Prof. J. Brych
Secretary: Mr. J. F. Thimus

Total number of Ordinary Members - 36
Total number of Supporting Members - 10

ORDINARY MEMBERS

10657 BOLLE
Quai Banning, 6
4000 LIEGE

501 * + BONNECHERE, F.
Quai Banning, 6
4000 Liege

7694 + BRYCH, J.
Rue du Bois La-Haut, 30
7000 MONS

10658 BUTTIENS
Barrikadenplein, 13
1000 BRUXELLES

10659 CARPENTIER
Olmenstraat 8
8020 OOSTKAMP

13008 CAUFRIEZ
Rue Augile, 33b
6401 PETIGNY

10660 CRUCIFIX
Rue des 4 Arbres, 36
5160 LUSTIN

220 * + DE BEER, M.E.
Keizerlijkeplaats,3
9300 AALST

605 + DELMER, A.
Avenue Colonel Daumerie, 16
1160 BRUXELLES

7695 + DESPERT, M.
Quai Roi Albert
4000 LIEGE

478 * + FAGNOUL, A.
Quai Banning, 6
4000 LIEGE

7696 + FUNCKEN, M.
Rue de la Science, 31
1040 BRUXELLES

609 + GILSOUL, G.
Rue Joseph Gos, 11
1420 BRAINE-L'ALLEUD

13101 GOUCHE
Rue du Joncquois, 43
7000 MONS

7697 * GROSJEAN, P.
Avenue du Rond Point. 10
1330 RIXENSART

10661 HUERGO, P.
Avenue Adolphe Buyl,87
1050 BRUXELLES

6304 * + LOUSBERG, E.
Avenue de l'Espinette, 25
1348 LOUVAIN-LA-NEUVE

1960 MONJOIE, A.
Avenue Champs La-Haut, 4
4920 EMBOURG

13105 NEERDAELD
Rue du Joncquois, 43
7000 MONS

10662 * NUYENS, J.
Avenue des Noisetiers, 5
1170 BRUXELLES

13009 * PAQUAY, J.
Rue Gretry, 196
4020 LIEGE

13102 PAREZ
Rue du Joncquois, 43
7000 MONS

10675 PIRSON
Rue du Moulin, 320
4020 LIEGE

10663 RINALDI
Allee des Maronniers, 2 Bte 25
5800 GEMBLOUX

6307 ROISIN, V.
* Rue du Cornet, 43
1040 BRUXELLES

13103 SAROT
Rue du Joncquois, 43
7000 MONS

6308 SCHROEDER, C.
\+ Les Forges, 60
5340 GESVES

10664 SCHWACHHOFER
* Rue du Parc, 47 Bte 3
6000 CHARLEROI

1961 STASSEN, P.
\+ Quai Marcellis, 25
4020 LIEGE

10665 THIMUS, J.F.
Rue du Monty, 3
6333 CORROY-LE-CHATEAU

7700 VAN BELLINGEN, R.
Rue Bidaut, 8
4800 VERVIERS

13010 VAN DAMME
Rue de la Station, 3
9600 RENAIX

13011 VAN DUYSE, H.
\+ Allee du Sainval, 23
4920 EMBOURG

13104 VAN EYLEN
Rue du Joncquois, 43
7000 MONS

10666 VAN IMPE
Doorsteekstraat, 9
9420 MERE

610 VAN WAMBEKE, A.
* Rue Knapen, 45
\+ 1040 BRUXELLES

SUPPORTING MEMBERS

10667 CENTRE DE RECHERCHES ROUTIERES
Boulevard de la Woluwe, 42
1200 BRUXELLES

10677 CIE INTERNATIONALE DE PIEUX ARMES
FRANKIGNOUL
Rue Gretry 196
4020 LIEGE

10676 COMISSION GENERALE D'ENTREPRISES
ELECTRIQUES ET INDUSTRIELLES
Place du Trone, 1
1000 BRUXELLES

10678 FORAKY
Place des Barricades, 13
1000 BRUXELLES

580 INIEX - INSTITUT NATIONAL DES
INDUSTRIES EXTRACTIVES
Rue du Chera
4000 LIEGE

10668 INISMA A.S.B.L.
Rue de Malplaquet, 32
7000 MONS

10669 INSTITUT GEOTECHNIQUE DE L'ETAT
Tramstraat, 44
9710 ZWIJNAARDE

10670 S.B.B.M.
Avenue Galilee, 5
1030 BRUXELLES

10679 SOCIETE DE TRACTION ET ELECTRICITE
Rue de la Science, 31
1040 BRUXELLES

10671 UNION MINIERE, S.A.
Rue de la Chancellerie, 1
1000 BRUXELLES

BOLIVIA

Total Number of Corresponding Members - 1

CORRESPONDING MEMBERS

8198 QUINTANA, C.E., Dr - Engr
Calle Hoyos 85
POTOSI

BRAZIL

ASSOCIACAO BRASILEIRA DE MECANICA DOS SOLOS (ABMS)

Caixa Postal 7141, 01000 SAO PAULO, SP

Secretary: Prof. A. D. F. Napoles Neto

Total Number of Ordinary Members - 84
Total Number of Supporting Members - 12

ORDINARY MEMBERS

13556 ABRAHAO, R.A., Dr - Geologist
\+ Av. Paulista, 2064 - 9º
01310 - SAO PAULO, SP

8036 ABRAO, P. C., Dr - Geologist
Av. Paulista, 688 - 17º
01310 - SAO PAULO, SP

13553 ACEVEDO, J.H.P., Dr - Geologist
Av. 9 de Julho, 4939 Bloco A - 11º
01407 - SAO PAULO, SP

13548 ALVES FILHO, A., Mr - Engr
Av. 9 de Julho, 4939 - 6º Bloco A
01407 - SAO PAULO, SP

8019 ARAUJO, E.S. de, Mr - Engr
R. Sa Ferreira, 44 - apto. 809
Copacabana
20000 - RIO DE JANEIRO, RJ

436 BARATA. F.E., Prof - Engr
* R. Prudente de Moraes, 331 - apto 101
20000 - RIO DE JANEIRO, RJ

1647 BARROS, F.P. de, Dr - Geologist
* Av. Paulista, 2064 - cj.86
01310 - SAO PAULO, SP

8012 BARTORELLI, A., Dr - Geologist
R. Fradique Coutinho, 623 - apto 71B
05416 - SAO PAULO, SP

7508 BERBERIAN, D., Prof - Engr
* Consulting Engineer
Trecho 3 nº 1370
70000 - BRASILIA, DF

8014 BEZERRA, D.M., Mr - Engr
Av. Paulista, 2064 - 9º
01310 - SAO PAULO, SP

7509 BJORNBERG, A.S., Dr - Geologist
* Chief Geologist
\+ Av. 9 de Julho, 4939 - Bloco A-11º
01407 - SAO PAULO, SP

3827 BOCK, E.I., Mr - Engr
* Av. Atlantica, 3102 - apto. 1201
20000 - RIO DE JANEIRO, RJ

3131 BOGOSSIAN, F., Mr - Engr
* R. Alvaro Ramos, 83 - ZC 082 - Botafogo
\+ 20000 - RIO DE JANEIRO, RJ

13551 BRANCO, J.J.R., Mr - Engr
R. Araguari, 1705 s/201/202
30000 - BELO HORIZONTE, MG

8038 BUSCH, R.G., Mr - Engr
Av. Agami, 22 - Indianopolis
04522 - SAO PAULO, SP

6699 CAMARGO, F.P., Dr - Geologist
Estrada da Gavea Pequena, 952/114 C
20000 - RIO DE JANEIRO, RJ

8039 CELERI, R. de O., Mr - Engr
\+ Av. Paulista, 2064 - sede II - 9º
01310 - SAO PAULO

13554 COELHO, L.F.M., Mr - Engr
IPT - DMGA - Cidade Universitaria
Caixa Postal 7141
01000 - SAO PAULO, SP

296 COSTA NUNES, A.J. da, Prof - Engr
* Consulting Engineer
\+ R. Barao de Sao Felix, 202
20000 - RIO DE JANEIRO, RJ

13547 CURY Jr., A., Mr - Engr
IPT - DMGA - Caixa Postal 7141
01000 - SAO PAULO, SP

8032 DONDICI RUIZ, M., Mr - Engr
Alameda Franca, 84
01422 - SAO PAULO

8022 ECKSCHMIDT, H.R., Mr - Engr
\+ Largo do Arouche, 24-5º
01219 - SAO PAULO, SP

8029 EIGENHEER, L.P.Q.T., Mr - Engr
Av. 9 de Julho, 4939 - Bloco A-6º
01407 - SAO PAULO, SP

13557 ESTON, S.M. de, Mr - Engr
IPT - DMGA - Cidade Universitaria
Caixa Postal 7141
01000 - SAO PAULO, SP

6697 FARJALLAT, J.E.S., Dr - Geologist
Caixa Postal 7141
01000 - SAO PAULO, SP

13546 FERNANDES, A.C., Mr - Engr
IPT - DMGA - Caixa Postal 7141
01000 SAO PAULO, SP

8026 FIGUEIREDO, J.T.N. de, Mr - Engr
R. Domingos Ferreira, 210 - apto.702
20000 - RIO DE JANEIRO, RJ

8013 FONSECA, A.M.M. da C.C., Mrs - Engr
R. Mary Pessoa, 54 - Gavea
20000 - RIO DE JANEIRO, RJ

6819 FRANCISS, F.O. Mr - Engr
Consulting Engineer
R. Araujo Porto Alegre, 70
20000 - RIO DE JANEIRO, RJ

8027 FRANCO, J.O. de J., Mr - Engr
Av. 9 de Julho, 4939 - Bloco A - 11º. andar
01407 - SAO PAULO, SP

6695 + FUJIMURA, F., Mr - Engr
Caixa Postal 7141
01000 - SAO PAULO, SP

7510 * GAIOTO, N., Prof - Engr
U. Teacher
Av. 9 de Julho, 4939 - Bloco A-11º
01407 - SAO PAULO, SP

8034 + GANDOLFI, N., Dr - Geologist
Escola de Engenharia de Sao Carlos - USP
13560 - SAO PAULO, SP

13552 * GONCALVES, J.L.A., Mr - Engr
R. D - Casa 6
15600 - FERNANDOPOLIS, SP

8140 * + GOTLIEB, M., Mr - Engr
Consulting Engineer
Alameda Itu, 448
01421 - SAO PAULO, SP

8024 GRANDIS, I., Mr - Engr
Av. Paulista, 688 - 17º
01310 - SAO PAULO, SP

6701 + GUIDICINI, G., Dr - Geologist
Instituto de Pesquisas Tecnologicas
Cidade Universitaria
Caixa Postal 7141
01000 - SAO PAULO, SP

8043 HASUI, Y., Dr - Geologist
Instituto de Geociencias USP
Caixa Postal 20889
01000 - SAO Paulo, SP

8041 HENNIES, W.T., Mr - Engr
R. Antonio de Andrade Rebello
05692 - SAO PAULO, SP

8040 * + HSU, S.J.C., Mr - Engr
Consulting Engineer
Largo do Arouche, 24 - 10º
01219 - SAO PAULO, SP

8033 + INFANTI Jr., N., Dr - Geologist
Av. 9 de Julho, 4939 - Bloco A-11º andar
01407 - SAO PAULO, SP

6700 JARDIM, N.S., Dr - Geologist
IPT - Caixa Postal 7141
01000 - SAO PAULO, SP

3843 * + KANJI, M.A., Prof - Geologist
Vice-President
GH Engenharia SC Lda.
Av. Brig. Faria Lima, 560-11º
01452 SAO PAULO, SP

8042 * + LACERDA, W.A., Prof - Engr
U. Teacher
R. Cardoso Jr., 5-apto.202
20000 - RIO DE JANEIRO, RJ

8018 * LEME, C.R. de M., Mr - Engr
Av. Paulista 2064 - 9º
01310 - SAO PAULO, SP

7512 LIMA, R. de S., Mr - Engr
a/c Promon Engenharia S.A.
Av. Rio Branco, 103 - 10º andar
20000 - RIO DE JANEIRO, RJ

13562 MAFFEI, C.E.M., Mr - Civil Engr
Av. 9 de Julho, 4939-9º
01407 - SAO PAULO, SP

13560 MANO, V.T., Dr - Geologist
IPT-DMGA - Cidade Universitaria
Caixa Postal 7141
01000 - SAO PAULO, SP

13563 MASSARENTE, G., Mr - Engr
CESP-Av. Paulista 2064 - 11º andar
01310 SAO PAULO, SP

8031 MATOS, M.M. de, Mr Engr
R. Barao de Capanema, 98 - apto.43
01411 - SAO PAULO, SP

13550 MELO, F.C.F. de, Mr - Engr
IPT - DMGA - Caixa Postal 7141
01000 - SAO PAULO, SP

7515 MELLO, F.M. de, Mr - Engr
ENGE-RIO, Engenharia e Consultoria S.A.
R. Mexico, 3 - 9º andar - Centro
20000 - RIO DE JANEIRO RJ

435 * + MELLO, V.F.B. de, Prof - Engr
Consulting Engineer
R. Frederico Chopin, 190
01454 - SAO PAULO, SP

8020 MENEZES, G., Mr - Engr
Av. Paulista, 688-17º
01310 - SAO PAULO, SP

6696 + MIDEA, N.F., Mr - Engr
R. Joaquim Machado 153-57
05050 - SAO PAULO, SP

13564 + MIOTO, J.A., Dr - Geologist
IPT-MGA - Caixa Postal 7141
01000 SAO Paulo, SP

13567 MORIMOTO, S., Mr - Engr
CESP - Vila dos Operadores,
Casa nº 8
86160 PORECATU, PR

13566 MORITA, L., Mr - Engr
CESP - Av. Paulista 2064 - 11º andar
01310 SAO PAULO, SP

8015 MORUZZI, C., Mr - Engr
Av. Paulista, 2064
01310 - SAO PAULO

8035 MOURA FILHO, O., Mr - Engr
R. Zaira, 85
01252 - SAO PAULO, SP

3829 * + NAPOLES NETO, A.D.F., Mr - Engr
Consulting Engineer
Caixa Postal 11326 - Pinheiros
01000 - SAO PAULO

13555 NETO, P.P., Mr - Engr
Av. 9, 162
15600 - FERNANDOPOLIS, SP

6693 + NIEBLE, C.M., Mr - Engr
R. Francisco Teixeira Nogueira 128
V. Romana
01000 - SAO PAULO, SP

8025 + NOGAMI, J.S., Mr - Engr
R. Bela Cintra, 321, aptº 33º
01415 - SAO PAULO, SP

6698 OLIVEIRA, A.M.dos S., Dr - Geologist
Caixa Postal 7141
01000 - SAO PAULO, SP

8021 * + OLIVEIRA, H.G. de, Mr - Engr
Consulting Engineer
R. Afonso Celso, 235
03119 - SAO PAULO, SP

8023 PACHECO, I.B., Mr - Engr
Av. Paulista, 2064 - 9º
01310 - SAO PAULO, SP

13561 PASTORE Jr, W.P., Dr - Geologist
Al. Ribeirao-Preto, 515 aptº 133
01331 - SAO PAULO, SP

13565 PAULA, J.R. de, Dr - Geologist
IPT - DMGA - Caixa Postal 7141
01000 - SAO PAULO, SP

3831 * PEREIRA, M.B., Mr - Geologist
R. Senador Simonsen, 54 -
Jardim Botanico
20000 - RIO DE JANEIRO, RJ

8037 POCCIOTTI, P.C., Dr - Geologist
Av. Paulista, 688-17º
01310 - SAO PAULO, SP

8142 + RE, G., Mr - Engr
Al. Ministro Rocha Azevedo, 976 - aptº 61
01410 - SAO PAULO, SP

8017 REN, C., Dr - Geologist
R. Real Grandeza, 219 - Bloco A-7º
20000 - RIO DE JANEIRO, RJ

8028 REDAELLI, L.L., Mr - Engr
Caixa Postal 18463 - Aeroporto
01000 - SAO PAULO, SP

13549 RUZZI, E.E., Mr - Geom.
R. Jacarei, 39 aptº 1412
01319 - SAO PAULO, SP

13558 SATO, S.T., Mr - Engr
R. Joaquim Antunes, 981 aptº 76
05415 - SAO PAULO, SP

6692 + SARAGIOTTO, L.F.R., Dr - Geologist
Caixa Postal 7141 - IPT
01000 - SAO PAULO, SP

13559 + SHAYANI, S., Dr - Geologist
R. Afonso Celso, 235
04119 - SAO PAULO, SP

7521 * SILVEIRA, E.B.S., Prof - Engr
Consulting Engineer
GH Engenharia S.C. Lda.
Av. Brig. Faria Lima 560-11º
01452 - SAO PAULO, SP

7523 + SILVEIRA, J.F.A., Mr - Engr
R. Natingui, 389
05443 - SAO PAULO, SP

8016 SORREGOTTE, C., Dr - Geologist
Av. Paulista, 2064 - 9º
01310 - SAO PAULO, SP

3842 * TEIXEIRA, A.H., Mr - Engr
Consulting Engr
Av. Paulista, 688 - 17º
01310 - SAO PAULO, SP

437 * + VARGAS, M., Prof - Engr
Consulting Engineer
Largo do Arouche, 24 - 7º
01210 - SAO PAULO, SP

8030 * WERNECK, M.L.G., Mr - Engr
R. Antonio Basilio, 434, aptº 102
Tijuca
20000 - RIO DE JANEIRO, RJ

SUPPORTING MEMBERS

7549 CENTRAIS ELECTRICAS DE SAO PAULO S.A. - CESP
Av. Paulista, 2086
01310 - SAO PAULO, SP

7524 ENGEVIX S.A. ESTUDOS E PROJETOS DE ENGENHARIA
R. Senador Pompeu, 46/60 - Caixa Postal 1920
20000 - RIO DE JANEIRO, RJ

7693 EPAMINONDAS MELO DO AMARAL FILHO
C.P. 20 728
R. Funchal 220
Vila Olimpia
SAO PAULO

7525 ESCRITORIO TECNICO J.C. DE FIGUEIREDO FERRAZ LTDa.
Alameda Santos, 1827 - 17º
01419 - SAO PAULO, SP

7967 FURNAS - CENTRAIS ELETRICAS S.A.
Divisao de Biblioteca
R. Real Grandeza, 219 - 2º Bloco B s/201
20000 - RIO DE JANEIRO, RJ - Botafogo

7581 GEOTECNICA S.A.
Av. Paulista, 688 - 17º
01310 - SAO PAULO, SP

7580 HIDROSERVICE ENGENHARIA DE PROJETOS LTDA.
R. Afonso Celso, 235
01419 - SAO PAULO, SP

7819 INSTITUTO DE PESQUISAS TECNOLOGICAS
Caixa Postal 7141 - Cidade Universitaria
01000 - SAO PAULO, SP

7526 PROMON ENGENHARIA S.A.
Av. 9 de Julho, 4939
01407 - SAO PAULO, SP

7330 SONDOTECNICA ENGENHARIA DE SOLOS
Largo dos Leoes 15
RIO DE JANEIRO, RJ

7531 TAMS ENGENHARIA LTDA.
Av. Afonso Pena, 981 - 2º
Caixa Postal 1269
30000 - BELO HORIZONTE, MG

7112 TECNOSOLO ENGENHARIA E TECNOLOGIA DE SOLOS E MAT. S.A.
R. Barao de Sao Felix, 202
20000 - RIO DE JANEIRO, RJ

BULGARIA

Total Number of Corresponding Members - 1

CORRESPONDING MEMBERS

7955 GANEV. I.C., Prof - Engr
blvd Christo Smirnenski 40
1421 SOFIA

CANADA

CANADIAN NATIONAL COMMITTEE ON ROCK MECHANICS

c/o Department of Geological Sciences, University of Saskatchewan, SASKATOON, SASKATCHEWAN S7N OWO

Secretary: Dr M. S. King

Total Number of Ordinary Members - 114
Total Number of Supporting Members - 11

ORDINARY MEMBERS

13444 AMES, D.W., Mr - Geotech. Engr
* 221 Colonial Court
SUDBURY, ONT. P3A 4X3

7591 ARCHAMBAULT, G., Prof.
Dept des Sc. Appliquees
U. of Quebec
930 Est. Rue Jacques-Cartier
CHICOUTIMI, QUE. G7H 2B1

600 AUDY, R., Mr
* Consultant
\+ Piette Audy Bertrand Lemieux
Fugere Leblond
2570 Chemin Ste-Foy
STE-FOY, QUE. G1V 1T9

7531 AZIZ, Z., Mr - Engng Geologist
Abbdl Inc.
85 Ste Catherine W.
MONTREAL, QUE. H2X 3P4

222 BAAR, C.A., Dr
\+ Consultant
5923 Trollinger St N.E.
CALGARY, ALTA. T2K 3W5

6171 BALLIVY, G., Prof.
* Dept of Civil Engineering
Universite de Sherbrooke
SHERBROOKE, QUE. J1K 2R1

2423 BARRON, K., Dr
Manager
Coal Mining Research Centre
9915 - 108 St.
EDMONTON, ALTA. T5K 2C9

13445 BARSOTTI, C., Mr - Mining Engr
Mines Engin. Dept
Inco Metals Ltd
COPPER CLIFF, ONT. POM 1NO

13446 BENSON, R.P., Dr
Consultant
Klohn Leonoff Consultants Ltd
10180 Shellbridge Way
RICHMOND, B.C. V6X 2W7

3239 BIELENSTEIN, H.U., Dr
Research Scientist
CANMET
3303 33rd St. N.W.
CALGARY, ALTA. T2L 2A7

13447 BLACK, G.D., Mr - Geol.Tech.
175 Hilda Ave., Apt 1110
WILLOWDALE, ONT. M2M 1V8

7592 BLAGONA, N., Mr
Consultant
Geotechnical Associates Ltd
P.O. Box 5895
ST JOHN'S, NFLD A1C 5X4

6823 BLOOMFIELD, W.G., Mr - Mining Engr
410-132-21 Macdonald Drive
FORT McMURRAY, ALTA. T9H 4H2

7593 BOURBONNAIS, J., Mr
* Consultant
\+ 201 - 3210 Forest Hill
MONTREAL, QUE. H3V 1C7

7594 BRAY, R.C.E.
Consultant
21 Sydenham Street
KINGSTON, ONT. K7L 3G8

6709 BROWN, J.D., Prof.
* Dept of Civil Engineering
\+ N.S. Tech. College, P.O. Box 1000
HALIFAX, N.S.

6711 CALDER, P.N., Prof.
Dept of Mining Engineering
Queen's University
KINGSTON, ONT. K7M 1E1

7596 CAMPBELL, D.S., Mr - Engr
Chief Engineer
Iron Ore Company of Canada
100 Retty St
SEPT-ILES, QUE. G4R 4L5

13448 CAPELLE, J.F., Mr - Engr
\+ Chief Engineer
Roctest Ltee.
665 Pine
ST LAMBERT, QUE. J4P 2P4

2444 COATES, D.F., Dr
* Director General
\+ CANMET
555 Booth St
OTTAWA, ONT. K1A OG1

6715 COHEN, A., Mr - Engng Geologist
* c/o Ont. Hydro Geotech. Engng Dept
+ 700 University Avenue
TORONTO, ONT. M5G 1X6

2446 COMEAU, W., Mr - Geologist
7580 Milan Blvd.
BROSSARD, QUE.
J4Y 1H5

6202 CRUDEN, D.M., Prof.
* Dept of Civil Engineering
University of Alberta
EDMONTON, ALTA. T6G 2G7

13449 CURRAN, J.H., Prof.
Dept of Civil Engineering
University of Toronto
TORONTO, ONT. M5S 1A4

7585 DAS, B., Dr
Mining Researcher
Western Office-MRL CANMET
3303 33rd St N.W.
CALGARY, ALTA. T2L 2A7

13450 DAVIES, K.G., Mr
Manager
Cominco Ltd
TRAIL, B.C.

13451 DICK, R.C., Mr - Geotech. Engr
* 1558 Burrill Ave
NORTH VANCOUVER, B.C. V7K 1L9

7600 DUFOUR, R., Prof.
Ecole Polytechnique, Dept Genie Min
C.P. 6079 Succ. A
MONTREAL, QUE H3C 3A7

13452 DURAND, M., Prof.
Dept Sciences de la Terre
U. of Que., C.P. 8888 Succ. A
MONTREAL, QUE. H3C 3PB

7601 EMERY, J.J., Prof.
+ Dept of Civil Engineering
McMaster University
HAMILTON, ONT. L8S 4L7

2185 FRANKLIN, J.A., Dr
* Consultant
Franklin Geotechnical Engineering Ltd
58 Guided Court
Rexdale ONTARIO M9V 4K6

13453 GLADWIN, D.B., Mr - Mining Engr
Klohn Leonoff Consultants Ltd
10180 Shellbridge Way
RICHMOND, B.C. V6X 2W7

7605 GOLD, L.W., Dr
* Research Manager
Div. of Building Research
Nat. Res. Council of Canada
OTTAWA, ONT. K1A OR6

13454 GORE, I.W., Mr - Geotechnical Engr
* 28 Delia Court
+ SUDBURY, ONT. P3E 4T8

2470 GRAY, W.M., Dr
* Post Office
TANTALLON, N.S. BOJ 3JO

13455 GREEN, D.H., Mr - Mining Engr
Central Canada Potash Co. Ltd
P.O. Box 1500
COLONSAY, SASK. SOK OZO

13456 HAMBLEY, D.F., Mr - Mining Engr
+ Denison Mines Ltd
Box B 2600
ELLIOT LAKE, ONT. P5A 2K2

3241 HEDLEY, D.G.F., Dr
Research Scientist
Elliot Lake Lab CANMET
P.O. Box 100
ELLIOT LAKE, ONT. P5A 2J6

3242 HERGET, G., Dr
Research Scientist
Elliot Lake Lab CANMET
P.O. Box 100
ELLIOT LAKE, ONT. P5A 2J6

13457 HEYSTEE, R.J., Mr - Geotech. Engr
* 111-260 Wellesley St. E.
+ TORONTO, ONT. M4X 1G6

2256 HOEK, E., Dr
Consultant
Golder Associates Ltd
224 West 8th Ave.
VANCOUVER, B.C. V5Y 1N5

6245 HUCKA, V.J., Prof.
* Mines et Metallurgie
Universite Laval
Cite Universitaire, QUE. G1K 7P4

13458 IMRIE, A.S., Mr - Engng Geologist
* 3816 Puget Drive
+ VANCOUVER, B.C. V6L 2V1

13459 ISAACS, R.M., Mr - Geotech. Engr
+ 14424 Northeast 14th Place
BELLEVUE, WASH. 98007, USA

13460 JEREMIC, M., Prof.
11639 72 Ave
EDMONTON, ALTA. T6G OB9

13461 KAISER, P.K., Prof.
* Dept of Civil Engineering
+ Univ. of Alberta
EDMONTON, ALTA, T6G 2G7

13462 KENNEY, T.C., Prof.
* Dept of Civil Engineering
Univ. of Toronto
TORONTO, ONT. M5S 1A4

2485 KING, M.S., Prof.
* Dept of Geological Sciences
\+ University of Sask.
SASKATOON, SASK. S7N OWO

7612 KIPNIS, N., Mr
Director
Mineral Lands & Mines Division
95 Bonaventure Ave
ST JOHN'S, NFLD A1C 5T7

7613 KRZYWICKI, H.R., Mr
Geotech. Specialist
Fondex Ltee
123 Jean Proulx
HULL, QUE. J8Z 1T4

13464 KWAN, W., Mr - Geotech. Engr
* c/o Alf Seto & Associates Ltd
34 Bentley Ave
OTTAWA, ONT. K2E 6T8

3243 LACHANCE, L., Mr - Engng Geologist
* A/S Tecsult International Ltee
\+ 85 W.St Catherine
MONTREAL, QUE. H2X 3P4

568 LADANYI, B., Prof.
* Ecole Polytechnique
\+ C.P. 6079 Succ.A
MONTREAL, QUE. H3C 3A7

2486 LAJTAI, E.Z., Prof.
* Dept of Geol., Univ. of N.B.
\+ P.O. Box 4400
FREDERICTON, N.B. E3B 5A3

502 LECOMTE, P., Dr
\+ Chef d'etudes
Hydro-Quebec, Place Dubois
855 Est Rue Ste Catherine 17E
MONTREAL, QUE. H2L 4P5

13463 LEE, C.F., Dr
* Manager
\+ Geotech. Engng Dept.
Ontario Hydro
700 University Ave
TORONTO, ONT. M5G 1X6

13464 LETENDRE, R., Mr - Engng Geologist
2612 Rue French
MONTREAL, QUE. H1L 4R9

13465 LOHSE, H., Mr - Geologist
* 1580 Mississauga Valley Blvd.
\+ Unit 2005
MISSISSAUGA, ONT. L5A 3T8

13469 MACKINTOSH, A.D., Mr - Mining Engr
Cominco (Potash) Ltd.
VANSCOY, SASK. SOL 3J0

13470 MAHYERA, A., Mr
Manager
Joy Manufacturing Co. (Canada)
Box 1150
CAMBRIDGE, ONT. N1R 6C9

13471 MANSSAN, N.L., Mr - Engr
Ontario Hydro
700 University Ave H15 F19
TORONTO, ONT. M5G 1X7

13472 MARCH, R.G., Mr
Chief Inspector of Mines
Mineral Lands and Mines Division
49-55 Elizabeth Ave
ST JOHN'S, NFLD A1C 5T7

6186 MASSIERA, M., Prof.
* Dept de Genie
\+ Univ. de Moncton
MONCTON, N.B. E1A 3E9

13466 McCREATH, D.R., Dr
Consultant
Acres Consulting Services Ltd
5259 Dorchester Rd
NIAGARA FALLS, ONT. L2E 6W1

13467 McEWAN, S., Mr - Mining Engr
Potash Corp, Saskatchewan
ROCANVILLE, SASK. SOA 3LO

13468 McKAY D.A., Mr - Engng Geologist
c/o Ontario Hydro
800 Kipling Ave
TORONTO, ONT. M8Z 5S4

7618 MORASH, B.J., Mr
Pit Superintendant
P.O. Box 877
KIRKLAND LAKE, ONT. P2N 3K7

6190 MORIN, J.P., Prof
* Dept de Genie Civil
\+ Universite de Sherbrooke
SHERBROOKE, QUE. J1K 2R1

13473 MRAZ, D., Mr
Manager
IMC (Canada) Ltd
ESTERHAZY, SASK. SOA OXO

7620 NANTEL, J., Prof.
Goodwin Hall
Queen's University
KINGSTON, ONT. K7L 3N6

7621 NGUYEN, D., Dr - Engr
Service Geologie-Geotechnique
Hydro-Quebec
855 Est. Ste-Catherine
MONTREAL, QUE. H2L 4M4

6202 OLIVER, P.H., Mr
Superintendant Rock Mech.
Mines Engineering
Inco Ltd
COPPER CLIFF, ONT. POM 1NO

13474 PAKALNIS, V., Mr - Engr
260 Cedar Street
SUDBURY, ONT. P3B 3X2

13475 PALMER, J.H.L., Dr
* Res. Scientist
Geotech Sec., Div. of Bldg Res.
National Res. Council
OTTAWA, ONT. K1A OR6

7646 PAPANTONOPOULOS, C., Mr - Engr
3-2670 Barclay
MONTREAL, QUE. H3S 1J6

13476 PARKES, D.E., Mr
Res. Tech.
Galt Advanced Development Center
Joy Manufacturing Co. (Canada)
Box 1150
CAMBRIDGE, ONT. N1R 6C9

13477 PASCOLI, L.S., Mr
Contractor
Canadian Longyear Ltd
Box 113
11 Creditstone Rd
Unit 13
CONCORD, ONT. L4K 1B6

6193 PATCHING, T.H., Prof.
Dept of Mineral Engng.
Univ. of Alberta
EDMONTON, ALTA, T6G 2G6

7623 PATTON, F.D., Dr
+ Consultant
2-265 25th St
WEST VANCOUVER, B.C. V7V 4H9

6895 PITEAU, D.R., Dr
* Consultant
+ D.R. Piteau & Associates Ltd
708-100 Kapilano
WEST VANCOUVER, B.C. V7T 1A2

13478 POULIN, B., Mr
President
Labo S.M. Inc.
76 Sud, 12E Ave
SHERBROOKE, QUE. J1G 2V4

13479 POWERS, G.J., Mr
Technologist
Franklin Trow Assoc. Ltd
43 Baywood Rd
REXDALE, ONT. M9V 3Y8

2524 RICE, H.R., Prof.
172 Hanna Road
TORONTO, ONT. M4G 3N8

3245 ROY, M. Prof.
* Dept de Genie Civil
+ Universite Laval
QUEBEC, QUE. G1K 7P4

13480 RUSSELL, D.J., Prof.
Dept of Geology
Univ. of Windsor
WINDSOR. ONT. P9B 3P4

7624 RUSSEL, M.H., Mr
Manager
Union Carbide Canada Ltd
311 - 8925 - 51st Avenue
EDMONTON, ALTA. T6E 5J3

13481 SAGE, R., Mr - Engr
CANMET
555 Booth St
OTTAWA, ONT. K1A OG1

6210 SAMSON, L., Mr
* President
+ Terratech Ltee
275 Benjamin-Hudon
MONTREAL, QUE. H4N 1J1

6194 SCHWARTZ, A., Mr - Mining Engr
Box 338
MANITOUWADGE, ONT. POT 2CO

6195 SCOTT, J.D., Prof.
* Dept of Civil Engineering
+ University of Ottawa
OTTAWA, ONT. K1N 9B4

13482 SEGO, D.C., Prof.
* Dept of Civil Engineering
+ Univ. of Alberta
EDMONTON, ALTA. T6G 2G7

7626 SEYCHUK, J.L., Mr - Engr
* H.Q. Golder & Associates Ltd
+ 3151 Wharton Way
MISSISSAUGA, ONT L4X 2B6

13483 SINGH, K., Dr
Superintendant Mines Engng
73 Normandy Court
SUDBURY, ONT. P3A 2E9

13484 SINH, D.L., Mr - Engr
76-12 Ave. Sud
SHERBROOKE, QUE. J1G 2V4

6726 SMITH, J.D., Mr
Consultant
Smith Engng Assoc. Ltd
110 Lakeshore Blvd.
KINGSTON, ONT. K7M 6R5

2536 SODERLUND, H., Mr
Manager
Algoma Ore Division
Algoma Steel Corp. Ltd
WAWA, ONT. POS 1KO

2538 SOLES, J.A., Dr - Mineralogist
Mineral Sciences Lab CANMET
405 Rochester Street
OTTAWA, ONT. K1A OG1

117 SOLYMAR, Z.V., Mr - Geotech. Engr
285 Mayfair Drive
BEACONSFIELD, QUE. H9W 1S2

13485 STEWART, J.W., Mr
Consultant
Dolmage, Mason & Stewart Ltd
314-355 Burrard Street
VANCOUVER, B.C. V6C 2G8

7629 STIMPSON, B., Prof.
Dept of Mineral Engineering
University of Alberta
EDMONTON, ALTA. T6G 2E2

13486 STIRBYS, A., Mr - Engr
1101 - 2221 Sherobee Road
MISSISSAUGA, ONT. L5A 2H5

7957 * + THOMSON, S., Prof.
Dept of Civil Engineering
University of Alberta
EDMONTON, ALTA. T6G 2G7

13487 + UDD, J.E., Prof.
Dept Min. and Metall. Engineering
McGill University
3480 University St
MONTREAL, QUE. H3A 2A7

7631 * + VANDINE, D.F., Prof.
Dept of Geological Sciences
Queen's University
KINGSTON, ONT. K7L 3N6

13488 VLADUT, T., Mr
4-225 Kensington Ave
MONTREAL, QUE. H3Z 2G9

7958 VONGPAISAL, S., Dr - Engr
Falconbridge Nickel Mines Ltd
Edison Annex
FALCONBRIDGE, ONT. POM 1SO

7633 WALTON, T.R., Mr
Manager
701-240 Duncan Mill Road
DON MILLS, ONT. M3B 3H4

7885 WEIR-JONES, J., Dr
Consultant
Weir-Jones Engng Consultant Ltd
2215 Birch Street
VANCOUVER, B.C. V6H 2S9

7959 * WHITE, W.L., Mr - Geotech. Engr
Nolan White & Associates Ltd
P.O. Box 5453, 215 Blackmarsh Road
ST JOHN'S, NFLD A1C 5W4

7886 WILSON, J.C.
Director
School of Engineering
Laurentian University
SUDBURY, ONT. P3E 2C6

13489 YU, T.R., Dr
Research Engineer
543 Fay Crescent
TIMMINS, ONT. P4N 7K8

2555 YU, Y.S., Dr
Research Scientist
Mining Research Laboratories EMR CANMET
555 Booth Street
OTTAWA, ONT. K1A OG1

2556 ZAHARY, G., Mr
Manager
Elliot Lake Laboratory CANMET
P.O. Box 100
ELLIOT LAKE, ONT. P5A 2J6

SUPPORTING MEMBERS

147 ACRES CONSULTING SERVICES LTD
5259 Dorchester Rd
P.O. Box 1001
NIAGARA FALLS, ONT. L2E 6W1

13490 ALGOMA ORE DIVISION
Algoma Steel Corporation Ltd
WAWA, ONT. POS 1KO

6846 CANADIAN LONGYEAR LTD
Box 113
11 Creditstone Rd
Unit 13
CONCORD, ONT. L4K 1B6

13491 COMINCO LTD
TRAIL, B.C.

7707 D.R. PITEAU & ASSOCIATES Ltd
708 - 100 Kapilano
WEST VANCOUVER, B.C. V7T 1A2

13492 FRANKLIN TROW ASSOCIATES LTD
43 Baywood Road
REXDALE, ONT. M9V 3Y8

13493 GEOTECHNICAL ASSOCIATES Ltd
P.O. Box 5893
ST JOHN'S, NFLD A1C 5X4

13494 GOLDER ASSOCIATES LTD
224 West 8th Ave
VANCOUVER, B.C. V5Y 1N5

7638 MINES GASPE
MURDOCHVILLE, QUE.

6587 THE LIBRARY, SERIALS SECTION
Carleton University
Colonel By Drive
OTTAWA 1

7641 WEIR-JONES ENGIN. CONSULT. LTD
2215 Birch Street
VANCOUVER, B.C. V6H 2S9

CHILE

SOCIEDAD CHILENA DE MECANICA DE SUELOS E INGENIERIA DE FUNDACIONES (SOCHIMSYF)

Instituto de Ingenieros de Chile, San Martin 352, Casilla 487, SANTIAGO 1

President: Mr. E. Rowe
Secretary: Mr. J. H. Troncoso

Total Number of Ordinary Members -- unknown
Total number of Corresponding Members - 6

CORRESPONDING MEMBERS

10688 CABELLO C., J., Mr - Civil Engr
Researcher CIMM
Casilla 170, Correo 10
SANTIAGO

10689 HERRERA, C., Mr -- Mining Engr
Researcher CIMM
Casilla 170, Correo 10
SANTIAGO

10595 KRSTULOVIC, G., Prof.- Mining Engr
U. Prof., Rock Mech. Div. Head CIMM
Centro de Investigacion Minera y Metalurgica
Avda. Parque Institucional 6500
Casilla 170
SANTIAGO 10

10691 MARCHANT, A., Mr - Mining Engr
Division Head
Codelco, Division Chuquicamata
CHILE

10690 MORALES, R., Mr - Mining and Civil Engr
Researcher CIMM
Casilla 170, Correo 10
SANTIAGO

10692 PALMA C.,R., Mr - Mining Engr
Mining Div. Head
Codelco, Division Salvador
Mina Salvador
CHILE

CHINA

CHINESE NATIONAL GROUP FOR ROCK MECHANICS

Institute of Geophysics, Academia Sinica, BEIJING

Chairman: Prof. Tan Tjong-kie

Total Number of Ordinary Members - 10
Total Number of Corresponding Members - 2

ORDINARY MEMBERS

10708 ELL Bingjun, Mr - Engr
The Water Conservancy and Hydroelectric Power Scientific Research Institute
BEIJING

10700 GU Dezhen, Prof.
U. Prof.
Institute of Geology, Academia Sinica
BEIJING

10704 KANG Wenfa, Mr
Assist Prof.
Institute of Geophysics, Academia Sinica
BEIJING

10706 LIU Baoshen, Dr
Assoc. Prof.
Changsha Institute of Mining and Metallurgy
Ministry of Metallurgy Industry
CHANGSHA

10707 NIU Xizhou, Mr - Engr
Deputy Director
The Mining Research Institute
Academy of Scientific Research of Coal Mining Industry
BEIJING

10699 TAN Tjong-kie, Prof.
U.Prof., Scientific Director
Institute of Geophysics, Academia Sinica
BEIJING

10703 TSAO Loan, Mr - Engr
Deputy Chief Engr
Yangtze Valley Planning Office
WUHAN

10701 WANG Sijing, Mr
Assoc. Prof.
Institute of Geology, Academia Sinica
BEIJING

10702 ZHU Keshan, Prof.
U. Prof.
The Chongqing Architectural Institute
CHONGQING

10705 ZHU Weishen, Dr
Assoc.Prof., Deputy Director
Institute of Rock and Soil Mechanics, Academia Sinica
WUHAN

CORRESPONDING MEMBERS

13197 HUNG Ju-Jiang, Prof.
National Taiwan University
Department of Civil Engineering
TAIPEI, TAIWAN

10632 LING Tse-chi
Power Development Department
Taiwan Power Company
76 Hoping East Road Section 1
TAIPEI, TAIWAN

COSTA RICA

Total Number of Corresponding Members - 1

CORRESPONDING MEMBERS

8141 LAPORTE M., G., Prof. - Civil Engr
Oficina Control de Materiales
Instituto Costarricense de Electricidad
Apartado 10032
SAN JOSE

CZECHOSLOVAKIA

NARODNI SKUPINA PRO MECHANIKU HORNIN

Federalni ministerstvo paliv a energetiky odbor mezinarodni spoluprace
Vinohradska 8, 120 70 PRAHA 2

Chairman: Prof. L. Siska
Secretary: Mr A. Skrabis

Total Number of Ordinary Members - 30

11055 DLOUHY, J., Mr - Engr
Vedeckovyzkumny uhelny ustav
716 07 OSTRAVA-RADVANICE

13541 DOLEZALOVA, M., Mr. - Engr
Hydroprojekt
Taborska 31, p.p. 10
140 43 PRAHA 4 - NUSLE

13542 DOLEZEL, V., Mr - Engr
Ustav geologie a geotechniky CSAV
V Holesovickach 41
182 09 PRAHA 8 - LIBEN

6691 * DROZD, K., Mr - Engr
Stavebni geologie
Gorkeho nam. 7
113 09 PRAHA 1

3226 * DVORAK, A., Dr - Engr
Stavebni geologie
Gorkeho nam. 7
113 09 PRAHA 1

11061 * FEDA, J., Mr - Engr
UTAM CSAV
Vysehradska 49
128 49 PRAHA 2

11057 FIRLA, V., Mr - Engr
Generalni reditelstvi OKD, koncern
nam. Rijnove revoluce 6
728 30 OSTRAVA 1

11072 * GROMMA, B., Mr - Engr
IGHP, organizace pro inzenyrskou
geologi a hydrogeologii
Rajecka cesta
010 01 ZILINA

11067 HATALA, J., Mr - Engr
Vysoka skola technika
Svermova 5
040 01 KOSICE

11063 HAVIR, J., Mr - Engr
Vychodoceske uhelne doly, k.p.
542 34 MALE SVATONOVICE, OKR.TRUTNOV

11071 HUCKO, A., Mr - Engr
Uhelne a lignitove bane, koncorn
K.I. Savina 10
971 71 BRIEVIDZA

11054 JANAS, P., Mr - Engr
Vedeckovyzkumny uhelny ustav
716 07 OSTRAVA-RADVANICE

11069 KOCI, M., Mr - Engr
Banicky ustav SAV Kosice
Solovjevova 45
040 00 KOSICE

13543 KONECNY, P., Mr - Engr
Ustav geologie a geotechniky CSAV
Hladnovska 9
710 00 OSTRAVA 2

7087 + MATULA, M., Prof. - Engr
Prirodovedecka fakulta UK
Zadunajska 15
886 02 BRATISLAVA

11062 NEVLUD, J., Mr - Engr
OKR Dulni pruzkum a bezpecnost, k.u.o.
739 21 PASKOV

11059 + NOVOSAD, S., Mr - Engr
Geotest, k.p.
Gottwaldova 2036
709 00 OSTRAVA-MARIANSKE HORY

13544 OBORNIK, M., Mr - Engr
Banske projekty, k.u.o.
Pri starej pracharni 10
894 14 BRATISLAVA

11065 PICHLER, Mr - Engr
Vyzkumny ustav pro hnede uhli SHD
Budovatelu 2830
434 37 MOST

11060 PROCHAZKA, J., Mr - Engr
Geotest, k.p.
trida kapitana Jarose 28
600 00 BRNO

13545 RUZICKA, J., Mr - Engr
Uranove doly Pribram, k.p.
Konetopy
261 13 PRIBRAM

11068 SEKULA, F., Mr - Engr
Banicky ustav SAV Kosice
Solovjevova 45
040 00 KOSICE

492 SISKA, L., Prof.- Engr
Vysoka skola banska
trida Vitezneho unora
708 00 OSTRAVA-PORUBA

11058 SKALA, V., Mr - Engr
Ustav geotechniky a geologie CSAV
V Holesovickach 41
180 00 PRAHA 8-LIBEN

11053 SKRABIS, A., Mr - Engr
Urad predsednictva vlady CSSR
nabrezi kapitana Jarose
125 09 PRAHA - MALA STRANA

11064 STANEK, J., Mr - Engr
Vyzkumny ustav pro hnede uhli SHD
Budovatelu 2830
434 37 MOST

11070 TAVODA, O., Prof. - Engr
Prirodovedecka fakulta UK
Gottwaldovo nam. 19
886 02 BRATISLAVA

11066 TRANCIK, P., Mr - Engr
Vysoka skola technika
Svermova 5
040 01 KOSICE

11056 VAVRO, M., Prof.- Engr
Vysoka skola banska
trida Vitezneho unora
708 00 OSTRAVA-PORUBA

3225 ZAJIC, J., Mr - Engr
Stavebni geologie
Gorkeho nam. 7
113 09 PRAHA 1

ECUADOR

SOCIEDAD ECUATORIANA DE MECANICA DE SUELOS Y ROCAS

Chile 409, Casilla 3442, GUAYAQUIL

President: Mr. A. Silva Sanchez
Secretary: Mr. L. Palomeque Freile

Total Number of Ordinary Members - 48

ORDINARY MEMBERS

13310 AGUIRRE M.,C., Mr - Civil Engr
Pilotes Franki S.A. Km. 1 1/2
Av. C.J. Arosemena
GUAYAQUIL

13311 AGUIRRE R.,H., Mr - Consulting Engr
P. Moncayo 2602
GUAYAQUIL

11079 ALCIVAR A.,J., Mr - Civil Engr
Padre Solano 1317
GUAYAQUIL

13312 ANDRADE J.,N., Mr - Civil Engr
L. de Garaycoa 1709
GUAYAQUIL

13313 ANDRADE N.,C., Mr - Civil Engr
Cdla. Alborada Mz. CC V-3 (3ra. Etapa)
GUAYAQUIL

13314 ARMAS F.,T., Mr - Civil Engr
Av. C.J. Arosemena, Km. 2 1/2
GUAYAQUIL

10191 AYON J.,H., Prof. - Engng Geologist
Professor
Brasil 905 y L. de Garaycoa
GUAYAQUIL

13315 BLUM G., E., Prof. - Consulting Engr
Professor
Cuenca 426
GUAYAQUIL

10194 * CAMPOSANO L.,J., Mr - Civil Engr
Rumichaca 828
GUAYAQUIL

10195 CASARES C.,M., Mr - Civil Engr
Boyaca 642
GUAYAQUIL

10196 * CASTRO A.,T., Mr - Consulting Engr
Balsamo 707 (Urdesa)
GUAYAQUIL

13316 CASTRO V.,P., Mr - Civil Engr
CEDEGE
P. Icaza y Pedro Carbo
GUAYAQUIL

11083 CHAVEZ M.,M., Prof. - Engng Geologist
Professor
CEDEGE
P. Icaza 220
GUAYAQUIL

11082 CORAL N.,E., - Civil Engr
Cdla. Miraflores Calle 3ra.
No. 419 y Av. Miraflores
GUAYAQUIL

10199 * CORONEL J.,R., Mr - Civil Engr
Boyaca 1122
GUAYAQUIL

13317 CRUZ C.,B., Mr - Civil Engr
CEDEGE
Pedro Carbo y P. Icaza
GUAYAQUIL

13318 ECHEVERRIA C.,C., Mr - Civil Engr
G. Cordova 504
GUAYAQUIL

10202 * ESPINDOLA R., Mr - Civil Engr
Casilla 4610
GUAYAQUIL

13319 FORAY P., Prof. - Engr
Professor
Tamayo 943
QUITO

13320 GONZALEZ A.,R., Mr - Civil Engr
Machala 2110 y Manabi
GUAYAQUIL

10203 * GUZMAN F.,G., Mr - Civil Engr
Chimborazo 413, ler
Piso alto
GUAYAQUIL

11086 HERBOZO A.,E., Mr - Civil Engr
Professor
INERHI, Escobedo 914
GUAYAQUIL

13321 ICAZA E.,A., Mr - Civil Engr
Andres Marin 318
GUAYAQUIL

13322 LARA M.,O., Mr - Consulting Engr
Professor
Av. del Ejercito 303
GUAYAQUIL

13323 LASSO M.,E., Mr - Civil Engr
P. Aguirre 456
GUAYAQUIL

10206 LOOR L.,M., Mr - Consulting Engr
* Tulcan 1102 y Luque
GUAYAQUIL

10208 MANCHENO O.,A., Mr - Civil Engr
* Eng. Constructor
Orellana 211
GUAYAQUIL

11089 MANRIQUE P.,E., Mr - Civil Engr
Pilotes Franki S.A.
Av. C.J. Arosemena Km. 1 1/2
GUAYAQUIL

10209 MARIN N.,L., Prof. - Consulting Engr
* Professor
Colon 404 y Chile
GUAYAQUIL

10210 MARTIN V.,C., Mr - Civil Engr
* Yaguachi (14 ava) 1718 y Portete
GUAYAQUIL

10211 MARURI D.,R., Mr - Consulting Engr
* CIMENTACIONES S.A.
Av. Juan Tanca Marengo Km. 6,
GUAYAQUIL

10214 MEDINA C.,M., Prof. - Consulting Engr
* Professor
CEDEGE
Pedro Carbo y P. Icaza
GUAYAQUIL

10212 MENOSCAL V.,R., Mr - Consulting Engr
* Luis Urdanete 1600-A,
GUAYAQUIL

10213 MORENO L.,V., Prof. - Consulting Engr
Professor
Cdla las Acacias Mz. B-3 V-10
GUAYAQUIL

11091 MOYA C.,R., Mr - Civil Engr
INECEL, Cdla Kennedy Calle H ≠ 507-A
entre la Calle 4ta. y 5ta. Oeste
GUAYAQUIL

13324 NIAMA Z.,C., Mr - Consulting Engr
Pedro Moncayo 135 (Barrio El Dorado)
QUITO

10216 NUQUES C.,J., Mr - Consulting Engr
* Av. C.J. Arosemena Km. 2 1/2
GUAYAQUIL

11094 PESANTES P.,J., Mr - Civil Engr
Tulcan 1102 y Luque
GUAYAQUIL

13325 QUINONEZ C.,L., Mr - Civil Engr
Cdla. Los Esteros Mz. 16-A2 V-66
GUAYAQUIL

10220 RENALDI R.,C., Mr - Civil Engr
* Balsamo 400 (Urdesa)
GUAYAQUIL

11099 SERRANO C.,R., Mr - Civil Engr
INECEL, 9 de Octubre y P. Carbo
GUAYAQUIL

12000 SILVA S.,A., Prof. - Civil Engr
Professor
INERHI Escobedo 914
GUAYAQUIL

13326 SORIANO Z.,J., Mr - Consulting Engr
Cdla. La Pradera 1, Mz. D-6 V-2
GUAYAQUIL

13327 TERREROS C. de Ve., Mrs - Civil Engr
* Cdla Nueva Kennedy
Calle 7ma. 103 entre la D y la E
GUAYAQUIL

12003 VELASCO F.,A., Mr - Civil Engr
INECEL
Cdla. Kennedy Calle H
≠ 507-A entre la Calle 4ta. y 5ta.
- Oeste
GUAYAQUIL

10229 YAMBAY R.,M., Mr - Civil Engr
* Machala 2222 y Huancavilca
GUAYAQUIL

10231 YEPEZ C.,G., Mr - Consulting Engr
* Dolores Sucre 609 y Oriente
GUAYAQUIL

10230 YUNES F.,S., Mr - Civil Engr
* Calderon 103 y Malecon
Edf. El Libertador
GUAYAQUIL

FINLAND

SUOMEN KALLIOMEKANIIKKATOIMIKUNTA

Teknillinen Korkeakoulu, Vuoriteollisuusosasto, Louhintatekniikan Laboratorio,
02150 ESPOO 15

Chairman: Prof. R. Matikainen
Secretary: Dr P. Särkkä

Total Number of Ordinary Members - 71

ORDINARY MEMBERS

6598 AARREVAARA, E., Mr - Consultant
* Pulmusuontie 45
20360 TURKU 36

6599 ANTTIKOSKI, U., Mr - Civil Engr
* Helsingin kaupungin Geoteknillinen toimisto
Yrjönkatu 21 B A
00100 HELSINKI 10

7533 AULANKO, H., Mr - Mining Engr
Vuoriharjuntie 35
02320 ESPOO 32

6642 BÄCKSTRÖM, C-F., Mr - Mining Engr
08700 VIRKKALA

6643 EDELMAN, N., Prof
U. Prof
Abo Akademi
20500 TURKU 50

6601 EKLUND, P., Mr - Civil Engr
* Forsellesintie 20 D
02700 KAUNIAINEN

6645 GAAL, G., Dr
U. ass. Prof.
Niipperintie 86
02970 ESPOO 97

6646 GARDEMEISTER, R., Dr - Engng Geologist
* Viherkallionkuja 3 F 40
\+ 02710 VIHERLAAKSO

6603 HAAHTI, K., Mr - Mining Engr
Karl Forsström Oy
25640 FÖRBY

3821 HAKALEHTO, K., Dr
Manager
Kahmankaari 1 A 3
33310 TAMPERE 31

6604 HAKAPÄÄ, A., Mr - Mining Engr
71470 ORAVIKOSKI

6605 HAKOLA, A., Mr - Mining Engr
Outokumpu Oy
83500 OUTOKUMPU

6606 HEIKKILÄ, I., Mr - Mining Engr
\+ Rautaruukki Oy
88200 OTANMÄKI

6607 HERMONEN, O., Mr
Manager
Tarmontie 40 as 2
90530 OULU 53

7534 HINTIKKA, P-J., Mr - Mining Engr
* Tornitaso 7 A 41
02120 ESPOO 12

6608 HOLKKO, J., Mr - Civil Engr
* Metsätontunkuja 3 E 20
02200 ESPOO 20

7535 HOLM, C., Mr
Manager
Tallvägen 1 C 51
10210 INKOO

13045 HOLOPAINEN, P., Mr - Civil Engr
Kolmaslinja 31 A 16
00530 HELSINKI 53

6816 JAKSOLA, V., Mr - Consultant
Parkatti-Salonen Ky
Töölönkatu 3
00100 HELSINKI 10

6647 JOHANSSON, S., Mr - Engng Geologist
* Valimontie 23 E
\+ 00380 HELSINKI 38

6817 JUHOLA, M.O., Mr
* U. ass. Prof.
Lounaisväylä 2 A 4
00200 HELSINKI 20

13046 JÄÄSKELÄINEN, H., Mr - Civil Engr
* Geotek Oy
Valimontie 23
00380 HELSINKI 38

13047 KAJAN, M., Mr - Mining Engr
Harjukatu 3 C 51
00500 HELSINKI 50

4233 KALLA, J., Mr
* Manager
\+ Vyökatu 2 A
00160 HELSINKI 16

6609 KAURANNE, K., Prof.-Geologist
* Satukuja 1 F 35
\+ 02230 ESPOO 23

6648 KOIVIKKO, L., Mr - Mining Engr
58220 LOUHI

13048 KOPPINEN, J., Mr - Civil Engr
Soukankuja 8 A 49
02360 ESPOO 36

6650 KORPELA, K., Prof.
* U.Prof
\+ Kalkkipaadentie 2 G
00340 HELSINKI 34

6610 KUJANPÄÄ, J., Mr - Geologist
Outokumpu Oy
Kemin kaivos
94100 KEMI 10

6611 KUPIAS, P., Mr - Mining Engr
Outokumpu Oy
83500 OUTOKUMPU

7536 KURPPA, R., Mr
Manager
Kuunsäde 10 C 73
02210 ESPOO 21

13049 LAHTINEN, R.A., Mr - Civil Engr
* YIT
Ratamestarinkatu 12
00520 HELSINKI 52

6649 LAINE, T.K., Mr - Civil Engr
Haukiverkko 13 C 9
02170 ESPOO 17

13050 LAPPALAINEN, P., Mr - Engng Geologist
YIT
Ratamestarinkatu 12
00520 HELSINKI 52

6613 LAPPALAINEN, V., Dr
* U. ass. Prof.
\+ Kotitontuntie 17 C
02200 ESPOO 20

13051 LINDHOLM, J., Mr - Civil Engr
Rakennushallitus
PL 333
00530 HELSINKI 53

6614 LINDHOLM, O., Mr - Geologist
Pihatie 16
90650 OULO 65

1967 MAIJALA, P., Prof.
U. Prof.
Teknillinen korkeakoulu
Vuoriteollisuusosasto
02150 ESPOO 15

6615 MANUNEN, T., Mr - Mining Engr
Outokumpu Oy
86900 PYHÄKUMPU

6616 MARKKANEN, P., Mr - Geologist
* Aarnivalkeantie 2 D
\+ 02100 ESPOO 10

1966 MATIKAINEN, R., Mr - Mining Engr
\+ Kuhatie 23 A
02170 ESPOO 17

4058 MUSTALA, J., Mr - Geologist
\+ Outokumpu Oy
PL 27
02201 ESPOO 20

4148 NATUKKA, A., Mr - Civil Engr
* Vaisalantie 10 A
02130 ESPOO 13

6617 NIINI, H., Dr - Geologist
* Koukkusaarentie 7 C 329
\+ 00980 HELSINKI 98

13052 PARTIO, E., Mr - Geologist
\+ Maa ja Vesi Oy
Itälahdenkatu 2
00210 HELSINKI 29

13053 PATRIKAINEN, P., Mr - Geologist
* VTT
\+ Geoteknillinen osasto
02150 ESPOO 15

4059 PELTOLA, E., Dr - Geologist
Outokumpu Oy
83500 OUTOKUMPU

6618 PIHKO, E., Mr
Manager
Kuparipera 6
83500 OUTOKUMPU

7537 PUSKALA, R., Mr - Civil Engr
* YIT
Ratamestarinkatu 12
00520 HELSINKI 52

13054 RAUDASMAA, P., Mr - Geologist
\+ Laajakorvenkuja 2 as 24
01620 VANTAA 62

13055 RIEKKOLA, R., Mr - Civil Engr
Kimmeltie 26 D 31
02100 ESPOO 10

4149 RUOPPA, A., Mr
* Manager
Katajaharjuntie 22 G
00200 HELSINKI 20

7538 RÄSÄNEN, T., Mr - Civil Engr
Maa ja Vesi Oy
Itälahdenkatu 2
00210 HELSINKI 21

7539 RÖNKKÖ, K., Mr - Geologist
+ Eliaksentie 11 as 24
87200 KAJAANI 20

6615 SAANIO, V., Mr - Civil Engr
Vuorikummuntie 12 A
00390 HELSINKI 39

7540 SAARI, K., Mr - Civil Engr
* VTT
Geotekniikan laboratorio
02150 ESPOO 15

4060 SALO, U., Mr - Mining Engr
Vaahtokuja 5 E 57
01600 VANTAA 60

6619 SALONEN, A., Mr
Consultant
Kivikkotie 8-10
01560 MAANTIEKYLÄ

13056 SAMMALISTO, J., Mr - Mining Engr
Outokumpu Oy
KTR
83500 OUTOKUMPU

4147 SOLOVJEW, N., Mr
* Manager
Katajaharjuntie 20 A 3
00200 HELSINKI 20

6620 SUOMINEN, V., Mr - Geologist
Geologinen tutkimuslaitos
02150 ESPOO 15

6621 SÄRKKÄ, P., Dr - Mining Engr
Teknillinen korkeakoulu
Vuoriteollisuusosasto
02150 ESPOO 15

6622 TOIVONEN, K., Mr - Civil Engr
Otakallio 4 B 15
02150 ESPOO 15

6623 TUISKU, T., Mr - Civil Engr
* Ehrensvärdintie 5 A
00150 HELSINKI 15

4061 ULVELIN, E., Mr - Mining Engr
Impivaarantie 2 B 8
02880 VEIKKOLA

6624 UUSINOKA, R., Dr - Geologist
+ Putousrinne 1 D 27
01600 VANTAA 60

6625 VALTAKARI, U., Mr
Manager
Paraisten Kalkki Oy
21600 PARAINEN

6626 VILLIKKA, J., Mr - Mining Engr
Huhtiniemenkatu 30
53810 LAPPEENRANTA 81

6627 VORNANEN, E., Mr
Manager
Rautaruukki Oy
88200 OTANMÄKI

6652 VUOLIO, R., Mr
+ Manager
Oy Finnrock Ab
Lapinlahdenkatu 21
00100 HELSINKI

13057 VÄHÄSARJA, P., Mr - Engng Geologist
* Honkavaarankuja 1 L 82
+ 02710 ESPOO 71

FRANCE

COMITE FRANCAIS DE MECANIQUE DES ROCHES

Ecole Nationale Superieure des Mines, 60 Boulevard Saint Michel, 75272 PARIS cedex 06

President: Prof. E. Tincelin
Secretary: Dr C. Louis

Total Number of Ordinary Members - 213
Total Number of Supporting Members - 31

ORDINARY MEMBERS

13494 AHU, J.
Bureau VERITAS
2, Avenue Hoche
74008 PARIS

7262 * ALLARD, P.
CETE Division Laboratoire
Zone Industrielle
13290 LES MILLES

13495 ANDRE JEHAN, R.
6 rue de l'Ancienne Comedie
75006 PARIS

7263 * ARATHOON
I.T.B.T.P.
9, rue de La Perouse
75016 PARIS

4208 * ARNOULD, M.
Ecole des Mines
60 Boulevard Saint Michel
75272 PARIS cedex 06

7264 d'ARRAS, A.
Societe USINOR
B.P. 4-177
59307 VALENCIENNES

1144 * BARBEDETTE
36, Boulevard de la Republique
92100 BOULOGNE BILLANCOURT

13496 BARDES
S.N.C.F.
Department des Ouvrages d'Art
122, rue des Poissonniers
76876 PARIS cedex 18

1145 * BARON, G.
8, Avenue des Pommerots
78400 CHATOU

3250 * BARREYRE, M.
Societe Nat. des Petroles d'Aquitaine
Centre de recherches
Allees de Morlaas
64000 PAU

1148 * BEAUJOINT
E.D.F. - D.T.G.
37, rue Diderot
38000 GRENOBLE

1149 BEAUMONT, C.
MINATOME S.A.
69-73 rue Dutot
75738 PARIS cedex 15

7266 BELIN
CERCHAR
Chef du departement d'etudes Mine
B.P. 2
60550 VERNEUIL EN HALATTE

1150 BELLIER, J.
5 rue de Villeneuve
92380 GARCHES

13497 BEREST, P.
Directeur du Sous-Sol
Ministere de L'Industrie
101 rue de Grenelle
75700

1152 * + BERNEDE, J.
10 rue des Dardanelles
75017 PARIS

1153 * BERTHIER
C.E.S.T.
4, rue Wilfried Kilian
38000 GRENOBLE

1154 * BERTHIER, J.
Direction SETRA
46, Avenue Aristide Briand
92223 BAGNEUX

7269 * BIAREZ, J.
Ecole Centrale des Arts et Manufactures
Grande Voie des Vignes
92290 CHATENAY MALABRY

13498 * BOEHLER, J.P.
Institut de Mecanique de Grenoble
B.P. 53 - Centre de Tri
38041 GRENOBLE

1155 BOLELLI, E.
* B.R.G.M.
+ Chambre regionale de commerce et d'industrie
3 Boulevard de Chezy
B.P. 667
35008 RENNES cedex

1156 BOLLE, G.
* CITRA
13, Avenue Morane Saulnier
78140 VELIZY VILLACOUBLAY

1157 BOLLO, M.
* 106, Boulevard Pereire
75017 PARIS

7271 BONIN
* Ciments Lafarge
B.P. 8
07220 VIVIERS

1158 BORDET, C.
* E.D.F. - Service geologique et geotechnique
+ 3, rue de Messine
75008 PARIS

13499 BOUCHER, M.
Gaz de France
23, rue Philibert Delorme
75840 PARIS cedex 17

13500 BOUCLY, P.
Gaz de France
Centre de recherches gazieres
361, Avenue du President Wilson
B.P. 33
93212 LA PLAINE SAINT DENIS

7272 BOURGUIGNE, M.
* Cabinet d'Expertises Techniques
B.P. 60
30104 ALES cedex

13501 BOUVEL
Soletanche Entreprise
6 rue de Watford
92005 NANTERRE Cedex

7273 BRISAC
* Total Exploration Production
39-43, quai Andre Citroen
75739 PARIS Cedex 15

7274 BRUGMANN, B.J.
* Beethoven str. 11
CH-3073 GUMLIGNE/BE (SUISSE)

1160 CAMBEFORT, H.
* 112 ter rue Marcadet
75018 PARIS

1161 CAPDECOMME, L.
* Laboratoire de Mineralogie
+ 39, Allees Jules Guesde
31400 TOULOUSE

3453 CARPENTIER, L.
* S.N.C.F.
1, rue Olchanski
75016 PARIS

7276 CASTANY, G.
* 3, Residence du Petit Chambord
+ 92340 BOURG LA REINE

7277 CAUSERO, L.
* Laboratoire regional des Ponts et Chaussees
+ B.P. 8
Rue Grande Haie
54510 TOMBLAINE

1163 CHADEISSON, R.
* Soletanche
6 rue de Watford
B.P. 11
92005 NANTERRE cedex

7821 CHAMBON, C.
* Ecole des Mines
Parc de Saurupt
54000 NANCY

4210 CHAMBRON
* Chef de la Division des Ouvrages d'Art de la S.N.C.F.
122, rue des Poissonniers
75876 PARIS cedex 18

3254 CHAMPETIER de RIBES
* L.C.P.C.
+ 58, Boulevard Lefebvre
75732 PARIS cedex 15

13502 CHARPENTIER, J.P.
Laboratoire de Mecaniques des Solides
Ecole Polytechnique
Route de Saclay
91128 PALAISEAU cedex

7822 CHAYE d'ALBISSIN
* Faculte des Sciences
Laboratoire de Geol. dynamique
Tour 26
11 Quai Saint Bernard
75005 PARIS

13503 CHEYLAN, G.
Cie Nationale d'amenagement de la region du Bas Rhone et du Languedoc
685 routes d'ARles
30001 NIMES cedex

1165 CHOLET, H.
* 10 Allee des Lilas
Domaine de Grandchamp
78230 LE PECQ

4211 CLERC RENAUD, A.
* Geostock
Tour Aurore
92080 PARIS defense

7280 COLOMBET, G.
Coyne et Bellier
5 rue d'Heliopolis
75017 PARIS

13570 COMBE, R.
* Coyne et Bellier
5 rue d'Heliopolis
75017 PARIS

13504 COME, B.
* Societe Geostock
Tour Aurore - Cedex 05
92080 PARIS la Defense

1167 COMES, G.
* E.D.F.
Service geologique et geotechnique
3 rue de Messine
75008 PARIS

13505 CORNET, F.-H
Institut de Physique du Globe
Tour 14
4, Place Jussieu
75230 PARIS cedex 05

3257 DAGNAUX, J.-P.
* C.P.G.F.
77, Avenue Victor Hugo
92500 RUEIL MALMAISON

13506 DARCY, J.-P.
* C.N.E.T.U.
B.P. 48
109, Avenue de S. Allende
69672 BRON

3258 DAVOINE
* Ecole des Mines
158 bis, cours Fauriel
42100 SAINT ETIENNE

1169 DAYRE, M.
* Institut de recherches interdis-
ciplinaires en Geologie Mecanique
(I.R.I.G.M.)
B.P. 53 - Centre de Tri
38041 GRENOBLE cedex

13507 DEGAND
C.E.A. - Centre de Marcoule
Documentation
B.P. 106
30200 BAGNOLS SUR CEZE

4213 DEJEAN, M.
* Laboratoire de Mecaniques des Terrains
Ecole des Mines
Parc de Saurupt
54000 NANCY

13508 DELMER
* 13 rue de Jenner
1040 BRUXELLES (BELGIQUE)

7282 DELOS, J.-P
* Sce Reservoir Engineering
39-43 quai Andre Citroen
75739 PARIS cedex 15

1170 DEMOUY, F.
* 26, rue Dagorno
75012 PARIS

3260 DERLICH
* C.E.A. - D.A.M.
Service du Personnel
B.P. 1
78140 VILLACOUBLAY

4214 DESSENNE, J.-L.
* B.R.G.M.
Domaine de Luminy
Route Leon Lachamp
13009 MARSEILLE

7283 DUBOIS, D.
* Geostock
Tour Aurore
92080 PARIS defense

13509 DUC
Laboratoire de Mecanique des Solides
Ecole Polytechnique
Route de Saclay
91128 PALISEAU cedex

7284 DUCHESNE, M.
* Ecole des Mines
60, Boulevard Saint Michel
75272 PARIS cedex 06

3262 DUCLAUX, F.
* 15, rue de Villiers
91880 BOUVILLE

1172 DUFFAUT, P.
* 130, rue de Rennes
\+ 75006 PARIS

7285 DUNGLAS, J.
* Parc de Tourvoie
92160 ANTONY

1173 DUPEUBLE
S.I.F.
* Directeur du Bureau d'Etudes
11, Avenue du Colonel Bonnet
75016 PARIS

13510 ESTEULLE, F.
Coyne et Bellier
5, rue d'Heliopolis
75017 PARIS

13511 FABRE, D.
* I.R.I.G.M.
\+ B.P. 53 - Centre de Tri
38041 GRENOBLE cedex

1174 FARRAN, J.
292, rue Saint Martin
75141 PARIS cedex

13512 FAVAT
32 rue du Lac
La Clairiere
78120 RAMBOUILLET

7287 FENOUX, G.
* Soletanche Entreprise
6 rue de Watford
B.P. 511
92005 NANTERRE cedex

7288 FERRIEUX, H.
* 12, rue de la Paroisse
78000 VERSAILLES

7289 FILLIAT, G.
* Professeur
\+ 4, rue de Cice
75006 PARIS

1176 FINE, J.
* Chef du C.E.M.R.
Ecole des Mines
60 Boulevard Saint Michel
75272 PARIS cedex 06

7290 FOURMAINTRAUX, D.
* L.C.P.C.
\+ 58, Boulevard Lefebvre
75732 PARIS cedex 15

4217 FOURNIER, G.
* 53, rue Georges Pezieres
\+ 66000 PERPIGNAN

13513 FRANCK, J.
* Coyne et Bellier
5, rue d'Heliopolis
75017 PARIS

1181 GARNIER, J.-C.
* E.D.F. - SEPTEN
Tour EDF-GDF
Cedex nº 8
92080 PARIS la Defense

7291 GAUDIN, B.
* CETE
Zone Industrielle
13290 LES MILLES

7824 GAUTIER, J.B.
* 53 rue Ganneron
75018 PARIS

13514 GAZEL, J.
B.R.G.M. - D.S.G.N.
B.P. 6009
45018 ORLEANS cedex

4218 GIROUD, J.-P.
* Laboratoire de Mécanique des Sols
Universite de Grenoble
Cedex 53
38041 GRENOBLE

4219 GLARD, Y.
* Camp de Menthe
Petite route des Milles
13100 AIX EN PROVENCE

1182 GOGUEL, J.
* 100 rue du Bac
\+ 75007 PARIS

3265 GOLIGER, J.
* Service M.D.
Centre de Recherches du Bouchet
91710 VERT LE PETIT

7293 GRIVEAUX, B.
* Docteur en Geologie
\+ Laboratoire regional des Ponts et Ch.
Avenue de l''amiral Chauvin
49130 LES PONTS DE CE

4220 GROLIER
* Laboratoire de Petrologie
Faculte des Sciences
45100 ORLEANS

7294 GROS, J.C.
* Faculte des Sciences
Laboratoire Ragiogeol. et Mecanique des Roches
Avenue des Facultes
33405 TALENCE

7295 GUELLEC
* L.C.P.C.
58, Boulevard Lefebvre
75732 PARIS cedex 15

1184 HABIB, P.
* Laboratoire de Mecanique des Solides
\+ Ecole Polytechnique
Route d'Orsay
91120 PALAISEAU

7296 HAFFEN, M.
* Soletanche Entreprise
\+ 6 rue de Watford
B.P. 511
92005 NANTERRE cedex

7825 HARLE, D.
* Ing. civil des Mines de Paris
Directeur Tech. Nitrochimie
61 rue Galilee
75008 PARIS

13515 HENRI, J.P.
* U.D.I.
B.P. 36
59620 VILLENEUVE D'ASCO

13516 HOMAND, F.
Assistante en Geotechnique
E.N.S.G. - Centre de Geomecanique
B.P. 452
54001 NANCY cedex

1186 HOUPERT, R.
* E.N.S.G.
\+ B.P. 452
54000 NANCY

3266 HUET, C.
* Centre technique des Tuiles et Briques
200 Avenue de la Liberation
92140 CLAMART

1182 JANOD, A.
* E.D.F.
3 rue de Messine
75008 PARIS

7298 JAUGEY, P.
* Societe Generale d'Entreprise
\+ Departement Travaux Publics
Internationaux
21, rue du Pont des Halles
94150 CHEVILLY LARUE

4221 JOSIEN, J.P.
* Ecole des Mines
Laboratoire de Mecanique des Terrains
Parc de Surupt
54000 NANCY

4222 JOUANNA, P.
* Faculte des Sciences
Place Eugene Bataillon
34000 MONTPELLIER

13517 KERISEL, J.
* 28 Avenue d'Eylau
\+ 75016 PARIS

13518 KONGRAD, J.
Campenon Bernard Cetra
42, Avenue de Friedland
75363 PARIS cedex 18

7826 KUTKAN, E.
* Geosciences
\+ 1 rue de Lorraine
92600 ASNIERES

7829 LAREAL, P.
* I.N.S.A.
20, Avenue Albert Einstein
69100 VILLEURBANNE

13519 LAUNAY, D.
* 19, rue du Censier
75005 PARIS

1190 LAURENT, D.
* B.R.G.M.
\+ P.O. Box 1492
JEDDAH (ARABIE SAOUDITE)

1194 LEBRETON
* 9, rue de Milan
75009 PARIS

1195 LEGRAND, B.
* Ingenieur des Mines
11, Impasse des Chevrefeuilles
84100 ORANGE

3267 LEGRAND, J.
* Directeur du CETE
\+ Rue de l'Europe
B.P. 241
13605 AIX EN PROVENCE

7300 LEMOINE, Y.
* C.P.G.F.
77, Avenue Victor Hugo
92500 RUEIL MALMAISON

1196 LE ROHELLEC
* C.E.B.T.P.
12, rue Brancion
75015 PARIS

7302 LETHUAIRE, R.
* Lenoir et Mernier
14, rue Pasteur
ESCH S/ALZETTE
(GRAND DUCHE DU LUXEMBOURG)

1197 LE TIRANT, P.
* 13, rue des Pimeveres
92500 RUEIL MALMAISON

1198 LETOURNEUR
* 6, rue Louise Michele
\+ 38100 GRENOBLE

7303 LETULLIER, A.
* ELF-ERAP
7, rue Nelaton
75015 PARIS

1199 LEVEQUE, P.
* 50, rue Gounod
\+ 33400 TALENCE

1201 LEVY, G.
* 60 Avenue Anatole France
93140 BONDY

7304 L'HERITEAU, G.
* 50, rue Gerard
75013 PARIS

1204 LINNIKOFF, N.
* 6, rue Alcide Delapierre
\+ 92370 CHAVILLE

1205 LONDE, P.
* Coyne et Bellier
\+ 5, rue d"Heliopolis
75017 PARIS

1206 LOUIS, C.
* 11, Avenue Bosquet
75007 PARIS

13520 LOURDEL, Y.
Mines de Potasse d"Alsace S.A.
11 Avenue d'Altkirch
68055 MULHOUSE cedex

13521 LUNARDI, P.
COGEFAR-BASTIONI
Di-P-Nuova 21
MILAN (ITALIE)

1207 LUONG, M.P.
* Laboratoire de Mecanique des Solides
17, rue Descartes
75005 PARIS

13522 LUPIAC
President de l'AFTES
RATP Travaux Neufs
21, Boulevard Bourdon
75004 PARIS

4225 MALBECQ
* Christensen
Place de la Gare
78320 LA VERRIERE

7305 MALCOR
* Conseil General des Ponts et Chaussees
Ministere de l"Equipement
244, Boulevard Saint Germain
75007 PARIS

1209 MANDEL, J.
* Laboratoire de Mecaniques des Solides
17, Rue Descartes
75005 PARIS

6306 MARCHAND, P.
* CNARBRL
685 route d'Arles
30000 NIMES

3269 MARCHAND
* S.N.C.F.
229, rue du Faubourg Saint Honore
75008 PARIS

1211 MARINIER, P.
* 23, rue des Laitieres
94300 VINCENNES

13523 MARTINET, A.
Direction de l'Equipement E.D.F.
3, rue de Messine
75008 PARIS

13524 MARTORANO, A.
* Corso V. Emanuelle, 104
93100 CALTANISETTA (ITALIE)

3271 MASO, J.Cl.
* I.N.S.A.
Department de Genie Civil
Avenue de Rangueil
31400 TOULOUSE

4226 MASSE
* E.R.A.P. - Division Exploitation
7, rue Nelaton
75015 PARIS

7307 Van MASSENHOVE, H.
* 13, rue Laugier
75017 PARIS

7308 MASURE, P.
* B.R.G.M. - AGE
B.P. 6009
45018 ORLEANS cedex

7309 MAUMENE, J.M.
* Ciments Lafarge
Laboratoire Central
B.P. 8
07220 VIVIERS

1213 MAURY, V.
* Geostock
+ Tour Aurore
92080 PARIS defense

1214 MAYER, A.
* 51, rue Raynouard
75016 PARIS

7310 MAZIER, G.
* SOCOSOR
5, Avenue Dorion
77210 AVON

7311 MENEROUD, J.P.
* Laboratoire de Nice
+ 56 Boulevard de Stalingrad
06300 NICE

13525 MERCKLE, G.
* 761 Le Bottet
Appartement 116
69140 RILLIEUX

7312 MICHEL, G.
* Societe du Canal de Provence
13100 LE THOLONET

7313 MLADYENOVITCH, V.
* 10, Avenue Lamartine
78170 LA CELLE SAINT CLOUD

7827 MONTEAU, R.
* La Pauline, 12 c
13009 MARSEILLE

1219 MORLIER, P.
* I.U.T. de Bordeaux
Department Genie Civil
33405 TALENCE

13526 MOUNIER
Service de Documentation du CERCHAR
33 rue de la Baume
75008 PARIS

13527 MULLET, G.
20 Allee Chatrian
93340 LE RAINCY

1222 NGUYEN, M.
* Laboratoire de Mecanique des Solides
17, rue Descartes
75005 PARIS

1224 OLLIVIER MARTIN, D.
* 7, rue Desrenaudes
75017 PARIS

1225 PANET, M.
* L.C.P.C.
+ 58, Boulevard Lefebvre
75732 PARIS Cedex 15

PAQUET, J.
* Professeur - EUDIL
+ Universite de Lille 1
B.P. 36
59650 VILLENEUVE D'ASCQ

1228 PAUTRE, A.
* La Cedraie
+ Route de la Corniche
74290 VEYRIER DU LAC

13528 PEYREGNE
Societe Telemac
2, rue Auguste Thomas
92600 ASNIERES

1231 PERAMI, R.
* Laboratoire de Mineralogie
39, Allees Jules Guesde
31400 TOULOUSE

13529 PFISTER, P.
* Directeur de la Societe Sol Expert
International
6 rue de Watford
92000 NANTERRE

13530 PHILIPPONNAT, G.
* C.E.B.T.P.
12, rue Brancion
75737 PARIS cedex 15

13531 PIGUET, J.-P.
Laboratoire de Mecanique des Terrains
Ecole des Mines
54000 NANCY

7314 PIRAUD, J.
* B.R.G.M.
\+ B.P. 6009
45018 ORLEANS cedex

1234 PLICHON, J.N.
* E.D.F.
3, rue de Messine
75008 PARIS

1235 PLOUVIEZ, P.
* SITUMER SETEC
Tour Gamma - Batiment D
58, quai de la Rapee
75583 PARIS cedex 12

1236 POINSOT, P.
* Nobel-Bozel - Tour Nobel
8, Avenue du General de Gaulle
92800 PUTEAUX

7315 POIROT, R.
* 5, rue Jean Fernel
60600 CLERMONT

7316 POST, G.
* Ingenieur
5, rue Henri Regnault
92210 SAINT CLOUD

3274 POTTIER, M.
* Gaz de France - DETN
23, rue Philibert Delorme
75017 PARIS

7317 POULET, M.
* 2 bis chemin de Versailles
92500 RUEIL MALMAISON

7322 POULET, M.
I.F.P.
B.P. 18
92502 RUEIL MALMAISON

7320 QUIBLIER
* I.F.P.
B.P. 18
92502 RUEIL MALMAISON

3275 RADENKOVIC
* Ecole Polytechnique
Laboratoire de Mecanique des Solides
Route d'Orsay
91128 PALAISEAU cedex

4229 RAFFOUX, J.F.
* Chef du groupe Terrains du CERCHAR
B.P. 2
60550 VERNEUIL EN HALATTE

3276 RAT
* L.C.P.C. - Annexe d'Orly
\+ Cidex A 155
94396 ORLY GARE

4230 RAUD, J.
* 2, rue du Marais
78220 VIROFLAY

3278 REITH, J.-L.
* C.N.E.T.U.
109, chemin Saint Jean
69672 BRON

7828 REYMOND, M.Ch.
79, avenue Niel
75017 PARIS

13532 RICARD, A.
* L.C.P.C.
58, Boulevard Lefebvre
75732 PARIS cedex 15

13533 RISS, J.
* Laboratoire de Petrologie
Faculte des Sciences
45045 ORLEANS

1238 RIVET, J.
* 10, Avenue Charles de Gaulle
77230 LE PECQ

3280 ROCCA, M.
* Geostock
Tour Aurore
92080 PARIS la Defense

3281 ROCHET, L.
* C.E.T.E. - Div. Laboratoire
\+ Ponts et Chaussees
109, Chemin Saint Jean
69672 BRON

3282 ROQUEJEOFFRE, M.
43, Boulevard Garrisson
82000 MONTAUBAN

1241 SABARLY, F.
* Geoconseil
\+ 31, Avenue de Versailles
78170 LA CELLE SAINT CLOUD

1242 SADRAN, G.
* Societe Lafarge
28, rue Emile Menier
75016 PARIS

13534 SALEMBIER, M.
* Coyne et Bellier
5, rue d'Heliopolis
75017 PARIS

3284 SANTA LUCIA, D.
C.E.T.E. Div. Laboratoire
Ponts et Chaussees
109, Chemin Saint Jean
69672 BRON

3285 SARCIA, A.J.
* C.E.A.
+ Office de Gestion des dechets
31-33 rue de la Federation
75732 PARIS cedex 15

1243 SARDA, J.P.
* 13 bis rue Daniel Casanova
92500 RUEIL MALMAISON

13535 SCHLOSSER
+ Societe Terre Armee
Tour Horizon
52 quai National
92806 PUTEAUX

1244 SCHWEITZER, R.
* Charbonnages de France
9, Avenue Percier
75008 PARIS

450 SIRIEYS, P.
* I.N.S.A.
Avenue de Rangueil
31077 TOULOUSE cedex

1245 SIVIGNON
* R.A.T.P.
Direction des Travaux neufs
21, Boulevard Bourdon
75004 PARIS

13536 SOUBROUILLARD, G.
* Mines de Potasse d'Alsace
11, rue d'Altkirch
B.P. 399
68055 MULHOUSE cedex

4231 SOVICHE
* E.R.A.P.
7, rue Nelaton
75015 PARIS

3286 STRUILLOU, R.
* Ecole des Mines
+ 60, Boulevard Saint Michel
75272 PARIS cedex 06

7324 TALBERT, J.Cl.
* 19300 EGLETONS

1246 TCHENG, Y.
* C.E.B.T.P.
12, rue Brancion
75015 PARIS

1247 TEMPERE, M.
* E.R.A.P.
Division Exploitation
7, rue Nelaton
75015 PARIS

1248 THENOZ, B.
* Departement Genie Civil
Inst. Nal. Sciences Appliquees
156, rue de Rangueil
31400 TOULOUSE

7325 THOMAS, A.
* E.N.S.G.
+ B.P. 452
54000 NANCY

13537 THOMAS, J.
* S.N.C.F.
Departement des Ouvrages d'Art
122, rue des Poissonniers
75876 PARIS cedex 18

13538 TIJANI
C.E.M.R.
Ecole des Mines
35, rue Saint Honore
77300 FONTAINEBLEAU

1249 TINCELIN, E.
* Ecole Nationale Superieure des Mines
60, Boulevard Saint Michel
75272 PARIS cedex 06

1250 TISOT, M.
* E.N.S.G.
+ B.P. 452
54000 NANCY

13539 TOULEMONT, M.
* Laboratoire Regional de l'Equipement
B.P. 34
93350 LE BOURGET

3287 TOURENCQ, C.
* L.C.P.C.
+ 58, Boulevard Lefebvre
75732 PARIS cedex 15

7327 VERDIER, J.
* Coyne et Bellier
+ 5, rue d'Heliopolis
75017 PARIS

7328 VERSMEE, P.
* 33, Quai Orloing
69009 LYON

387 VIGNAUD, R.
* 81, rue Laugier
75017 PARIS

1253 VIGNEAUX, M.
* Faculte des Sciences
351, Cours de la Liberation
33400 TALENCE

1254 VOUILLE, G.
* Ecole des Mines
35 rue Saint Honore
77300 FONTAINEBLEAU

1255 WEBER, P.
* E.N.T.I.M.A.
Centre d'Etude des Materiaux
Avenue de Clavieres
30107 ALES

SUPPORTING MEMBERS

13393 BUREAU DE RECHERCHES GEOLOGIQUES
ET MINIERES
Departement Genie Geologique
B.P. 6009
45018 ORLEANS cedex

13394 CENTRE D'ETUDES DU BATIMENT ET DES
TRAVAUX PUBLICS (CEBTP)
12, rue Brancion
75015 PARIS

13395 CENTRE D'ETUDES ET RECHERCHES DES
CHARBONNAGES DE FRANCE (CERCHAR)
Secretariat Technique
B.P. 27
60100 CREIL

13576 CENTRE D'ETUDE DES TUNNELS
109 Avenue S. Allende
69672 BRON cedex

13396 CHRISTENSEN DIAMONDPRODUCTS Cie
La Verriere
78320 LE MESNIL SAINT DENIS

13399 CIE ROYALE ASTURIENNE DES MINES
5, rue Delandre
33000 BORDEAUX CAUDERAN

13398 COMPAGNIE DE PROSPECTION GEOPHYSIQUE
FRANCAISE
77, Avenue Victor Hugo
92500 RUEIL MALMAISON

6847 COYNE ET BELLIER
5, rue d'Heliopolis
75017 PARIS

13387 EDF-DTG
37, rue Diderot
38000 GRENOBLE

13386 EDF - IGECO
68, rue du Fg Saint Honore
75008 PARIS

13388 ENTREPRISE LEON BALLOT
155, Boulevard Haussmann
75008 PARIS

13397 ETUDE ET DEVELOPPEMENT DE SAMIFER
19, rue de Metz
B.P. 11
54150 BRIEY

13385 FRANCE EXPLOSIFS
Tour Gamma D
197, rue de Bercy
75583 PARIS cedex 12

13389 INSTITUT FRANCAIS DU PETROLE
DIR-DOC-INF (D 30)
1 et 4 rue du Bois Preau
92500 RUEIL MALMAISON

13390 LABORATOIRE CENTRAL DES PONTS ET
CHAUSSEES
58, Boulevard Lefebvre
75732 PARIS cedex 15

13392 MECASOL
50, rue Gerard
75013 PARIS

13391 MINES DE POTASSE D'ALSACE
B.P. 399
68100 MULHOUSE

13572 PENNAROYA
1, Boulevard de Vaugirard
75015 PARIS

13574 PONTS ET CHAUSSEES CETU
Centre de Documentation
109 Av. Salvador Allende
B.P. 48
69672 BRON cedex

13577 REGIE AUTONOME DES TRANSPORTS
PARISIENS
Go Documentation
53 ter quai des Grands Augustins
75271 PARIS cedex 06

13578 SAINRAPT ET BRICE
3, Place Paul Verlaine
75013 PARIS

13579 SERVICE DES MINES DE PARIS
247 rue du Bercy
75012 PARIS

13580 SERVICE DE PRODUCTION HYDRAULIQUE
3, rue de Messine
75008 PARIS

13581 SERVICE TECHNIQUE DE L'ENERGIE
ELECTRIQUE ET DES GRANDS BARRAGES
9, rue de Milan
75009 PARIS

13575 SIMECSOL
115, rue Saint Dominique
75007 PARIS

13582 SOCIETE DU CANAL DE PROVENCE ET DE
L'AMENAGEMENT DE LA REGION
PROVENCALE
10, Boulevard d'Athenes
13001 MARSEILLE

13583 SOCIETE NATIONALE DES CHEMINS DE FER
Division des Installations Fixes
Division des Ouvrages d'art
75876 PARIS cedex 18

13573 SOCIETE DE RECHERCHES GEOPHYSIQUES
81, rue Laugier
75017 PARIS

13540 SOLETANCHE
6, rue de Watford
B.P. 511
92005 NANTERRE cedex

6849 SOL-EXPERT INTERNATIONAL
6, rue de Watford
B.P. 515
92005 NANTERRE cedex

13384 SONDAGES, INJECTIONS, FORAGES - BACHY
11, Avenue du Colonel Bonnet
75016 PARIS

GERMANY, FEDERAL REPUBLIC OF

DEUTSCHE GESELLSCHAFT FÜR ERD-UND GRUNDBAU e.V. (DGEG)

Kronprinzenstrasse 35a, 4300 ESSEN 1

Secretary: Dr -Ing. K.H. Idel

Total Number of Ordinary Members - 236
Total Number of Supporting Members - 9

ORDINARY MEMBERS

226 + ABRAHAM, K.H., Dipl.-Ing.
Siemens AG
Unternehmensbereich
Energieversorgung
Werner-von-Siemens-Str. 50
8520 ERLANGEN

225 ADLER, R.E., Prof.
Geologisches Institut
Adolf-Römer-Str. 2A
3392 CLAUSTHAL-ZELLERFELD

2912 * ALPMANN, B., Dr -Ing.
i.T.Ü. insaat Fak. zemin
Mek. Kürsüsü
Taksim-ISTAMBUL (TÜRKEI)

7990 AMBATIELLO, P., Dipl.-Ing.
Salzbergwerk Berchtesgaden
Bergwerkstr. 83
8240 BERCHTESGADEN

13081 ASDECKER, H., Dipl.-Ing.
Franken-Consult
Carl-Schüller-Str. 27
8580 BAYREUTH

10816 AXHAUSEN, K., Dipl.-Ing.
Neureuther Str. 20
8000 MÜNCHEN 21

6629 * BAUDENDISTEL, M., Dipl.-Ing.
Tulpenstr. 20
7501 BRUCHHAUSEN

2914 BAUMANN, Dipl.-Ing.
Fleichhauserstr. 21
7000 STUTTGART 80

2915 * de BEER, E. Prof.
Rijksinstitut voor Grondmechanica
44 Tramstraat
B-9710 ZWIJNAARDE (BELGIEN)

2916 + BENZ, H., Prof.
Silberburgstr. 157
7000 STUTTGART

7205 * BERGER, W., Dipl.-Ing.
Hessenlauweg 14
7000 STUTTGART 80

10625 BISBOS, C., Dipl-Ing.
Giurdas Verlag
Sergiu Patriarchou 4
ATHEN (GRIECHENLAND)

10594 * BLINDE, A., Prof.
Breslauer Str. 66 a
7500 KARLSRUHE

458 + BOCK, H., Dr
Department Civil Engineering
James Cook University
Qld. 4811 (AUSTRALIA)

7833 BORBEREKY, H., Prof.
Swietelsky Bau-GmbH & Co KG
Museumstr. 7
A-4010 LINZ (AUSTRIA)

8194 * BORM, G., Dr rer. nat.
Institut für Bodenmech. u.
Felsmechanik,
Universität Karlsruhe
Richard-Willstätter-Allee 10-12
7500 KARLSRUHE

10626 BORST, W.G., c.i.
F.C. de Weger Int. BV.
William Boothln. 20
ROTTERDAM (HOLLAND)

8195 + BOUCEK, B., Dr-Ing.
Institut für Grundbau, Bodenmech.,
Felsmechanik u. Verkehrswasserbau
TH Aachen
Mies-van-der-Rohe-Strasse
5100 AACHEN

2920 BRAND, W., Dr-Ing.
Deutscher Platz 7
5060 BENSBERG

7834 * BRÄUNER, G., Dr-Ing.
Steinkohlenbergbauverein
Frillendorfer Str. 351
4300 ESSEN

10630 BRETZ, H., Dipl.-Ing.
Naupliastr. 74 A
3000 MÜNCHEN 90

7972 BUCHNER, Dr
\+ Alfons Bayerer Str. 14/1
8400 REGENSBURG

7212 BÜHL, H.H.
\+ Bühl & Faubel GmbH
Haeckelstr. 14
A-1235 WIEN 23 (AUSTRIA)

245 BUTTMANN, W., Dipl.-Ing.
\+ Lüderitzstr. 20
6430 BAD HERSFELD

254 CHRISTMANN, W., Dipl.-Ing.
* Pettenkoferstr. 21
\+ 8000 MÜNCHEN 15

2922 DAHM, H., Dr
\+ Stadtvermessungsamt
Deutschlandhaus
4300 ESSEN

7835 DAVID, K., Dr-Ing.
* Boiestrasse 11
2300 KIEL

10578 DEMAN, F., Dr-Ing.
WPW Beratende Ingenieure
Feldmannstr. 72-74
6600 SAARBRÜCKEN

7204 DENZER, K., Dr
\+ Hauptstrasse 3
8101 UNTERAMMERGAU

9009 DEUTSCH, R., Dipl.-Geol.
\+ Ruhrtalsperrenverein Essen
Abt. Geologie
Birkenfeld 8
5952 ATTENDORN

13199 DICKFELD, L., Dipl.-Geol
Neuhasstr. 10
6370 OBERURSEL

13196 DISTELMEIER, H., Dipl.-Ing.
Bilfinger/Berger
Bauaktiengesellschaft
Carl-Reiss-Platz 1-5
6800 MANNHEIM

464 DÖSCHER, H.D., Dipl.-Ing.
* Am Berg 32
7500 KARLSRUHE 41

7207 DRÄGERT, K., Dr
* Lilienweg 1
3500 KASSEL

156 DREYER, W., Prof.
Technische Universität Clausthal
Adolf-Römer-Strasse 2a
3392 CLAUSTHAL-ZELLERFELD

8196 DÜLLMANN, H., Dipl.-Ing.
* Lehrstuhl für Ingenieur-geologie
\+ TH Aachen
Kopernikusstr. 6
5100 AACHEN

195 DÜRO, F., Dipl.-Ing.
\+ Geologisches Landesamt NW.
De-Greiff-Strasse 195
4150 KREFELD

6628 EGGER, P., Dr-Ing.
* Eidgen. Techn. Hochschule Lausanne
Av. de Provence 22
CH - 1007 LAUSANNE (SCHWEIZ)

13082 EICHHOLZ, K., Dr-Ing.
Westf. Berggewerkschaftskasse
Institut für Markscheidewesen
Herner Str. 45
4630 BOCHUM

7993 EIGENBROD, K.D., Dipl.-Ing.
* Luegplatz 2
4000 DÜSSELDORF 11

2932 EISSELE, K., Dr
* Geologisches Landesamt
\+ Baden-Württemberg
Albertstr. 5
7800 FREIBURG

13083 ERALP, H., Dr-Ing.
* Poyraak Sokak
\+ Astorya Aot. Kat. 2 D 4, Nisantasi,
ISTANBUL (TÜRKEI)

10579 ERDEMGIL, M., Dr-Ing.
Middle East Technical Univers.
Dept. of Civil Eng.
ANKARA (TÜRKEI)

189 von ESBECK-PLATEN
* Salzgitter Consult GmbH
\+ Postfach 411169
3321 SALZGITTER 41

2973 ETSCHEL, H., Ing.
E+M Bohr-GmbH
August-Mohl-Strasse 38
8670 HOF (SAALE)

57 EVERLING, G., Dr-Ing.
\+ Bergbauforschung
Postfach 130140
4300 ESSEN

13084 FECKER, E., Dipl.-Ing.
\+ Institut für Bodenmechanik
und Felsmechanik
Universität Karlsruhe
Kaiserstr. 12
7500 KARLSRUHE

7994 FEDINGER, H., Dipl.-Ing.
* Rhein-Ruhr-Ingenieur GmbH
Burgwall 5
4600 DORTMUND

8145 FEESER, V., Dipl.-Ing.
\+ Institut für Geologie und
Paläontologie
TU Braunschweig
Postfach 3329
3300 BRAUNSCHWEIG

13085 FIALA, F., Dipl.-Ing.
Thyssen Bergbautechnik
Schwanthalerstr. 73
8000 MÜNCHEN

13086 FLECK, H., Dr-Ing.
Gotthardstr. 68
8000 MÜNCHEN 21

2925 FRANCHINI, F., Dr-Ing.
* Direktor "Consonda S.p.A."
Via Generale Albricci 7
MAINLAND

6158 FRANK, A., Dipl.-Ing
Held & Franke Bau AG
Aschauer Str. 21
8000 MÜNCHEN 90

8002 FRIEDRICH, H., Dipl.-Ing.
* Seydlitzstr. 48
4000 DUSSELDORF

7837 FRIEDRICH, K., Dr
\+ Hess. Landesamt für Bodenforschung
Leberberg 9
6200 WIESBADEN

10595 FRÖHLICH, B., Dipl.-Ing.
Inst. f. Boden u. Felsmechanik
Universität Karlsruhe
Postfach 6380
7500 KARLSRUHE

7438 GARTUNG, E., Dipl.-Ing.
* Grundbauinstitut der Landes-
gewerbeanstalt Bayern
Gewerbemuseumsplatz
8500 NURNBERG

10580 GELL, K., Dipl.-Ing.
Promenadenstr. 7
5100 AACHEN

2927 GOSSWEILER, H., Dipl.-Ing.
* AG. H. Hatt-Haller
Bärengasse 25
CH-8001 ZÜRICH (SCHWEIZ)

179 GRAUPNER, A., Dr -phil.
\+ Am Kastanienberg 6
6903 NECKARGEMUND

7986 GRIMME, K., Dipl.-Ing.
* Niedersächsisches Landesamt
\+ für Bodenforschung
Postfach 230153
3000 HANNOVER 23

2928 GROB, H., Prof.
* ETH Zürich Lehrstuhl für Strassen
und Untertagebau
Clausiusstr. 33
CH-8006 ZÜRICH (SCHWEIZ)

2930 GRUHLE, H.D., Oberbaurat
* Schlossblick 83
5900 SIEGEN

8202 GUDEHUS, G., Prof.
* Institut für Bodenmechanik und
Felsmechanik, Universität Karlsruhe
Richard-Willstatter-Allee
7500 KARLSRUHE

10817 HAARDT, H., Oberbaurat
U-Bahnbauamt
Stadt Essen
4300 ESSEN 1

78 HABETHA, E., Prof.
\+ U - Bahnbauamt
Stadt Essen
4300 ESSEN 1

7987 HACKL, E., Dipl.-Ing.
* Ingenieurkonsulent f. Bauwesen
Geoconsult Hackl-Golser
Franz-Josef-Str. 20A
A-5020 SALZBURG (AUSTRIA)

10656 HANISCH, J., Dr -Ing.
Institut für Bautechnik
Reichpietschufer 72-76
1000 BERLIN 30

7838 HARSCH, W., Dr
* Geotechnisches Institut
Gartenstr. 13
CH-3007 BERN (SCHWEIZ)

6153 HARTMANN, F., Dr -Ing.
* Bahnhofstr. 39
2241 HEMMINGSTEDT

2933 HAUSS, K.D., Dipl.-Ing.
Bundesanstalt f. Wasserbau
Abt. Erd- u. Grundbau
Kussmaulstr. 17
7500 KARLSRUHE 21

72 HEITFELD, K.H., Prof.
* Rue du Calvaire 16
\+ B-4671 MORESNET (BELGIEN)

6159 HEIWOLT, G., Dipl.-Ing.
* Bundesministerium der Verteidigung
Trierer Strasse 78
5300 BONN

2931 HELLERER, H.O., Dipl.Geol.
+ Mittenwalder Str. 37a
8000 MÜNCHEN 70

328 HENKE, K.F., Prof.
* Amtl. Forschung u. Material
+ Prüfungsanstalt f.d. Bauwesen
Otto-Graf-Institut
Pfaffenwaldring 4
7000 STUTTGART 80

3801 HENNING, D., Dr -Ing.
* Rheinische Braunkohlenwerke
502 FRECHEN

6695 HERBST, T., Dr -Ing.
+ Dyckerhoff & Widmann AG
Lessingstr. 9
8000 MÜNCHEN 15

6309 HILMER, K., Dr -Ing.
* Landesgewerbeanstalt
Bayern, Grundbauinstitut
Gewerbemuseumsplatz 2
8500 NURNBERG

8003 HINTEREGGER, H., Dipl.-Ing.
+ Konradstr. 16
8000 MÜNCHEN 13

169 HOEPPENER, R., Prof.
+ Ruhr-Universität Bochum
Geolog. Institut
Buscheystrasse
4630 BOCHUM

7437 HOFMANN, H., Dr -Ing.
* Wasserstr. 58a
4630 BOCHUM

4136 HOLLMANN, F., Dr -Ing.
+ Westf. Berggewerkschaftskasse
Herner Strasse 45
4630 BOCHUM

10682 HOLZAPFEL, O., Ing. grad.
Klingenstr. 28
8501 ECKENTAL

166 HORN, A., Prof.
* Hochschule der Bundeswehr
Werner Heisenberg Weg 39
8014 Neubiberg
MÜNCHEN

7973 HORNEMANN, D., Dipl.-Ing.
* Kütingsgarten 27
4300 ESSEN 14

10631 HOSANG, K.H., Dipl.-Ing.
Junkerstr. 9
5100 AACHEN

13087 HUBER, H., Dr techn.
Verbundgesellschaft
Am Hof 6A
A-1010 WIEN (AUSTRIA)

627 HÜLSMANN, K.H., Dipl.-Ing.
+ Westf. Berggewerkschaftskasse
Herner Str. 45
4630 BOCHUM

7840 JAGAU, H., Dr -Ing.
Hochschule für Technik der
Freien Hansestadt Bremen
Langemarkstr. 116
2800 BREMEN

3802 JÄGER, B., Dipl.-Geol.
+ Geologisches Landesamt NW
De-Greiff-Str. 195
4150 KREFELD

84 JELINEK, R., o. Prof.
* Technische Universität München
Paul-Gerhardt-Allee 2
8000 MÜNCHEN 60

80 JOHN, K.W., o.Professor
* Geologie III - Geotechnik
+ Ruhruniversität Bochum
Postfach 2148
4630 BOCHUM

2935 JÖRGER, H., Dipl.-Ing.
Tiroler Str. 41
6000 FRANKFURT/MAIN

8146 JÜNGLING, H., Dipl.-Ing.
* Institut f. Verkehrswasserbau
+ Grundbau u. Bodenm. TH Aachen
Mies-van-der-Rohe-Strasse
5100 AACHEN

7213 KAHL, R.-W., Dr -Ing.
+ Blohm & Voss AG
P.O. Box 737
BANDAR ABBAS (IRAN)

157 KALTER, A., Dipl.-Ing.
* Techniques Louis Menard
Effnerstr. 111
8000 MÜNCHEN 81

158 KANY, M., Dr -Ing.
* Landesgewerbeanstalt
Bayern- Grundbauinstitut
Gewerbemuseumsplatz 2
8500 NÜRNBERG

13098 KATZENBACH, R., Dipl.-Ing.
Institut für Bodenmechanik und
Grundbau
TH Damstadt, Petersenstr. 13
6100 DARMSTADT

274 KEIL, K., Prof.
* Gasselstiege 433
+ 4400 MÜNSTER i.W.

10577 KILCHERT, M., Dipl.-Ing.
Inst. f. Grundbau u. Bodenmech.
TU Berlin
Strasse des 17. Juni 135
1000 BERLIN 12

7991 KISTENBRÜGGER, K., Dipl.-Ing.
* Vodestr. 52
4630 BOCHUM

13088 KLAPPERICH, H., Dipl.-Ing.
TU Berlin
Inst. für Grundbau u. Bodenmech.
Strasse des 17. Juni 135
1000 BERLIN 12

153 KLÖSS, M., Dr -Ing
+ Brunnenstr. 19
5240 BETZDORF/SIEG

148 KNAUSENBERGER, H., Dipl.-Ing.
+ Zeunerstr. 13
4300 ESSEN 1

2937 KNICKELMANN, F.-J., Dipl.-Ing.
+ Meisenweg 15
4355 WALTROP

10814 KOCK, F., Ing. Grad.
Ingenieurbüro für Bauingenieurwesen
Geibelstr. 10
4470 MEPPEN

3804 KOENIG, H.-W., Dr -Ing. E.h.
* Moorenstr. 33
+ 4300 ESSEN 1

6696 KOENZEN, J.P., Dr -Ing.
* Institut f. Bodenmech. u. Felsmechanik
+ Universität Karlsruhe
Richard-Willstatter-Allee
7500 KARLSRUHE

64 KÖRNER, H., Dr -Ing.
* Bayrisches Geologisches Landesamt
+ Prinzregentenstr. 28
8000 MÜNCHEN 22

229 KRAMER, J., Bauing.
* Erdbaulaboratorium
Prof. Dr.-Ing. Schmidbauer Nachf.
Susannastr. 31
4300 ESSEN 1

6693 KRAUTER, E., Dr phil.
* Geolog. Landesamt Rheinl. Pfalz
+ Flachsmarkt 9
6500 MAINZ

7841 KÜHN, G., Dipl.-Geol.
* Argelandstr. 86
+ 5300 BONN

8004 KÜHN, G., Dr rer. nat.
* Oberbergstr. 39 IV
6200 WIESBADEN-MEDENBACH

10651 KUHNHENN, K., Dr -Ing.
Douglasweg 2
7517 WALDBRONN

7209 KUTTER, H.K., Prof.
+ Inst. für Geol. Ruhr-Universität
Universitätsstrasse
4630 BOCHUM

4137 KUTZNER, C., Dr -Ing.
* Marktplatz 9
6231 SCHWALBACH

6155 LACKINGER, B., Dipl.-Ing.
* Institut f. Bodenmechanik
Universität Innsbruck
Technikerstr. 13
A-6020 INNSBRUCK (AUSTRIA)

2938 LACKNER, E., o. Prof.
* Lindenstr. 1a
2820 BREMEN 70

252 LANGER, M., Prof.
* Bundesanstalt für Geowissenschaften
+ u. Rohstoffe
Postfach 51 01 53
3000 HANNOVER 51

13089 LEICHNITZ, W., Dipl.-Ing.
Lehrstuhl für Felsmechanik
Kaiserstr. 12
7500 KARLSRUHE

144 LEINS, W., Prof.
* Technische Hochschule Aachen
Mies-van-der-Rohe-Strasse
5100 AACHEN

65 LEUSSINK, H., o. Prof.
* Strähler Weg 45
7500 KARLSRUHE-DURLACH

67 LINK, H., Regierungsbaum. a.D.
+ Lahmeyer International GmbH
Lyoner Str. 22
6000 FRANKFURT/MAIN 71

2939 LOCHER, H.-G., Dipl.-Bauing.
Losinger AG
Könizstr. 74
CH-3001 BERN (SCHWEIZ)

7842 LOEM, K., Dipl.-Ing.
* Planungsbüro Obermeyer
Speyerer Str. 3
6000 FRANKFURT/MAIN

554 LOKAU, H., Dipl.-Ing.
* Karl Bauer KG
Wittelsbacher Str. 5
8898 SCHROBENHAUSEN

8142 LÖTTGEN, O., Dipl.-Ing.
* Lehrstuhl für Strassenwesen
Erd- u. Tunnelbau RWTH Aachen
Mies-van-der-Rohe-Strasse
5100 AACHEN

374 MAGAR, K., Dr
* Ing. Büro f. Grundbau
+ Neubergstr. 19
8700 WÜRZBURG

8197 MARKL, S.W., Dipl.-Ing.
* E. Heitkamp GmbH
Fraunhoferstr. 11
8033 MARTINSRIED

9010 MARR, G., Ing.
Ingenieurbüro f. Bauwesen
Alpenroderstr. 1
5238 HACHENBURG

7843 MATUSCHKA, W.S., Dipl.-Ing.
* Heidestr. 119
5840 SCHWERTE

10582 MAYER, G., Dipl.-Ing.
Inst. Grundbau u. Baubetrieb
TU Berlin
Strasse des 17. Juni 135
1000 BERLIN 12

139 MEISSNER, H., Dipl.-Ing.
* Gänsegasse 4
6091 WEILBACH

6631 MEISTER, D., Dr.-Ing.
* Bundesanstalt für Bodenforschung
Stillweg 2
3000 HANNOVER-BUCHHOLZ

8000 MELZER, Dr.-Ing.
* Batelle-Institut e.V.
Am Romerhof 35
6000 FRANKFURT/MAIN 90

2941 MENNERICH, H.H.
+ Geotechnik H.H. Mennerich
Gabelsbergerstr. 5
3000 HANNOVER

138 MERKEL, G., Dipl.-Ing.
* Rhein-Ruhr-Ingenieur-GmbH
Burgwall 5
4600 DORTMUND

13090 MIETENS, D., Dr rer. nat.
+ Casilla 11069
LIMA 14 (PERU)

10581 MOSBACH, K.-H.
H. Gerke-Dr.-Ing. W. Meihorst
Kleefelder Str. 29
3000 HANNOVER

10628 MÜHLHAUS, H.-B., Dipl.-Ing.
Institut für Bodenmechanik u.
Felsmechanik, Universität Karlsruhe
Richard-Willstätter-Allee
7500 KARLSRUHE

131 MÜLLER, G., Dr
* Interfels GmbH
Hilgenstiege 82
4442 BENTHEIM

130 MÜLLER, H., Dipl.-Ing.
* Schafir-Mugglin AG
Boglerenstr. 46
CH-87 Küsnacht
ZURICH (SCHWEIZ)

2943 MÜLLER, L., Prof.
* Abt. Felsmechanik, Universität
+ Karlsruhe
Richard-Willstätter-Allee
7500 KARLSRUHE

7844 MÜLLER-KIRCHENBAUER, H., Prof.
* TU Berlin
Strasse des 17. Juni 135/Bauing. Flügel
1000 BERLIN 12

7440 NATAU, O., Prof.
* Institut f. Bodenmechanik u.
Felsmechanik
Universität Karlsruhe
Richard-Willstätter-Allee
7500 KARLSRUHE

6631 NAUJOKS, L., Dipl.-Ing.
* Wersener Str. 45
4500 OSNABRÜCK

304 NEFF, H.K., Dipl.-Ing.
* Königsberger Str. 9
+ 6303 HUNGEN/OBERHESSEN

3806 NEUBER, H., Dr -Ing.
+ Geologisches Landesamt NW
De-Greiff-Str. 195
4150 KREFELD

8198 NUSSBAUMER, M., Dipl.-Ing.
* Am Langenbühl
7253 RENNINGEN

13007 OBENAUER, P.W., Dipl.-Ing.
J.S. Bachstr. 4
6906 Leimen - St.
ILGEN

125 OPPERMANN, W., Dipl.-Berging.
+ Wilhelmsplatz 4
Heimatmuseum
3370 SEESEN

13091 PACHEN, H., Dipl.-Ing.
Lessinglaan 79
UTRECHT

415 PAHL, A., Dr
* Bundesanstalt f. Bodenforschung
Stilleweg
3000 HANNOVER-BUCHHOLZ

2944 PARAKENINGS, H., Dipl.-Ing.
* Geotest GmbH
Koloniestrasse
4040 ST. PETER STÜRZELBERG

6157 PAULY, E., Dr
Hessisches Landesamt für
Bodenforschung
Leberberg 9
6200 WIESBADEN

2945 PENTZIN, W., Dipl.-Ing.
* Hinschenfelder Stieg 1
2000 HAMBURG 70

2946 PFISTERER, E., Dr
* Schluchseewerk AG
Rempartstr. 16
7800 FREIBURG i Br.

8198 NUSSBAUMER, M., Dipl.-Ing.
* Am Langenbühl
7253 RENNINGEN

13007 OBENAUER, P.W., Dipl.-Ing.
J.S. Bachstr. 4
6906 Leimen - St.
ILGEN

125 OPPERMAN, W., Dipl.-Berging.
+ Wilhelmsplatz 4
Heimatmuseum
3370 SEESEN

13091 PACHEN, H., Dipl.-Ing.
Lessinglaan 79
UTRECHT

415 PAHL, A., Dr
* Bundesanstalt f. Bodenforschung
Stilleweg
3000 HANNOVER-BUCHHOLZ

2944 PARAKENINGS, H., Dipl.-Ing.
* Geotest GmbH
Koloniestrasse
4040 ST. PETER STÜRZELBERG

6157 PAULY, E., Dr
Hessiches Landesamt für
Bodenforschung
Leberberg 9
6200 WIESBADEN

2945 PENTZIN, W., Dipl.-Ing.
* Hinschenfelder Stieg 1
2000 HAMBURG 70

2946 PFISTERER, E., Dr
* Schluchseewerk AG
Rempartstr. 16
7800 FREIBURG i. Br.

8199 PHILIPP, G., Dipl.-Ing.
Ber. Ing. f. Grundbau und Tunnelbau
Maximilianstr. 29
8000 MÜNCHEN 22

8144 PIERAU, B., Dipl.-Ing.
+ Institut f. Verkehrswasserbau
Grundbau u. Bodenmech. TH Aachen
Mies-van-der-Rohe-Strasse
5100 AACHEN

13092 PLISCHKE, B., Dipl.-Phys.
Institut für Grundbau, Bodenmech.
Felsmechanik u. Verkehrswasserbau
TH Aachen
Mies-van-der-Rohe-Strasse
5100 AACHEN

13093 PREIFER, H., Obering.
Schluchseewerk AG
Rempartstr. 12-16
7800 FREIBURG i. Br.

2947 PRINZ, H., Dr
+ Hessisches Landesamt f. Bodenforschung
Leberberg 9
6200 WIESBADEN

13099 RAAB, W., Dipl.-Geol.
+ Wachholtzstr. 15
3300 BRAUNSCHWEIG

10681 RAFEK, A.G., Geol.
Inst. f. Geologie
Lehrstuhl Geologie III Geotechnik
Ruhruniversität Bochum
Universitätstr. 150
4630 BOCHUM

13094 RAPP, R., Dipl.-Ing.
Ruhr-Universität
Postfach 102148
4630 BOCHUM

13208 RAUTENBERG, A., Dipl.-Geol.
Bonnersweg 83
4150 KREFELD 1

8200 REIK, G., Dipl.-Geol.
+ Institut für Bodenmechanik u.
Felsmechanik
Universität Karlsruhe
Kaiserstr. 12
7500 KARLSRUHE

2948 REINHARDT, M., Dr
+ Geologisches Landesamt NW
De-Greiff-Strasse 195
4150 KREFELD

7845 REIPSCHLÄGER, H.H., Bauing.
* Bonner Str. 4
2800 BREMEN

2949 REUM, H., Dr
+ Bundesanstalt f. Bodenforschung
Sven-Hedin-Strasse
3000 HANNOVER-BUCHOLZ

6692 REUTER, G.
+ Deutschlandhaus
4300 ESSEN 1

13191 de REUTER, W., Dipl.-Ing.
Ingenieur für Baustofftechnologie
Lindenstr. 1
4401 ALTENBERGE

335 REXHÄUSER, H., Prof.
+ Landesamt für Bodenforschung
Alfred-Bentz-Haus
3000 HANNOVER-BUCHHOLZ

2950 RIEDEL, H., Dr -Ing.
+ Wittener Str. 72
4630 BOCHUM

10634 RISCH, H., Dr
Merzenbachstr. 17
6610 LEBACH

7846 RISSLER, P., Dr -Ing.
* Ruhrtalsperrenverein
Kronprinzenstr. 37
4300 ESSEN 1

6694 RODATZ, W., Dr -Ing.
* Breckenheimer Str. 10
6238 HOFHEIM

3807 RÖHNISCH, A., o. Prof.
Universität Stuttgart
Keplerstr. 11
7000 STUTTGART

7442 RUMMEL, F., Dr
* Institut für Geophysik
Ruhr-Universitätsstrasse
4630 BOCHUM-QUERENBURG

9011 RYCHLOVSKY
Beton-u. Monierbau AG
Postfach 1704
4000 DÜSSELDORF

6148 SAVIDIS, S., Prof.
* TU Berlin Lehrstuhl für
Grundbau u. Bodenmechanik
Hardenbergstr. 35
1000 BERLIN 12

8143 SCHEIBE, H.-J., Dr
\+ Lahmeyer International GmbH
Lyoner Str. 22
6000 FRANKFURT/MAIN 71

6352 SCHETELIG, K., Dr
* Lahmeyer AG.
\+ Guiolettstr. 48
6000 FRANKFURT/MAIN

2927 SCHIKOR, G., Dr
\+ Geotechnisches Institut
Gartenstr. 13
CH-3007 BERN (SCHWEIZ)

10636 SCHLEINER, H., Bauing. (grad.)
Herkentrup 6
4401 HAVIXBECK

7210 SCHMIDT, M.W., Dipl.-Geol.
\+ Ges. f. Strahlen-u.
Umweltforschung GmbH. München
Institut für Tieflagerung
Berliner Str. 2
3392 CLAUSTHAL-ZELLERFELD

7211 SCHNEIDER, H.J., Dr -Ing.
\+ Bundesanstalt f. Geowissenschaften
u. Rohstoffe
Postfach 51 01 53
3000 HANNOVER 51

4138 SCHÖNE-WARNEFELD, G., Dr
\+ Westf. Berggewerkschaftskasse
Geologisches Institut
Herner Str. 45
4630 BOCHUM

2955 SCHRIMPF, P., Dipl.-Ing.
* Longyear Deutschland GmbH.
Geologisches Institut
Herner Str. 45
4630 BOCHUM

2956 SCHÜTZ, H., Prof.
* Staatliches Laboratorium für
Bodenmechanik Erd-und Grundbau
Pauluskirchstr. 7
5600 WUPPERTAL - BARMEN

2957 SCHWEIKERT, K., Dr -Ing.
Brieger Str.12a
7500 KARLSRUHE

2951 SEELMEIER, H., Prof.
\+ Technische Hochschule Graz
Burgring 12/11
A-8010 GRAZ (AUSTRIA)

7847 SEMPRICH, S., Dipl.-Ing.
* Institut für Verkehrswasserbau
Grundbau u. Bodenmech. TH Aachen
Mies-van-der-Rohe-Strasse
5100 AACHEN

238 SENTKO, M., Prof.
\+ Fachhochschule
Glevenbecker Weg 65
4400 MÜNSTER

4139 SIEDEK/KÜGLER, H., Ing.
* Telemannstr. 6
4000 DÜSSELDORF-BENRATH

2952 SIOR, G., Dr -Ing.
\+ Fritz-Reuter-Str. 28
6000 FRANKFURT/MAIN 1

13206 SOBISCH, H.-G.
Mühlenstr. 25
5000 KÖLN 90

2953 SOMMER, H., Dr -Ing.
* Stephanstr. 3
\+ 6100 DARMSTADT

98 SONNTAG, G., Prof.
* Arcisstr. 21
8000 MÜNCHEN 2

116 von SOOS, P., Dipl.-Ing.
* Technische Hochschule München
\+ Paul-Gerhardt-Allee 2
8000 MÜNCHEN 60

2954 SPANG, J., Dr -Ing.
\+ Waxensteinstr. 44
8135 SÖCKING B. STARNBERG

7992 SPANG, R.M., Dipl.-Geol.
* Uferstr. 17
\+ 5810 WITTEN-BOMMERN

7848 SPAUN, G., Dr
\+ Hauptstr. 27
8228 FREILASSING

8001 STAUPENDAHL, G., Dipl.-Ing.
* Hüttenweg 1,A
3392 CLAUSTHAL-ZELLERFELD

337 STEINFELD, K., Prof.
+ Alte Königstr. 3
2000 HAMBURG 50

319 STENZEL, G., Dipl.-Ing.
* Grundbauinstitut
Landesgewerbeanstalt Bayern
Gewerbemuseumsplatz 2
8500 NÜRNBERG

10815 STREIM, W., Dr, Dipl.-Geol.
Sozietät Grundbau+Ingenieurgeologie
Salzschlifer Str. 16
6000 FRANKFURT/MAIN

2958 THOMAS, G., Dipl.-Ing.
Pestalozzistr. 25
4330 MÜLHEIM/RUHR

6354 THÜME, W., Dipl.-Ing.
+ Werderstr. 69
2000 HAMBURG 13

10653 THUT, A., Dr -Ing.
Solexperts AG
Postfach 274
CH-8034 ZÜRICH (SCHWEIZ)

13220 TRISCHLER, J., Dr, Dipl.-Geol.
Baugrundinstitut
Bergstr. 104-106
6102 PFUNGSTADT

13095 TROPP, W., Dr
* Königsberger Str. 9
+ 6303 HUNGEN/OBERHESSEN

2959 TUBBESING, K., Dipl.-Ing.
+ P.Eng., Consulting Engineer
3725 Bluebonnet Road
NORTH VANCOUVER, B.C. (CANADA)

4141 UHLENBECKER, F.-W., Dr -Ing.
* Kali u. Salz GmbH Werk Hattorf
6433 PHILIPPSTAHL/WERRA

2960 ULRICH, R., Dr
+ Bayr. Geologisches Landesamt
Prinzregentenstr. 28
8000 MÜNCHEN 22

2961 VEDER, C., Prof.
* Technische Hochschule Graz
+ Rechbauerstr. 12
A - 8010 GRAZ (AUSTRIA)

7443 VOORT, H., Dr
+ Hilgenstiege 82
4442 BENTHEIM

2962 WAGNER, H., Prof.
TU Hannover Fachgebiet
Erd-Strassen- u. Tunnelbau
Callinst. 15 c
3000 HANNOVER

13096 WAIBEL, P., Prof.
+ Versuchsanstalt f. Erdbau
u. Bodenmechanik HTL Rankweil
Negrellistr. 50
A - 6830 RANKWEIL (AUSTRIA)

7849 WALDBRÖHL, H.-T., Dipl.-Ing.
+ Hch. Waldbröhl GmbH & Co
Stollen u. Tunnelbau KG.
Nordstr. 73
5300 BONN 1

2963 WALLAYS, M., Dipl.-Ing.
+ Pieux Franki S.A.
196 rue Grety
LIEGE (BELGIEN)

7850 WALLNER, M., Dr -Ing.
Bundesanstalt F. Geowissenschaften
u. Rohstoffe
Stilleweg 2
3000 HANNOVER 51

13221 WANNINGER, R., Dr -Ing.
Dieburger Str. 212
6100 DARMSTADT

2964 WEBER, P., Dr
+ Geologisches Landesamt NW
De-Greiff-Str. 195
4150 KREFELD

6156 WEINHOLD, H., Prof.
* Spezialtiefbau Aufschläger KG
Frans-Josef-Str. 11
8000 MÜNCHEN 13

7851 WEIRICH, K.W., Dipl.-Ing.
Wingertweg
CH - 8888 MELS (SCHWEIZ)

7852 WERNER, H.-U., Dr.-Ing.
* Dorsch Consult
Elsenheimerstr. 63
8000 MÜNCHEN 21

4140 WERNICK, E., Dipl.-Ing.
* Universität Karlsruhe
Kaiserstr. 12
7500 KARLSRUHE

13096 WESELOH, K., Dr -Ing.
* An der Drift 4
3006 GROSSBURGWEDEL

8201 WICHTER, L., Dipl.-Ing.
Institut für Boden- und
Felsmechanik Universität
Richard-Willstätter-Allee
7500 KARLSRUHE

2965 WIERCZEYKO, E.
Prakla Ges. für praktische
Lagerstättenforschung mbH.
Haarstr. 5
3000 HANNOVER

9012 WIETEK, B., Dipl.-Ing.
+ Ingenieurgemeinschaft
Lässer-Feizlmayer
Framsweg 16
A - 6020 INNSBRUCK (AUSTRIA)

13207 WILHELM, R., Dipl.-Geol.
Waldensberger Str. 17
6480 WÄCHTERSBACH 2

473 WILMERS, W., Dipl.-Geologe
+ Bodenprüfstelle beim Strassbauamt
Hessen - Mitte
Am Güterbahnhof
6330 WETZLAR

311 WITTKE, W., Prof.
* Direktor des Institut für
+ Verkehrswasserbau, Grundbau
u. Bodenmechanik, TH Aachen
Mies-van-der-Rohe-Str.
5100 AACHEN

7206 WÖLFEL, E., Dr -Ing.
+ Institut für Bautechnik
Reichpietschufer 72-76
1000 BERLIN 30

2968 WOLTERS, R., Dr.
* Geologisches Landesamt NW
+ De-Greiff-Str. 195
4150 KREFELD

2970 WÜSTENHAGEN, K., Dr.
+ Bundesanstalt für Bodenforschung
Hans-Stille-Weg
3000 HANNOVER-BUCHHOLZ

13097 ZACAS, M., Dipl.-Bauing.
Paisley Inst. of Technology
High Street
Paisley
Renfrewshire PA1 2BE
Scotland (GROSSBRITANNIEN)

SUPPORTING MEMBERS

277 BUNDESANSTALT FÜR STRASSENWESEN
Brühler Strasse 1
Postfach 51 05 30
5 KÖLN 51

247 C. BARESEL AG
Postfach 86
Nordbahnhofstrasse 135
7000 STUTTGART 1

358 FERDINAND ENKE-VERLAG
Postfach 1304
Hasenbergsteige 3
7000 STUTTGART 1

243 JOHANN KELLER GmbH
Hanauer Landstr. 334
6 FRANKFURT/MAIN

194 LAHMEYER INTERNATIONAL GmbH
Lyoner Strasse 22
6000 FRANKFURT/MAIN 71

185 LEHRSTUHL FÜR INGENIEURGEOLOGIE
UND HYDROGEOLOGIE
Lochnerstr. 4-20
5100 AACHEN

101 SCHLUCHSEEWERKE AG
Postfach 1460
Rempartstr. 16
7800 FREIBURG

8146 TRANS TECH PULICATIONS
Adolf Ey Strasse 5
3392 CLAUSTHAL-ZELLERFELD

3029 WAYSS & FREYTAG KG
Neue Mainzer Str. 59
6 FRANKFURT/MAIN

GREECE

HELLENIC SOCIETY OF SOIL MECHANICS AND FOUNDATION ENGINEERING

42, Patission St (Polytechneion), ATHENS 147

President: Prof. A. Loizos
Secretary: Mr. D Frankidakis

Total Number of Ordinary Members - 55

ORDINARY MEMBERS

4175 ANAGNOSTOPOULOS, A., Dr
* Lecturer
National Techn. Univ. of Athens
42 Patission Street
ATHENS

10931 ANDRONOPOULOS, B., Dr - Geologist
* 31 Kousidou Street, Zographou
ATHENS

10932 BISBOS, Chr., Mr - Civil Engr
32 Kalymnou Street
ATHENS 813

10933 CATRAKIS, S., Prof.
* Emer. U. Prof
25 Metsovou St
ATHENS

10934 CAVOUNIDIS, S., Dr - Soils Engr
18 Anagnostopoulou St
ATHENS 136

10935 COUMOULOS, D.G., Dr - Soils Engr
* 10 Krisila St
ATHENS 502

10936 CORONIOTIS, C., Mr - Civil Engr
14 Drosopoulou St
ATHENS

4180 DALLAS, S.A., Mr - Civil Engr
* 17 Benaki St
Kifissia
ATHENS

10937 DAOUTIS, J.S., Mr - Consulting Engr
* 4 Vassilissis Sophias Ave
ATHENS 138

4181 DEMIRIS, C., Prof.
* U. Prof.
15 Kapidaki St
ATHENS 218

10938 DEMOIROS, E., Mr - Civil Engr
* 11b Massalias St
ATHENS

10939 DEREKAS, L., Mr - Civil Engr
* 7 Ioanninon St
EDESSA

10940 DRETTAS, Gr., Dr
* U. Lecturer
36 Amalias Ave
ATHENS

6380 FITROLAKIS, N., Dr - Geologist
* U. Lecturer
+ 4-6 Micromanis St
ATHENS

10941 FRANKIDAKIS, D.J., Mr - Consulting Engr
* 4 Vassilissis Sophias Ave
ATHENS 138

10942 GASSIOS, E., Mr - Chemical Engr
* 4 Terpsichoris St, Cholargos
ATHENS

4184 GOFAS, Th., Mr - Civil Engr
* 58 Souidias St
ATHENS 140

10943 KAPPOS, A., Mr - Civil Engr
* 17-19 Kalvou St
ATHENS

10944 KARAVOKYRIS, G.J., Mr - Consulting Engr
* 4 Vassilissis Sophias Ave
ATHENS 138

10945 KARYDIS, P., Dr
* U. Lecturer
31 Spefsippou St
ATHENS 140

10946 KARYOTIS, S., Mr
* U. Lecturer
17 Stournara St
ATHENS

10947 KOCHILAS, C., Mr
* U. Lecturer
134a M. Boda St
ATHENS

4186 KOLONIAS, E., Mr - Civil Engr
* 4 Phaedonos St, Pangrati
ATHENS

10948 KONTOS, D.S., Mr - Civil Engr
* 12 Karneadou St
ATHENS 139

4188 KORYALOS, Th., Mr - Soils Engr
* 66 Sp. Merkouri St
ATHENS 516

10949 KOSTOPOULOS, S., Mr - Civil Engr
* 66 Char, Trikoupi St, Kifissia
ATHENS

10950 KOTTA, N., Miss - Civil Engr
* 51 Stournara St
ATHENS

10951 KOTTEAKOS, Gr.J., Mr - Civil Engr
* General Director, Ministry Public Works
7 Pezopoulou St
ATHENS 617

4189 KOTZIAS, P., Dr
* Soils Consultant
5 Isavron St
ATHENS 706

4192 LENTOUDIS, G., Mr - Civil Engr
* Nat. Techn. Univ. of Athens
42 Patission St
ATHENS

10952 LIAKOURIS, D., Mr - Geologist
* 7 Pilika, Amerikis Square
ATHENS

414 LOIZOS, A. - Prof.
* U. Prof.
Foundation Engng and Soil Mech.
Nat. Techn. Univ. of Athens
42 Patission St
ATHENS

10953 MANOELOGLOU, J., Mr - Civil Engr
* 41 Efroniou St, Pangrati
ATHENS

599 MANOS, O.G., Mr - Civil Engr
* 43 Akadimias St
ATHENS

4194 PAGONIS, N., Mr - Civil Engr
6 Argiroupoleos St, Hymettus
ATHENS

10954 PANTELOPOULOS, P., Mr - Mining Engr
* 55 Dafnis St, Voula
\+ ATHENS

10955 PAPACHARISSIS, N., Dr
* U. Lecturer
17 Sidirokastrou St
SALONICA

4441 PAPAGEORGAKIS, J., Prof. - Geologist
*
\+ Nat. Techn. Univ. of Athens
42 Patission St
ATHENS

4196 PAPAGEORGIOU, O., Mr - Civil Engr
* 4 Diochariou St
\+ ATHENS 812

10956 PAPANTONOPOULOS, C., Mr - Civil Engr
*
16 Vallianon St
ATHENS 805

413 PAPASPIROU, S., Dr
* U. Lecturer
Nat. Techn. Univ. of Athens
42 Patission St
ATHENS

10957 PARASKEAVAIDIS, E., Prof.- Geologist
* 8 Plomariou St, Amaroussion
ATHENS

10958 PAVLATOS, Chr., Mr - Civil Engr
* 68 Friderikis St, P.Psychikon
ATHENS

10959 PILITSIS, S., Mr
* U. Lecturer
Nat. Techn. Univ. of Athens
42 Patission St
ATHENS

4198 PIPPAS, D., Prof.
* Emer. U. Prof.
49 Agathoupoleos St
ATHENS

10960 SIDERIS, N., Mr - Civil Engr
* 111 Paleas Philippou St
SALONICA

10961 SOTIROPOULOS, E., Mr - Soils Engr
*
32-36 Ag. Fanouriou St
ATHENS

4207 STAMATOPOULOS, A., Mr
* Soils Consultant
1 Voulgaroktonou St
ATHENS 705

597 TASSIOS, Th., Prof.
U. Professor
Nat. Techn. Univ. of Athens
21 Parnithos St, P.Psychikon
ATHENS

10962 TSOUTRELIS, Ch.E., Prof.
* U. Prof.
Nat. Techn. Univ. of Athens
21 Parnithos St, P.Psychikon
ATHENS

10963 VALALAS, D., Prof.
* U. Prof.
Polytechnic School of Salonica University
SALONICA

10964 VARVITSIOTIS, S.P., Mr - Civil Engr
* 12 Karneadou St
ATHENS 139

4204 VASSILOPOULOS, E., Mr - Soils Engr
* 41 Efessou St, N. Smyrni
ATHENS

4205 VETTAS, P., Mr - Civil Engr
* 43 Akadimias St
ATHENS

4206 VOUZARAS, E., Mr
* U. Lecturer
72 Admitou St
ATHENS 201

HUNGARY

NEMZETKÖZI KÖZETMECHANIKAI TARSASAG

Magyar Tudomanyos Akademia, V. Roosevelt Ter 9, H - 1361 BUDAPEST, Pf. 6

President: Prof. F. Martos
Secretary: Dr P. Kertesz

Total Number of Ordinary Members - 10

ORDINARY MEMBERS

10205 ASSZONYI Cs., Dr - Mining Engr
Manager
Magyar Szenbanyaszati Tröszt
H-2803 TATABANYA, Pf 305

13200 BOBOK, E.
Researcher
NME Olajtermelesi Tanszek
H - 3515 - MISKOLC - EGYETEMVAROS

10206 BODONYI J., Dr - Mining Engr
Head of Branch
Banyaszati Kutato Intezet
H-1300 BUDAPEST, Pf. 115

10207 * GRESCHIK Gy., Dr - Civil Engr
Assistant Manager
Közlekedesi es Metro Epito Vallalat
Böszörmenyi ut 20-22
H-1126 BUDAPEST

10208 KAPOLYI L., Dr - Mining Engr
Deputy Minister
Nehezipari Miniszterium
Marko utca 16
H-1055 BUDAPEST

10209 + KERTESZ P., Dr - Civil Engr
Assistant Professor
Budapesti Müszaki Egyetem
H-1521, BUDAPEST

10210 * KEZDI A., Dr Prof - Civil Engr
Professor
Budapesti Müszaki Egyetem
H-1521, BUDAPEST

10204 MARTOS F., Dr Prof - Mining Engr
Director
Banyaszati Kutato Intezet
H-1300 BUDAPEST, Pf. 115

10211 MEISEL J., Dr Prof. - Geologist
Professor, Rector
Budapesti Müszaki Egyetem
H-1521 BUDAPEST

10212 * PETRASOVITS G., Dr Prof. - Civil Engr
Professor
Budapesti Müszaki Egyetem
H-1521 BUDAPEST

INDIA

INDIAN NATIONAL GROUP OF THE ISRM

Central Board of Irrigation and Power, Kasturba Gandhi Marg, NEW DELHI - 110001

Secretary: Mr C.V.J. Varma

Total Number of Ordinary Members - 27
Total Number of Supporting Members - 1

ORDINARY MEMBERS

12045 CHAUDHURI, B.K.
10/3 Barrackpur Trunk Road
CALCUTTA-56

12046 DATTA, H.N., Prof.
U. Prof.
Department of Mining Engineering
Indian Institute of Technology
KHARAGPUR

12047 GUPTA, V.
Sub Divisional Officer
PWD (I.B.) Haryana
House No. 23 - Sector - 18 A
CHANDIGARH

12048 * KARACHIWALA, Z.M.
314, Jamshedj Jilibhoy Road
BOMBAY - 400 008

12049 MAHAMOOD, V.K.
Head of Civil Engineering Department
Government Polytechnic
CANNANORE-7

12050 MOKHASHI, S.L.
Chief Research Officer
Central Water & Power
Research Station
P.O. Khadakwasla Research Station
POONA - 411024

12051 SAXENA, P.C.
Director
Central Water & Power
Research Station
Post Office, Khadakwasla
POONA-24

12052 SHAMBHU, R.
Lecturer in Mining Engineering
Institute of Technology
Banaras Hindu University
Varanasi - 221 005
UTTAR PRADESH

3154 SIKKA, D.R.
Chief Engineer (Irrigation)
Mahanadi Bodhghat
Hasdeo Major Multipurpose Project
M.P. Govt.
(RAIPUR) M.P.

12053 SINGH, D.P., Dr
Department of Mining Engineering
Institute of Technology
Banaras Hindu University
VARANASI - 221005, U.P.

12054 THOMAS, K.C., Dr
General Manager
Chukha Hydel Project
P.O. Chimakothi
BHUTAN

12055 The Director of Design
Central Design Office
Irrigation & Waterways Directorate
9, Brabourne Road
CALCUTTA 700001

12056 The Superintending Engineer
Central Design Organisation
Block - 10
2nd Floor
Sachivalaya
Gandhinagar
GUJARAT

12057 The Superintending Engineer
Central Design Organisation
Konkan Bhavan
7th Floor
NEW BOMBAY 400614

12058 The Director
Central Road Research Institute
(Council of Scientific &
Industrial Research)
Post Office C.R.R.I.
Mathura Road
NEW DELHI

12059 M/s. Asia Foundation and
Construction Private Limited
254-D Dr. Annie Besant Road
Band Box House
BOMBAY 400025

12060 The Director
Engineering Research Labs,
Himayatsagar
Hyderabad 500030
ANDHRA PRADESH

12061 The Chief Engineer, P.W.D.
Irrigation South
Palace Office Building
New Sayyaju Rao Road
MYSORE 570001

12062 The Joint Director
Kerala Engineering Research Institute
Peechi 680653
Trichur District
KERALA

12063 The Deputy Director
Maharashtra Engineering
Research Institute
Nasik
MAHARASHTRA

12064 The Chief Engineer
Nagarjunasagar Right Canals Unit
Andhra Pradesh Government of
Major Projects Department
Hill Colony 508202
ANDHRA PRADESH

12065 The Managing Director
Patel Engineering Co. Ltd,
Jogashwari Estate
Swami Vivekananda Road
Jogeshwari
BOMBAY 400060

12066 The Registrar
Punjab P.W.D.
Irrigation Branch
Hydel Building, Sector-18
GHANDIGARH

12067 The Executive Engineer, P.W.D.
Soil Mechanics and Research Division
Chepuak
MADRAS 600005

12068 The Chief Engineer (Projects)
Srisailam Hydro-Electric Project
Srisailam Project Dam West
Mahaboobnagar District
ANDHRA PRADESH

12069 The Chief Engineer
(Supa Dam Construction)
Mysore Power Corporation Ltd,
Ganeshgudi 581365
KARNATAKA

12070 The Director
U.P. Irrigation Research Institute
ROORKEE

SUPPORTING MEMBERS

13197 GUJARAT ENGINEERING RESEARCH INSTITUTE
Irrigation Department
RACE COURSE VADODARA 390007

IRAQ

Total Number of Corresponding Members - 1

CORRESPONDING MEMBERS

10596 FATTOHI, Z.R., Dr
Lecturer
University of Baghdad
College of Engineering
Petroleum & Minerals Engineering Dept
BAGHDAD

ISRAEL

ISRAEL SOCIETY OF ROCK MECHANICS AND ENGINEERING GEOLOGY

Faculty of Civil Engineering, Technion - Israel Institute of Technology, Technion City
HAIFA 32000

Secretary: Dr Y. Tsur-Lavie

Total Number of Ordinary Members - 27

ORDINARY MEMBERS

384 AISENSTEIN, B., Prof. - Geologist
* Consultant
\+ Alozorof 10
JERUSALEM

3853 AMIR, J., Mr - Civil Engr
* Consultant
P.O. Box 6511
TEL AVIV

11000 BARJOSEF, A., Dr - Geologist
Geologist
15 Mabarot St
HAIFA

11001 BRAUU, M. Dr - Geologist
Consultant
Geoprospect, P.O. Box 8720
JERUSALEM

7217 DENEKAMP, S.A., Dr - Engng Geologist
* Mineral Engineering Department
\+ Technion City
HAIFA

11002 DIMANT, E., Mr - Geologist
Geologist
Petroleum Services, P.O. Box 24229
TEL AVIV

11003 FAINSTEIN, G., Mr - Mining Engr
Engineer
Herzel St
MAGDIEL, HOD-HASHARON

3848 FLEXER, A., Prof. - Geologist
\+ U. Prof.
Department of Geography
Tel-Aviv University
TEL-AVIV

856 GEFFEN, S., Mr - Civil Engr
* Consultant
39 Hatarsi St
TEL AVIV

11004 GOLECKI, J.J., Prof.-Mathematician
U.Prof
Faculty of Civil Engineering
Technion City
HAIFA

11005 GOTLIEB, M., Mr - Geologist
\+ Geologist
Tahal, 54 Even Gavirol St
TEL-AVIV

11006 HARASH, A., Dr - Engng Geologist
\+ Chief Engineering Geologist
Tahal, 54 Even Gavirol St
TEL-AVIV

11007 HAYATI, G., Dr - Engng Geologist
* Chief Engineering Geologist
Soil and Road Lab.
Technion City
HAIFA

11008 ISRAELI, A., Mr - Engng Geologist
\+ Engineering Geologist
Tahal, 54 Even Gavirol St
TEL-AVIV

11009 JABLONKA, R., Mr - Mining Engr
Engineer
Mineral Engineering Department
Technion City
HAIFA

11010 KATZ, A., Mr - Geologist
Geologist
Petroleum Services, P.O. Box 24229
TEL-AVIV

1131 KATSIR, M., Mr - Civil Engr
* Consultant
44 Revivim St
TEL-AVIV

11011 KINSKY, J.F., Mr - Civil Engr
* Research Engineer
Soil and Road Lab.
Technion City
HAIFA

11012 LASMAN, N., Mr - Geologist
Consultant
Geoprospect, P.O. Box 3720
JERUSALEM

11013 SALTZMAN, U., Dr - Engng Geologist
Lecturer
Department of Geography
Tel-Aviv University
TEL-AVIV

11014 SEBEL, J., Mr - Mining Engr
Mining Engineer
Petroleum Services, P.O. Box 24229
TEL-AVIV

11015 SHA'AL, B.Z., Dr - Civil Engr
* Consultant
Geomechanics, P.O. Box 7238
HAIFA

3854 SHENBERG, A., Mr - Mining Engr
G. Manager
68 Aluf David St
RAMAT-GAN

11016 TSUR, Y., Mr - Engng Geologist
\+ Senior Engineering Geologist
Tahal, 54 Even Gavirol St
TEL-AVIV

3836 TSUR-LAVIE, Y., Dr - Mining Engr
* Mineral Engineering Department
\+ Technion City
HAIFA

11017 VERED-VEISS, Y., Mr - Mechanical Engr
Deputy General Manager
\+ Petroleum Services, P.O. Box 24229
TEL-AVIV

11018 ZEITLEN, J.G., Prof.-Civil Engr
* U. Prof.
Faculty of Civil Engineering
Technion City
HAIFA

ITALY

ASSOCIAZIONE GEOTECNICA ITALIANA

Viale Regina Margherita 183, 00198 Roma

Chairman: Dr M. C. Dolcetta
Secretary: Dr S. Martinetti

Total Number of Ordinary Members - 1002
Total Number of Supporting Members - 6

ORDINARY MEMBERS

3409 ABBRESCIA, G., Prof.-Engr
* Via G. Capruzzi 228
70124 BARI

10686 ACAMPORA, C., Dr
* C.so Garibaldi 185
80055 PORTICI (NA)

10687 ACCORSI, D.M., Dr - Engr
* c/o Palisud Srl
Via Blandamura 50
74100 TARANTO

10549 AGATI, R., Mr
* Via M. Rapisardi 53
90144 PALERMO

3411 AGOSTA, C., Dr - Engr
* Via XX Settembre 27/A
46019 VIADANA (MN)

9019 AGOSTINUCCI, A., Dr
* Via G. Pezzana 102
00197 ROMA

9020 AIRO'FARULLA, C., Dr - Engr
* Via A.il Magnanimo 30
90127 PALERMO

10688 AIROLDI, S., Mr
* Via Carso 45
24100 BERGAMO

3413 ALBANI, E., Dr - Engr
* Via Palatino 11
20148 MILANO

9021 ALBERTARIO, G., Mr
* P.za Vittorio Veneto
27049 STRADELLA (PV)

4024 ALBERT, L., Dr - Engr
* Via 1º Maggio 1/14
Zelo Foramagno
20068 PESCHIERA BORROMEO (MI)

3415 ALGRANATI, S.P., Dr - Engr
* Via N. Martelli 1/A
00197 ROMA

9022 ALLAIS, E., Dr - Engr
* Via Nepote 7
27100 PAVIA

7332 ALLARIA, N., Dr - Engr
* Via Val Sesia 24
20152 MILANO

10689 AMADINI, G., Mr
* Via Meropia 67
00147 ROMA

3416 AMBROSANIO, A., Dr - Engr
* Ufficio Speciale Acquedotto
Sussidiario di Palermo
Ente Acquedotti Siciliani
Via N. Turrisi 38
90138 PALERMO

9023 AMICI, D., Mr
* Via T. Zigliara 24
00168 ROMA

3417 AMOROSI, D., Dr - Engr
* Via Piave 55
71100 FOGGIA

6248 ANGELI, A., Dr
* Via Rasi Spinelli 243
47023 CESENA (FO)

9024 ANICETI, G., Mr
* c/o Farroni Lucio
Via Scalo Settebagni 81
00138 ROMA

10690 ANSOLDI, C.A., Dr - Engr
* Servizi Tecnici Internazionali
I.T.S. SpA
Via di Villa Sacchetti 11
00197 ROMA

10691 ANTONINI, F., Mr
* Via Toscanini 19
58022 FOLLONICA (GR)

9025 ANTONINI, T., Prof.
* Via Udine 15
34132 TRIESTE

3419 APPENDINO, M., Dr - Engr
* ENEL-CPCIE di Torino
Via Avogadro 30
10121 TORINO

7333 ARGIROFFI, U., Mr
* Via Vilafranca 42
90141 PALERMO

3423 ARREDI, F., Prof.-Engr
* Ist. Costruz. Idrauliche
Via Eudossiana 18
00184 ROMA

3424 ARRIGONI, E., Dr
* Via G.B. Borghesi 1
43100 PARMA

9026 ASCIONI, S., Mr
* Via P. Marconcelli 44
00149 ROMA

4025 AURELI, A., Prof.
* V.le Andrea Doria 55
95123 CATANIA

10528 AURELI, C., Dr - Engr
* L.go G. Pepe 13
67055 GIOIA DEI MARSI (AQ)

10464 AVERSA, A., Mr
* Via F. Turati
87036 RENDE (CS)

3426 BAISTROCCHI, E., Dr
* Via Vittor Pisani 8
47037 RIMINI (FO)

9027 BAJETTI, F., Dr - Engr
* Via Latina 61
00179 ROMA

6994 BAJETTI, S., Mr
* Via Girolamo da Carpi 1
00196 ROMA

3427 BALDACCI, R., Prof.-Engr
* Direttore Ist. Scienza Costruz.
Via Montallegro 1
16145 GENOVA

10692 BALDI, G., Dr - Engr
* Via T. Tasso 17
24020 TORRE BOLDONE (BG)

479 BALDOVIN, G., Dr - Engr
* Via Roncaglia 14/B
20146 MILANO

9028 BALLADORE, L., Mr
* Via Umberto 1 108
27032 FERRARA ERBOGNONE (PV)

6995 BALLIS, G., Prof.
* Via Carpaccio 3
34143 TRIESTE

7334 BALOSSI RESTELLI, A., Dr - Engr
* Via Montenapoleone 23
20121 MILANO

10693 BALZARETTI, E., Dr - Engr
* Via Don Minzoni 28
21053 CASTELLANZA (VA)

10467 BARBARANO, A., Dr - Engr
* Via G. Iannelli 45/C
80128 NAPOLI

10470 BARBON, G., Dr - Engr
* Via Cannaregio 5580
30121 VENEZIA

3429 BARBUJANI, F., Dr - Engr
* Via Giuoco del Pallone 23
44100 FERRARA

9029 BARDELLA, C., Prof.
Laboratorio di Geotecnica
* Via Gramsci 9
43100 PARMA

3430 BARDI, B., Dr - Engr
* Via S. Biagio dei Librai 46
80138 NAPOLI

9030 BARDIN, G., Dr
* Via Feltre 74
32100 BELLUNO

3431 BARDIN, R.
* Via Feltre 74
32100 BELLUNO

9031 BARDUCCI, C., Mr
* Via A. Gandiglio 60
00151 ROMA

6997 BARLA, G., Prof. - Engr
* Istituto Arte Mineraria -
Politecnico
\+ C.so Duca degli Abruzzi 24
10129 TORINO

542 BARONCINI, G., Dr - Engr
* Via Nevio 102/D
80122 NAPOLI

9032 BARONE, A., Dr - Engr
* Via Campo d/Vettovaglie 10
98100 MESSINA

10539 BARONE, E., Dr - Engr
* c/o ENEL C.P.C.I.E.
V. le Kennedy 5
80125 NAPOLI

3433 BASCHIERI, L., Prof. - Engr
* Ist. Costruz. Stradali
Via Diotisalvi 2
56100 PISA

10419 BASILE, L., Mr
* C.so Garibaldi 181
82100 BENEVENTO

9033 BASTA, F., Mr
* C.so V. Emanuele II 14
87050 CELLARA (CS)

10694 BATTAGLIO, M., Dr - Engr
* Politecnico di Torino
Corso Duca degli Abruzzi 24
10129 TORINO

10420 BEGHER, E., Dr - Engr
* Via Amba Alagi 27
39100 BOLZANO

9034 BELGASEM, M., Mr
* Via Bottini 13
27100 PAVIA

9035 BELLADONNA, V., Mr
* Via Prenestina 911
00155 ROMA

9036 BELLEGONI, M., Mr
* P.za Accademia 2
54033 CARRARA (MS)

3435 BELLINI, A., Dr - Engr
* Via Fogazzaro 9
20135 MILANO

6250 BELLONI, L., Dr
* V. le Teodorico, 21
20149 MILANO

10421 BELLONI, M., Dr
* via Basilicata 11
27100 PAVIA

9037 BELLOTTA, W., Dr - Engr
* Via Bagnoli 46
67051 AVEZZANO (AQ)

3436 BELLOTTI, R., Dr - Engr
* ENEL
Via Ornato 90/14
20162 MILANO

10422 BELOTTI, M., Mr
* In. Labo Srl
P.za Trento 6
20135 MILANO

9039 BENCIVENGA, M., Mr
* Via Luigi Pirandello 72 p.2/A
00137 ROMA

3438 BENINI, A., Prof.-Engr
* V. le Ippocrate 93
00161 ROMA

9040 BENUSSI, G., Dr - Engr
* Azienda Nazionale Autonoma
Strade-Compartimento della
Viabilita
39100 BOLZANO

3439 BEOMONTE, M., Dr - Engr
* Via Archimede 92
00197 ROMA

3440 BERARDI, G., Prof.-Engr
* Facolta d'Ingegneria
Via Montallegro 1
16145 GENOVA

9041 BERARDI, R., Dr - Engr
* 67010 SAN VITTORINO (AQ)

3441 BERNABINI, M., Prof.-Engr
* Istit. Geofisica Mineraria
Via Eudossiana 18
00184 ROMA

10423 BERNACCHI, G., Dr - Engr
* ENEL-CRIS
Via Ornato 90/14
20162 MILANO

10695 BERNARDO, J.L., Dr -Engr
* Via Bitinia 5
00179 ROMA

9042 BERNASCONI, G., Mr
* V. le Cattaneo 1
22100 COMO

7335 BERRY, P., Dr - Engr
* Via C. Colombo 436
00145 ROMA

3442 BERTACCHI, P., Dr - Engr
* ENEL-CRIS
Via Ornato 90/14
20162 MILANO

9043 BERTERO, M., Dr - Engr
* c/o Pali Trevisani
Via Dismano 58/9
47023 CESENA (FO)

3443 BERTI, P., Dr - Engr
* Via Solferino 7
20121 MILANO

10488 BERTINI, T., Dr
* Via Silvani A 69
00139 ROMA

10424 BERTOSA, Dr - Engr
* c/o Geotecneco
61047 S. LORENZO IN CAMPO (PS)

3444 BETTINI, A., Dr
* Via Paradiso 287
47023 CESENA (FO)

7336 BEVILACQUA, V., Dr - Engr
* V.Dir Centr. ENEL
Via G.B. Martini 3
00198 ROMA

10519 BIAGINI, S., Dr - Engr
* Via Trieste 45
60100 ANCONA

10425 BIFERI, G., Dr
* Via Nazionale 91
64026 ROSETO D'ABRUZZI (TE)

3446 BIFFI, E., Dr
* Via S. Vigilio 11
31030 DOSSO DI CASIER/TV

3447 BIGALLI, F., Dr - Engr
* Via California 9
20144 MILANO

9044 BIGGI, A., Mr
* Via Nisco 30
00179 ROMA

7000 BILOTTA, E., Dr - Engr
* Istituto di Geotecnica
Via Claudio 21
80125 NAPOLI

10696 BINAZZI, P.P., Dr
* Via S. Biagio in Cascheri 59
51100 PISTOIA

7337 BIONDI, P., Dr - Engr
* Via Crema 5
00182 ROMA

9045 BISCEGLIA, D., Mr
* C.so Manfredi 95
71043 MANFREDONIA (FG)

3448 BOCCI, V., Dr - Engr
* Facolta di Ingegneria
Via Montallegro 1
16145 GENOVA

9046 BOGETTI, F., Dr - Engr
* Via I. Pindemonte 1
20129 MILANO

10697 BOGNOLO, G., Dr - Engr
* Via E. Fermi 20
30027 S. DONA'DI PIAVE

9047 BOIDO, C., Dr - Engr
* Via Gorizia 6
20144 MILANO

9048 BOLDONI, G., Dr - Engr
* S.P.E.A.
Parco Comola Ricci 155
80122 NAPOLI

9049 BOLDRIGHINI, U., Mr
* Via Nazionale
22050 COLICO (CO)

9050 BOLOGNINI, M., Dr
* Via G. Aristide Sartorio 147
00147 ROMA

10466 BOMPAROLA, G., Mr - Geom.
* Via Medaglie d'Oro 10
87100 COSENZA

10426 BONFANTI, S., Dr - Engr
* Via G. Galilei 16/E
35043 MONSELICE (PD)

9052 BONIFACIO, A., Dr - Engr
* Via Domenico Fontana 190
80131 NAPOLI

10427 BORDUGO, I., Dr - Engr
* Via Julia 11
33170 PORDENONE

3450 BORELLI, F., Dr - Engr
* Via Codogno 10
20139 MILANO

9070 BORGIA, G.C., Dr
* c/o Geo - Sal
Via A. Pirro 3
84100 SALERNO

3451 BORRELLI, M., Dr - Engr
* Via Florimo 3
80128 NAPOLI

7001 BOSELLI, L., Dr - Engr
* Via S. Bernardino de Siena 15
52100 AREZZO

10698 BOTTARI, R., Dr - Engr
* Via Pereira 15
00136 ROMA

6252 BOTTI, E., Dr - Engr
* Via Ampere 77
20131 MILANO

3454 BRACCESI, M., Dr
* Via XX Settembre 45
51100 PISTOIA

3455 BRAGA, G., Prof.
* Via Cazzamali 3
+ 27100 PAVIA

7338 BRANCALEONI, F., Mr
* Via Gianfranco Albani 15
00165 ROMA

10480 BRASCHI, G., Dr - Engr
* Via Ponte Garate
27010 S. GENESIO (PV)

9071 BRAY, C., Mr
* C.so Siracusa 106
10137 TORINO

3457 BREZZI., L., Dr - Engr
* C.so Rosselli 80
10129 TORINO

3458 BRIGHENTI, G., Prof.-Engr
* Istituto Scienze Minerarie
+ V.le Risorgimento 2
40136 BOLOGNA

10699 BROGGINI, N., Dr
* c/o Radaelli & Castelloti Snc
Via XXV Aprile 10
20090 SEGRATE (MI)

7339 BROGINI, F., Dr
* Via Alpi 9
00198 ROMA

10700 BRUGALETTA, V., Dr - Engr
* Via del Faggio 21
97100 RAGUSA

7349 BRUGGI, V., Mr
* Via Marazzani 18
27029 VIGEVANO (PV)

3460 BRUNETTI, G., Dr
* Via Cassiopea lotto 296
Casina de Pecchi
20100 MILANO

3461 BRUNNER, F., Dr - Engr
* Via L. Manara 1
20122 MILANO

10701 BRUNO BOSSIO, E., Dr - Engr
* Via Alba 27
00182 ROMA

7348 BRUSA, R., Dr - Engr
* Via Hermada 3
21100 VARESE

10702 BRUZZI, D., Dr
* c/o Radaelli & Castellotti, Snc
Via XXv Aprile 10
20090 SEGRATE (MI)

10428 BUCCELLA, L., Mr
* Via Colli 34
65010 FARINDOLA (PE)

3462 BUCCHI, R., Prof. - Engr
* Via Piccoli 4
35100 PADOVA

10472 BUCCI, L., Mr
* P.za Chiesa 15
86070 CONCA CASALE (IS)

10703 BUONANNO, A., Mr
* Via F. D'Ovidio 121
00137 ROMA

9072 BUONO, G., Dr - Engr
* Italsonda
V.le De Filippis 228
88100 CATANZARO

6253 BURGHIGNOLI, A., Dr - Engr
* Via Novacella 14
00145 ROMA

10473 BUSCARNERA, M., Dr - Engr
* Via S. Agostino 15a
92019 SCIACCA (AG)

9073 CACCIAMANI, V., Mr - Geom.
* 67017 PIZZOLI (AQ)

3463 CACCONE, M., Dr - Engr
Via R. Benini 20
00191 ROMA

10429 CACIOPPO, G., Mr
* Via Umberto 189
90043 CAMPOREALE (PA)

3464 CADDEO, S., Dr
* Via Marchioro 12
36015 SCHIO (VI)

7340 CAFACCI, C., Dr - Engr
* Soc. Fondedile
Via G. Verdi 35
80133 NAPOLI

9074 CAFIERO, G., Dr - Engr
* c/o Ferrofir
Via Ombrone 2/G
00198 ROMA

3465 CALABRESI, G., Prof.-Engr
* Via G. Vaccari 32
\+ 00194 ROMA

9075 CALCERANO, G., Dr - Engr
* c/o Soc. Di Penta
Via Agrigento 5
00161 ROMA

9076 CALDARERA, E., Dr
* Via Vallombrosa 22
00135 ROMA

3466 CALVINO, F., Prof.-Engr
* Via Giordano Bruno 42/7
\+ 16146 GENOVA

10476 CAMPONESCHI, B., Dr
* Via Lanuvio 25
\+ 00179 ROMA

10529 CAMPONOVO, R.A., Dr - Engr
* Via Emilio Bossi 22
6830 CHIASSO (CH)

6254 CAMPONUOVO, G., Dr - Engr
* ISMES
V. le Giulio Gesare 29
24100 BERGAMO

3469 CANCELLI, A., Dr
* Via Sansovino 14
\+ 20133 MILANO

9077 CANDELISE, F., Dr - Engr
* Via Piano Lago 169
87050 MANGONE (CS)

9078 CANDINI, M., Dr
* Fondedile
Via G. Verdi 35
80133 NAPOLI

3470 CANNAVO, S., Prof - Engr
* Via Monte Senario 95
00141 ROMA

3471 CAPALDO, G., Dr - Engr
* P.za Amedeo 8
80121 NAPOLI

9079 CAPALDO, M., Dr
* c/o Geoconsult
Via Nomentana 859/F
00137 ROMA

10704 CAPATA, W., Dr - Engr
* Via Pietro Blaserna 19
00146 ROMA

7342 CAPORASO, G., Dr - Engr
ENEL/CPCIE
V.le Kennedy 5
80125 NAPOLI

10705 CAPOTONDI, G., Dr - Engr
* Via Giacomo Costamagna 90
00181 ROMA

213 CAPOZZA, F., Dr - Engr
* Via di Vigna Murata 9
+ 00143 ROMA

9080 CAPULLI, F., Mr
* 67010 S.VITTORINO (AQ)

10479 CAPUOZZO, S., Dr Engr
* Via P. Mascagno 12
80128 NAPOLI

7243 CARABELLI, E., Prof.-Engr
* Via Tasso 10
20123 MILANO

9081 CARDINALE, S., Dr
* Via del Giuba 22
00199 ROMA

10489 CARERE, M., Mr
* Via Miceli 47
87100 COSENZA

3472 CARGNEL, G., Prof.
* Via Fantuzzi 8
32100 BELLUNO

10430 CARLA, L., Dr - Engr
* Capo Sezione ANAS
Via Amba Alagi 24
39100 BOLZANO

7347 CARNEVALE, G., Dr - Engr
* Soc. Fondedile
Via G. Verdi 35
80133 NAPOLI

10556 CAROSSO, G., Mr
* Via Nicola Sardi 23
14030 ROCCHETTA TANARO (AT)

3475 CAROTENUTO, U., Dr - Engr
* Dir. INFRASUD
Via Chiatamone 57
80121 NAPOLI

10431 CARRABBA, E., Dr - Engr
c/o ICOS
Via L. Manara 1
20122 MILANO

10475 CARRAPETTA, C., Dr - Engr
* Via Cosseria 14
50129 FIRENZE

10478 CARUANA, R., Dr - Engr
* Via Licia 19
00183 ROMA

3478 CASARA, F., Dr - Engr
* Contra Piazza del Castello 16
36100 VICENZA

9083 CASCINI, L., Dr - Engr
* Pal. Giorcelli-Castiglione Scalo
87036 RENDE (CS)

10706 CASELLI, G., Dr
* Via Manlio di Veroli 5
00199 ROMA

10707 CASSAR, S., Mr
* Via Firenze 8
95129 CATANIA

10481 CASSARA, R., Dr
* Via Republica Argentina 114/B
25100 BRESCIA

3479 CASSINIS, C., Dr - Engr
* Via Chiana 38
00198 ROMA

10708 CASTELLOTTI, L., Dr
* c/o Radaelli & Castellotti Snc.
Via XXV Aprile 10
20090 SEGRATE (MI)

10709 CASTIGLIONE, A., Mr
* Viale della Regione 134
93100 CALTANISSETTA

7003 CATALANO, A., Dr - Engr
* Via Pompeo Neri 32
00191 ROMA

7345 CATALDI, C., Dr - Engr
* c/o ENEL - C.R.G.
Piazza Bartolo da Sassoferrato
56100 PISA

10484 CAVALERA, L., Dr - Engr
* Via Teulada 65
00195 ROMA

10550 CAVALLARO, A., Mr
* Via Ardizzone Gioeni 35
95125 CATANIA

7004 CAVALLO, R., Dr - Engr
* Via Levico 11
00198 ROMA

7358 CAZZOLA, A., Dr - Engr
* Lagaspi Village
P.O. Box 1059
MAKATI RIZAL (PHILIPPINES)

10520 CELENTANI, G.B., Dr - Engr
* ENEL-CPCIE
V.le Kennedy 5
80125 NAPOLI

10710 CELLA, R., Dr
* c/o Radaelli & Castellotti Snc.
Via XXV Aprile 10
20090 SEGRATE (MI)

10482 CELLO, G., Dr
* c/o ENEL CRIS
\+ Via Ornato 90/14
20162 MILANO

3482 CELOTTI, U., Mr - Geom.
* Via Mincio 22
20139 MILANO

9053 CERATI, G., Mr
* Via Liberta 3
20080 OSPEDALETTO LODIGIANO

7346 CERBINI, G., Dr
* C.so Buenos Aires 1
20124 MILANO

7347 CERCIELLO, C., Dr - Engr
* Soc. Fondedile
Via G. Verdi 35
80133 NAPOLI

7005 CERIANI, G., Dr
* ELC - Elettroconsult
Via Chiabrera 8
20151 MILANO

10830 CERUTTI, A., Mr
* Via Eseguiti 31
27043 BRONI

10711 CERUTTI, G., Mr
* Via Faa di Bruno 29
15029 SOLERO (AL)

10541 CESTARI, F., Mr - Geom.
* c/o Studio Geotecnico Italiano
Via Ripamonti 89
20141 MILANO

7348 CESTELLI GUIDI, C., Prof - Engr
* Ist. Tecnica Costruz.
Facolta di Architettura
Via Gramsci 53
00197 ROMA

7007 CHIANTINI, B., Dr - Engr
* Via Montanini 99
53100 SIENA

3484 CHIARI, A., Dr - Engr
* AP 209
SAN JUAN MAGUANA (REPUBLICA DOMENICANA)

7349 CHIAVARINO, A., Mr - Geom.
* Via Bergognone 31
20144 MILANO

9054 CHIEREGATO, A., Dr - Engr
* S.P.E.A. SpA
Via G. Ferraris 33
15100 LAESSANDRIA

3487 CHIESURA, G., Dr
* c/o "C.N.D. Controlli non distruttivi" srl
Via Piambellino 15
20146 MILANO

9055 CHISARI, A., Dr - Engr
* METROROMA
P.le del fiocco (Villa Borghese)
00197 ROMA

9056 CHITIS, W., Dr - Engr
* Soc. Fondedile
Via Verdi 35
80133 NAPOLI

9057 CIARLA, M., Mr
* Via L. Cattolica 6
00195 ROMA

10530 CIARNELLI, M., Dr - Engr
* Via F. Caracciolo 14
80122 NAPOLI

9058 CIARROCCA, P., Mr
* Via U. Lusena 5
00136 ROMA

3489 CICALA, I., Dr - Engr
Via Messina 36/A
90141 PALERMO

9059 CICCHELLI, C., Dr - Engr
* Via G. Sasso 9
65027 SCAFA (PE)

10713 CICOGNANI, M., Dr - Engr
* Via Valpolicella 19
00141 ROMA

9060 CILENTO, M., Mr
* Via Capo Palinuro 2
20142 MILANO

3490 CIOPPA, O., Dr - Engr
* Via Bevagna 68
00191 ROMA

10714 CIPOLLONI, B., Dr
* c/o Sondedile SaS
Via Crispi 17
64100 TERAMO

10432 CIRILLO, A., Mr
* Via G. Leopardi 203
80125 NAPOLI

10715 CIRILLO, P., Dr - Engr
* Via Cicerone 28
00193 ROMA

9061 CIRONE, P., Dr - Engr
* Via Amendola 38
87023 DIAMANTE (CS)

10474 CIRUZZI, F., Dr - Engr
* Vicolo del Vicario 138
00165 ROMA

9062 CIVIDINI, A., Dr - Engr
* Via A. da Rosciate
24100 BERGAMO

9063 CLAPS, G.A., Dr - Engr
* Via Roma 63
85047 MOLITERNO (PZ)

9064 COCCHI, R., Mr
* Via Tuscolana km. 16,800
00044 FRASCATI (Roma)

3491 COEN, L., Dr - Engr
* Via Gallia 10
00183 ROMA

9066 COLA, S., Mr
* Via G. Allievo 39
00135 ROMA

10716 COLABATTISTA, G., Dr - Engr
* Via 123° Fanteria 55
66100 CHIETI

9067 COLAFRANCESCHI, P., Mr
* Via Panaro 17
00199 ROMA

10717 COLETTA, E., Dr - Engr
* c/o S.G.I. - Sogene
Piazzale Agricoltura 24
00144 ROMA

9068 COLLESELLI, F., Dr - Engr
* Centro Geotecnico Veneto
Via Ognissanti 39
35100 PADOVA

10477 COLLU, A., Dr - Engr
* Via Colombo n. 11
18039 VENTIMIGLIA (IM)

7009 COLOMBO, C., Dr
* Soc. It. per la Strumentazione
Geotecnica
Via Roma 15
20090 SEGRATE (MI)

3493 COLOMBO, P., Prof - Engr
* Centro Geotec. Veneto
\+ Via Ognissanti 39
35100 PADOVA

6256 COLOSIMO, P., Dr
* Via G. Berneri 20
\+ 00169 ROMA

9069 COLTRO, R., Prof
* Istituto di Geologia
\+ C.so Italia 55
95129 CATANIA

7350 CONTARDI, M., Dr - Engr
* Via G. Verdi 18
80133 NAPOLI

3496 CONTESSINI, F., Dr - Engr
* Galleria Borella 3
20123 MILANO

10518 CONTI, M., Dr
* Via De La Oliva 18
25084 GARGNANO (BS)

7010 CONVERSINI, P., Dr
* Via Torino 1
\+ 06034 FOLIGNO

7011 CORRERO, R., Dr - Engr
* Via Nerva 160/C
80126 NAPOLI

7012 COSCIA, M., Dr - Engr
* Via Mafalda 517
98100 MESSINA

10718 COSENTINO, C., Dr - Engr
* Via Caserta Crocevia 25 Sc-C
89100 REGGIO CALABRIA

10719 COSTA, C.P., Dr - Engr
* Piazza Nettuno 4
95126 CATANIA

3498 COTECCHIA, V., Prof - Engr
* Ist. Geologia Applicata
\+ Via Re David 200
70125 BARI

10433 COTZA, R., Prof - Engr
* c/o Istituto Arte Mineraria
Facolta di Ingegneria
P.za d'Armi
09100 CAGLIARI

9085 COVELLI, L., Mr
* Via L. Settino 57
88074 CROTONE (CZ)

3499 COZZI, L., Dr - Engr
* Via T. Tasso 23
30172 MESTRE (VE)

10483 CRAVERO, M., Dr - Engr
* C.so. Orbassano 118
10136 TORINO

7013 CREA, G., Dr - Engr
* Ufficio Naz. Minerario Idrocar.
Via Nomentana 41
00161 ROMA

7014 CRESCENZI, S., Dr
* Via Yser 8
00198 ROMA

10434 CRESPELLANI ALLEGRETTI, T., Prof.-Engr
* Via Gen. Cagna 23
09100 CAGLIARI

10720 CRICCHI, D., Mr
* 02010 VINDOLI DI LEONESSA (RI)

7351 CRISERA, S., Mr
* Via Fortuna 1
00042 ANZIO (Roma)

3500 CROCE, A. Prof - Engr
* Via Ferdinando Russo 29
80123 NAPOLI

10493 CROCE, P., Mr
* Via Ferdinando Russo 29
80123 NAPOLI

3501 CUCCINIELLO, E., Dr
* Ist. Scienza Costruzioni
Ist. Universitario di Architettura
30125 VENEZIA

7352 CUCUZZA SILVESTRI, S., Prof
* Casella Postale 345
95100 CATANIA

10721 CUIULI, E., Dr
* c/o Geostudi
\+ Viale Colombo 67
88046 LAMEZIA TERME-NICASTRO (CZ)

3502 CUOCOLO, M., Dr - Engr
* Via Pacuvio 4
80122 NAPOLI

9086 CURTACCI, A., Mr
* Via Arischia 8
67100 L'AQUILA

3503 CUSMANO, G., Dr - Engr
* V. le Regina Margherita 9
90138 PALERMO

10722 CUTRI, A., Mr
* Via Veneto 68
89013 GIOIA TAURO (RC)

9087 CUTRUZZULA, B., Dr
* S.I.G. snc.
P.za Meucci 5/8
00146 ROMA

10723 CUTTINI, M., Dr
* Largo Pecile 26
33100 UDINE

10724 CUTTINI, U., Dr - Engr
* Largo dei Pecile 21
33100 UDINE

9088 DA CASTO, A., Dr
* Via Anime 1
25087 SALO'(B5)

9089 DALERCI, G., Prof - Engr
* C.so A. Gastaldi 1/26
16131 GENOVA

9090 D'AMATO, A., Dr - Engr
* c/o INCO
28 rue Mokrane Aous
El Mouradia
ALGERI (ALGERIA)

9091 D'AMATO, A., Mr
* Via Telemaco 22
90149 PALERMO

3505 D'AMATO, S., Dr - Engr
* Via Roma 228
80132 NAPOLI

10725 DAMIANI, B., Dr - Engr
* Via S. Maria 1
65026 POPOLI (PE)

9092 D'AMICO, A., Mr
* Via F. Scaduto 14
90144 PALERMO

4027 D'AMICO, R., Dr - Engr
* Via Giordano Bruno 9/A
45100 ROVIGO

10726 DAMOCINI, T., Mr
* Corso Rosse li, 48
10128 TORINO

7353 D'AMORE, A., Dr - Engr
* V. le Liberta is. 481
98100 MESSINA

3507 D'ANDREA, D., Dr - Engr
* Via M. Cammarano 25
80129 NAPOLI

3508 D'ANGELO, A., Dr - Engr
* C.so Re Umberto I 237
80138 NAPOLI

10727 D'ANGELO, M., Dr
* V. le Alcione 519
66023 FRANCAVILLA M. (CH)

10728 D'ANIELLO, A., Dr - Engr
* C.so Vittorio Emanuele 6
80122 NAPOLI

7354 DE CAROLIS, P., Dr - Engr
* V. le Pinturicchio 45
00196 ROMA

7018 DE DOMENICO, R., Dr - Engr
* Via Tommaso Gulli 40
20147 MILANO

9093 DE DONATO, O., Prof.
* Res. Parco - Milano 2
20090 SEGRATE (MI)

9094 DEFEZ, A., Prof.
* Riviera di Chiaia 127
80122 NAPOLI

6260 DE FORTIS, E., Mr
* Via Dante n. 58/A
90141 PALERMO

7355 DE LA GRENNELAIS, A., Dr - Engr
* ENEL/CPCIE
V. le Kennedy 5
80125 NAPOLI

3512 DEL FELICE, E., Dr - Engr
* Via Ragazzi del '99 10
23100 SONDRIO

7887 DEL FIACCO, C., Mr
* Via Ferdinando di Savoia 3
00196 ROMA

3511 D'ELIA, B., Dr - Engr
* Via della Tribuna di Tor de Specchi
\+ 18/A
00186 ROMA

9096 DELLI VENERI, G., Dr - Engr
* Via P. Castellino 88
80131 NAPOLI

10435 DE LIO, A., Mr
* C. so XX Settembre 148
87037 SAN FILI (CS)

10729 DEL RE, M., Mr
* Via degli Orti Giancolensi 5
00152 ROMA

7358 DE LUCA, E., Dr - Engr
* Via Giannetto Valli 95 vil.B/10
00149 ROMA

3514 DE MAGISTRIS, F., Dr - Engr
* Via Cornelio Nepote 25
00136 ROMA

10730 DE MICHELE, G., Mr
* Via E. Corcione 21
81031 AVERSA

9097 DENTE, G., Dr - Engr
* Via Tripoli 5
83100 AVELLINO

9065 DE PAOLI, B., Dr - Engr
* Via V. Emanuele II 28
24050 COVO (BG)

7359 DEPOLI, A., Dr - Arch
* V. le E. Caldara 29
20122 MILANO

3517 DERIU, M., Prof - Engr
* Ist. Petrografia
Via Gramsci 9
45400 PARMA

10731 DE ROBERTO, A., Dr
* Via Teognide III
00126 ROMA

9098 DE SIMONE, P., Mr
* P.za S. Francesco 3
84100 SALERNO

3518 DESIO, A., Prof - Engr
* Ist. di Geologia
P. Le Gorini 15
20133 MILANO

7360 DIAMANTI, L., Dr - Engr
* Geosonda
Via Girolamo da Carpi 1
00196 ROMA

3519 DIAMANTI, R., Dr - Engr
* Via R. Zandonai 11
00196 ROMA

10732 DI BARTOLOMEO, S., Dr - Engr
* Via Principe Amedeo 303
\+ 70122 BARI

9099 DI BLASIO, L., Dr - Engr
* Via Capaldo 7
80128 NAPOLI

10000 DI CAMILLO, O., Dr
* Via Brigata Acqui 9
38100 TRENTO

10521 DI CARLO, A., Mr
* Vic. Pasqualucci 2
61040 MONDAVIO (PS)

10001 DI COSTA, G., Mr
* P. za Monte Gennaro 26/12
00139 ROMA

10733 DI CRISTINO, F., Dr - Engr
* Via Squarcialupo 14
00162 ROMA

3521 DIETRICH, E., Dr - Engr
* Via Conca del Naviglio 35
20123 MILANO

3523 DI FILIPPO, M., Dr - Engr
* Via Carmine 58
84100 SALERNO

10734 DI FILIPPO, S., Dr - Engr
* Via Ugo Betti 8
00152 ROMA

10436 DI GIACOMO, G., Mr
* Via Ponte S. Antonio 9
65017 PENNE (PE)

10002 DI LORETO, G., Dr - Engr
* V. le 6. Mazzini 50
67039 SULMONA (AQ)

3524 DI MARTINO, R., Prof - Engr
* Parco Margherita 8
80121 NAPOLI

7361 DI PAOLA, M., Dr - Engr
* Via Palmerino 6
90129 PALERMO

10003 D'ISCHIA, A., Dr - Engr
* c/o ENEL Dir. Lav. Imp. Presenzano
Via Sterpara Prop. Lancini
86079 VENAFRO (IS)

10496 DI ROSARIO, V., Mr
* Salita Pandolfi 28
65100 PESCARA

7362 DI VINCENZO, S., Dr - Engr
* Via Ortigara 2
44100 FERRARA

6261 DOLCETTA CAPUZZO, M., Dr - Engr
* c/o ENEL - Dir. Generale
Via G.B. Martini 3
00198 ROMA

6262 DOLCIMASCOLO, F., Dr - Engr
* C.so Sicilian. 45
94100 ENNA

7363 DOMENICHINI, L., Mr
* Via Asiago 2
00195 ROMA

10547 D'OPPIDO, M., Dr
* Via F. lli Bisogno 27
83100 AVELLINO

155 DRIOLI, C., Prof - Engr
* Via Aniello Falcone 153
80127 NAPOLI

7364 DRUSCO, A., Mr
* Via Moneglia 3
00168 ROMA

10735 DUGNANI, G., Dr - Engr
* Edilsonda Fondazioni
Piazza N. Tommaseo 2/A
20123 MILANO

10736 ELHSSEINI, A., Mr
* Corso Sebastopoli 48
10 134 TORINO

3530 ERRANI, V., Dr - Engr
* Via Patin 10
35100 PADOVA

10509 ESPOSITO, A., Mr
* Rione Lauro 70
80185 NAPOLI

10004 ESPOSITO, L., Mr
* P. za Gesu e Maria 20
80135 NAPOLI

525 ESU, F., Prof.-Engr
* Via Proba Petronia 69
\+ 00136 ROMA

7365 EVANGELISTA, A., Dr - Engr
* Via M. Cammarano 25
80129 NAPOLI

7367 FAIELLA, D., Dr - Engr
* Via Floridiana 10
00135 ROMA

7026 FAITELLI, G., Dr
* Via Gregorio VII 154
00165 ROMA

10437 FALCHI DELITALA, G., Dr - Engr
* Piazza Vittoria 15
16121 GENOVA

4029 FALCIOLA, G., Dr - Engr
* Via Visconti di Modrone 2
20122 MILANO

10468 FALSETTI, E., Dr - Engr
* Via Luigi Lilio 19
00147 ROMA

4030 FANELLI, M., Prof - Engr
* ENEL/CRIS
Via Ornato 90/14
20162 MILANO

3531 FANTOMA, D., Dr - Engr
* ENEL
C.so del Popolo 111
30172 MESTRE (VE)

10005 FARAVELLI, F., Mr
* Fraz. Bagarello
27047 MONTECALVO
VERSIGGIA (PV)

10006 FARINA, G., Mr
* Via Indipendenza 83
88013 DASA' (CZ)

3532 FASSER, C., Dr
* Via Bezzecca 6
25100 BRESCIA

3533 FASSO', C., Prof - Engr
Ist. di Idraulica
P.za d'Armi
09100 CAGLIARI

10491 FAVA, A., Dr - Engr
* Via Iseo 33
20052 S. FRUTTUOSO MONZA (MI)

10007 FAVIA, G., Dr - Engr
* c/o Caterini
Via Montepulciano 26
00182 ROMA

10008 FAZIO, F., Dr - Engr
* Via C. Menotti - pal. F. Guido
Contrada Commenda
87030 ROGES (CS)

10009 FAZIO, L., Mr
* Via G. De Rada 51
87100 COSENZA

7027 FEDERICO, A., Dr - Engr
* Via Franco 22
\+ 73057 TAVIANO (LE)

10737 FERABOLI, G., Dr
* Via Veneto 20
86100 CAMPOBASSO

7368 FEROLDI, M., Mr
* Via Verdi 6
43100 PARMA

10010 FERRANTI, N., Mr
* Via Militello 58
00132 ROMA

10011 FERRARA, V., Prof.
* Via Veneto 62
\+ 95024 ACIREALE (CT)

10012 FERRO, F., Dr - Engr
* Via Quintino Sella 45
95129 CATANIA

3536 FIGUNDIO, N., Dr - Engr
* Via della Pineta 19
85100 POTENZA

10738 FIORAVANTI, E., Dr - Engr
* Via Ardeatina 524
00179 ROMA

10012 FIORE, U., Mr
* Via Venezia 42
27100 PAVIA

7028 FIORINI, P., Dr - Engr
* V. le Italia 615
19100 LA SPEZIA

10471 FLORA, R., Mr
* Via Napoli 62
85042 LAGONEGRO (PZ)

10438 FLORIO, D., Mr
* Via M. Schipa 41
84100 SALERNO

3558 FOCARDI, P., Dr
* Via N. Machiavelli 2
50015 GRASSINA (FI)

7369 FOFFO, G., Dr - Engr
* Via della Riserva Grande 237
00166 ROMA

10739 FONTANIVE, A., Mr
* c/o CNEN
Laboratorio Ingegneria Sito-RAD
Centro Studi Nucleari della Casaccia
Casella Postale 2400
00100 ROMA

10740 FORACI, M., Dr - Engr
* Consorzio di Bonifica
"Delia Nivolelli"
Via M. Rapisardi 30
91026 MAZARA DEL VALLO (TP)

10741 FORASASSI, A., Dr - Engr
* Via C. Marx 138
50047 PRATO (FI)

10013 FORMIGONI, G., Dr
* Via V. Veneto 14
20124 MILANO

7031 FORMIS, L., Dr - Engr
* Via Palizzi 139
80127 NAPOLI

10742 FORMISANO, G., Dr - Engr
* Via Palizzi 139
80127 NAPOLI

10014 FORNARO, C., Dr
* "Geotecnica Veneta"
Via N. Bixio 7/9
OLMO DI MAERNE (VE)

7032 FORTIS, A., Mr
* Via Monserrato 61
00186 ROMA

10439 FORTUNATO, M., Mr
* Via Petrarca 129/D
80100 NAPOLI

3542 FRANCALANCIA, J., Dr - Engr
* Cannaregio 358/A
30121 VENEZIA

10015 FRANCESCHETTI, C., Dr - Engr
* 38085 PIEVE DI BONO (TN)

3543 FRANCHINI, F., Dr - Engr
* Via D. Cornaghi 2
20129 MILANO

3545 FRANCO, A., Dr - Engr
* Via C. de Fabritiis 113
00136 ROMA

10743 FRASSILE, G., Mr
* Via Morgex 6
11100 VILLAIR (AO)

10016 FRASSONI, A., Dr - Engr
* Via Medici 3
24100 BERGAMO

10744 FRETTI, C., Mr
* Via Biava 11
24100 BERGAMO

10017 FUDA, P., Mr
* Via Trento 29
89048 SIDERNO MARINA (RC)

3546 FUGANTI, A., Dr
* Via Bolognini 50
+ 38100 TRENTO

10745 FUGAZZA, E., Dr
* Via del Babuino 46
00187 ROMA

10028 FUINA, A., Mr - Geom.
* Via Alto Adige 60
74100 TARANTO

469 FUMAGALLI, E., Prof - Engr
* ISMES
V. le Giulio Cesare 29
24100 BERGAMO

3547 FUNEL, C., Dr - Engr
* Parco Comola 125
80122 NAPOLI

10746 GABRIELE, G., Mr
* 67045 LUCOLI FRAZ. COLLIMENTO (AQ)

3548 GAGLIANI, E., Dr - Engr
* Via Renato Fucini 48
00137 ROMA

10747 GALANTI, R.R., Dr
* Vico Paino 2
93012 GELA

3549 GALATERI, C., Dr - Engr
* Via S. Filippo 24
80122 NAPOLI

10748 GALEOTTI, G., Mr
* Via Valentino Mazzola
01011 CANINO (VT)

3550 GALLO, G.C., Dr
* Via A. Ferrero 7
00196 ROMA

10018 GALLO, M., Mr
* Via Virgilio 22
80053 CASTELL.RE DI STABIA (NA)

10749 GALLO, N., Dr - Engr
* Via Amedeo 52
88073 COTRONEI (CZ)

10490 GAMBINI, F., Dr - Engr
* c/o SCAC
Corso Italia 1
20122 MILANO

10019 GARASSINO, A., Dr - Engr
* Via Mincio 24
20139 MILANO

3551 GARGANO, A., Dr - Engr
* Via Emerico Amari 32
90139 PALERMO

4031 GARZETTI, B., Mr - Geom.
* Via E. Maccani 150
38100 TRENTO

10020 GASPARONI, L.V., Dr
* Via di Casalpalocco is. 53
00124 ROMA

3552 GATTI, G., Dr - Engr
* L.go Murani 3
20133 MILANO

3553 GENEVOIS, R., Dr
* Ist. di Geologia - Citta Univer.
P. le delle Scienze
00185 ROMA

10531 GENTILINI, L., Dr - Engr
* Via Perini 113
38100 TRENTO

10494 GERARDI, T., Dr
* Via A. Trionfi 9
00159 ROMA

3555 GERMANI, P. Dr - Engr
* V.le della Tecnica 102
00144 ROMA

7370 GERMANINO, G., Mr
* V.le Mazzini 9
28047 OLEGGIO (NO)

3556 GHEZZI, L., Dr - Engr
* Via Lorenteggio 53
20146 MILANO

3557 GHIRARDINI, A., Dr - Engr
* Via Lovanio 11
00198 ROMA

3559 GIANGRECO, E., Prof - Engr
* Ist. Tec. Costruzioni
P.le Tecchio
80125 NAPOLI

10027 GIANNETTI, M., Mr
* Via Bolognese 81
50036 PRATOLINO (FI)

10029 GIODA, G., Dr - Engr
* Ist. Scienza e Tecnica delle
Costruzioni Politecnico
P.za L. da Vinci 32
20133 MILANO

10030 GIORDANO, A., Mr
* Via Settimio Mobilio 82
84100 SALERNO

10031 GIORGI, P., Dr
* Via Valerio Anziate 16
00136 ROMA

10750 GIOVAGNOLI, M., Dr - Engr
* Servizi Tecnici Internazionali
I.T.S. SpA
Via di Villa Sachetti 11
00197 ROMA

10032 GIOVANNINI, W., Mr
* Strada Ponte Po
15040 TICINETO PO (AL)

3560 GIOVANNUCCI, A., Dr - Engr
* Via Castello 3
05100 TERNI

10506 GIOVINAZZO, F., Dr - Engr
* Via Stradone 187
06060 MOIANO (PG)

10751 GIUNTA, S., Mr
* Via Magenta 28
97019 VITTORIA (RG)

7034 GIUSTINIANI, A., Dr - Engr
* Via della Mattonaia 17
50121 FIRENZE

7035 GNISCI, U., Dr - Engr
L.go Forano 18
00199 ROMA

3562 GOFFI, L., Prof - Engr
* Ist. Tecnica Costruzioni
C.so Duca degli Abruzzi 24
10128 TORINO

10021 GONSALVI, L., Dr
* Via Magenta 17
20075 LODI (MI)

10752 GORETTI, M., Dr - Engr
* Via Lorenzetti 14
52100 AREZZO

13219 GRANATA, E., Dr
* Via San Alessandro 14-L
24068 SERIATE (BG)

10753 GRANATA, R., Dr
* Via Puccini 15
21100 VARESE

10754 GRANDE, E., Mrs
* Via Mondovi 5
00185 ROMA

10022 GRASSO, F., Dr - Engr
* Fondedile
Via G. Verdi 35
80133 NAPOLI

3563 GRASSO, F., Dr
* Via San Quintino 42
\+ 10121 TORINO

10023 GRAZIOTTI, G., Dr
* Via Po
01100 VITERBO

10522 GRECUZZO, R., Mr
* Via Mineo 2
95125 CATANIA

10755 GRIPPO, T., Dr - Engr
* Via G. De Falco 16
80136 NAPOLI

10024 GRISOLIA, M., Dr - Engr
* Istituto di Geologia Applicata
Facolta di Ingegneria
Universita degli studi
Via Eudossiana 18
00184 ROMA

10025 GROSSI, F., Dr
* Via Bari 24/D
20136 MILANO

10026 GRUNFEELD, P.J., Dr - Engr
* Fondedile
Via G. Verdi 35
80133 NAPOLI

7036 GUARASCIO, M., Dr - Engr
* V.le XXI Aprille 21
00162 ROMA

3567 GUIDUCCI, M.F., Dr - Engr
* Via M. Adelaide 8
00196 ROMA

3568 GURRERI, P., Dr - Engr
* Via Gradoli 11
00189 ROMA

10756 INGRAVALLE, F., Dr - Engr
* Via Alcide de Gasperi 2
00165 ROMA

10033 ISMAIL, R., Dr - Engr
* P.O. Box 12477
TRIPOLI (LIBIA)

3573 JAMIOLKOWSKI, M., Dr - Engr
* Via Ripamonti 89
20139 MILANO

3574 JAPPELLI, R., Prof - Engr
Via Principe di Paterno 72
90144 PALERMO

3576 JELMONI, F.A., Prof - Engr
* Ist. Vie e Trasporti
P.za L. da Vinci 32
20133 MILANO

10034 LA CIVITA, E., Mr
* Via Staz. Introdacqua 28
67039 SULMONA (AQ)

7037 LA FIURA, G., Dr - Engr
* Via Principe di Paterno 76
90144 PALERMO

10035 LAGUARDIA, G., Dr - Engr
* P.za B. Bonaventura 10
85100 POTENZA

3577 LA MAGNA, A., Dr - Engr
* Facolta d'Ingegneria
Via Montallegro 1
16145 GENOVA

10757 LANCELLOTTA, R., Dr - Engr
* Via Napione 33
10124 TORINO

6265 LARGAIOLLI, T., Prof.
* Via Grazioli 116
\+ 38100 TRENTO

10036 LA ROSA, G., Mr
* Via Diana 100
97017 S. CROCE CAMERINA (RC)

10440 LAVORATO, A., Dr - Engr
* Via Paleocapa 18
24100 BERGAMO

10037 LEONI, A., Dr
* Via S. Filippo 4
03020 CASTELMASSIMO (FR)

10038 LIBANO, N., Dr - Engr
* Via Vincenzo di Marco 8
90143 PALERMO

10039 LIBERATORI, A., Dr
* c/o Ballanti
Via Fontanellato 44
00142 ROMA

3579 LIMONCELLI, B., Dr - Engr
* Via Flora 4/18
16146 GENOVA

430 LINARI, C., Dr - Engr
* CPCIE/ENEL
C.so del Popolo 111
30172 MESTRE (VE)

3580 LIZZI, F., Dr - Engr
* Via G. Verdi 35
80133 NAPOLI

10758 LOCASCIO, A., Dr - Engr
* Via Taranto 59
00182 ROMA

3582 LO CICERO, L., Dr - Engr
* Via Gozzano 4
20131 MILANO

3583 LODIGIANI, P., Dr - Engr
* Via Senato 8
20121 MILANO

10040 LO GIUDICE, F., Dr - Engr
* Via Minicucca 38
95037 S. GIOVANNI LA PUNTA (CT)

10041 LOJELO, L., Dr - Engr
* Via Cogoleto 112
00168 ROMA

10759 LOLLINO, G., Dr - Engr
Via S. Carlo 12
70012 CARBONARA DI BARI (BA)

10042 LOMBARDI, A., Dr - Engr
* Via Lavanga 66
04023 FORMIA

10043 LOMBARDI, G., Mr
* Via Nicolardi P. Avolio 7
80100 NAPOLI

10760 LOMI, A., Dr
c/o Radaelli Castellotti snc
Via XXV Aprile 10
20090 SEGRATE (MI)

10044 LOPPO, G., Dr - Engr
* Via di Passolombardo la Tr. X.
00044 VERMICINO (Frascati) Roma

3584 LORENZINI, G., Dr - Engr
* Via le Ludovico il Moro 11
27100 PAVIA

3586 LOSARDO, R., Prof - Engr
* Via S. Filippo 20/F
80122 NAPOLI

549 LOTTI, C., Prof - Engr
* Via del Fiume 14
00186 ROMA

10761 LOVISON, L.C., Dr
* Via Tunisi 6
\+ 35100 PADOVA

7373 LUCCIOLI, F., Dr - Engr
* Via XX Settembre 89
06034 FOLIGNO (PG)

6266 LUCINI, P., Prof - Engr
* Via Cammarano 19
80129 NAPOLI

10762 LUDA di CORTEMIGLIA, F., Mr
* Via Benso 16
10022 CARMAGNOLA (TO)

10046 LUNARDI, P., Dr - Engr
* c/o Cogefar
\+ Bastioni di Porta Nuova 21
20121 MILANO

3588 LUONGO, E., Dr - Engr
* Via I. Nievo 1
20145 MILANO

3589 LUPINO, R., Dr - Engr
* Via S. Martino ai Monti 10
\+ 00184 ROMA

10763 LUPO, F., Mr
* Via E. D'Angio 12
95125 CATANIA

10540 LUPOI, F., Dr - Engr
* Comando Prov. le Vigili d.Fuoco
Via Scarlatti 16
90134 PALERMO

6267 LUSCO, F., Dr - Engr
* Via Trapani 1/D
90141 PALERMO

10532 LUZZATI, G., Dr - Engr
* Via del Caravaggio 11
20144 MILANO

3591 MADONI, G., Dr - Engr
* Via Magenta 62
25100 BRESCIA

7039 MADONIA, B., Dr - Engr
* Via Carini 1
90144 PALERMO

3592 MAGNANI, G., Dr - Engr
* Via Pellizzari 27
20059 VIMERCATE (MI)

10047 MAIER, G., Prof - Engr
* Quart. Milano 2
Res. Campo 251
20090 SEGRATE (MI)

10764 MALERBA, F.S., Dr
* Via Gramsci 12
95030 GRAVINA DI CATANIA

10048 MALICE, L.A., Dr
* Via S. Tugini 73
\+ 71100 FOGGIA

3593 MALLANDRINO, G., Mr
* Via Piana di Gallo 34 (Mondello)
90151 PALERMO

7374 MANCINI, P., Dr - Engr
* Dir.Tec.E.G.O.
Via Cessati Spiriti 60
00179 ROMA

10049 MANFREDI, G., Mr
* Via Conte Orsini 8
80035 NOLA (NA)

7041 MANFREDINI, G., Dr - Engr
* Via Pietro da Cortona 8
\+ 00196 ROMA

7042 MANFREDINI, M., Prof - Engr
* Via Livorno 63
\+ 00162 ROMA

7043 MANISCALCO, A., Dr
* Via Luigi Pulci 52
00162 ROMA

3405 MANTOVANI, E., Dr - Engr
* CPCIE/ENEL
Via Avogadro 30
10121 TORINO

543 MARAZIO, A., Prof - Engr
* ENEL
Via G.B. Martini 3
00198 ROMA

10050 MARCELLINO, P., Dr - Engr
* Via Pier Bertinoro 15 5
47034 FORLIMPOPLI (FO)

10765 MARCHETTI, G., Mr
* Viale Fonderia 4
05100 TERNI

3598 MARCHETTI, S., Dr - Engr
* Via Bracciano 38
00189 ROMA

3599 MARCHI, E., Prof - Engr
* Istituto Idraulica
Via S. Giuliano 11
16145 GENOVA

3600 MARCHINI, S., Dr - Engr
* RODIO
Casella Postale 7
20077 MELEGNANO (MI)

10499 MARELLA, G., Dr - Engr
* Via De Rosa 29
85100 POTENZA

10051 MARGIOTTA, G., Mr
* Via M. di Villa Bianca 111
90143 PALERMO

7375 MARI, B., Dr - Engr
* Dir. Serv. Tec. Consorzio Bonifica
Piana di Sibari e Valle Media del Crati
87100 COSENZA

7044 MARINO, L.M., Dr - Engr
* Via Brunetto Latini 80
50133 FIRENZE

10492 MARIOTTI, G., Dr
* C.so Mazzini 9
60011 ARCEVIA (AN)

3795 MARTINELLI, D., Dr - Engr
* SORINCO Srl
Via Marisiga 97
32100 BELLUNO

10052 MARTINELLI, F., Dr - Engr
* Via Cantalupo in Sabina 1
00191 ROMA

545 MARTINETTI, S., Dr - Engr
* Via Asmara 58
\+ 00199 ROMA

10486 MARTINI, M., Dr - Engr
* Via di S. Maria d. Speranza 11
00139 ROMA

7045 MARTINO, D., Dr - Engr
* Via S. Croce in Gerusalemme 107
00185 ROMA

4036 MARTORANA, A., Dr
* C.so Emanuele 104
+ 93100 CALTANISSETTA

10053 MARZIALI, M., Mr
* Via 4 Giornate di Napoli 39
63019 S. ELPIDIO A MARE (AP)

3603 MARZOLO, F., Prof - Engr
* Prato della Valle 78
35100 PADOVA

3604 MASCARDI, C., Dr - Engr
* Via S. Sofia 8
30122 MILANO

10054 MASELLA, L., Dr - Engr
* Taverna del Postiere
S.S. 19 delle Calabrie km. 137
85045 LAURIA (PZ)

10487 MASSARA, G., Mr
* Via Pisacane 49
27024 CILAVEGNA (PV)

10055 MASSETTI, S., Mr
* Via R. Alexandri 27
00151 ROMA

7046 MASTRANTUONO, C., Dr - Engr
* V.le Molise 63
20137 MILANO

10766 MASTRODICASA, G.C., Mr
* Via Castellari 2
65027 SCAFA (PE)

3606 MASTROPIETRO, G., Dr - Engr
* V.le Luciana 9
20139 MILANO

3607 MATERNINI, M., Prof - Engr
* Ist. Strade e Trasporti
P. le Europe 1
34172 TRIESTE

10767 MATRICIANI, D., Dr - Engr
* Via Convento Vecchio 2
64037 CERMIGNANO (TE)

3608 MATTEOTTI, G., Prof - Engr
* Universita
Via Ognissanti 39
35100 PADOVA

10055 MATTIUCCI, S., Mr
* Via S. Attanasio 74
80141 NAPOLI

4037 MAUGERI, M., Prof - Engr
* C.so Umberto 208
+ 95024 ACIREALE (CT)

3609 MAURI, S., Mr - Geom.
* Via Tuberose 6
20146 MILANO

10056 MAURI, Z., Mr
* Via Don Bosco 28
28021 BORGOMANERO (NO)

6270 MAURO, M., Dr - Engr
* Via Viminale 66
00184 ROMA

10768 MAZZA, A., Dr - Engr
* Via D. Marzi 17
04015 PRIVERNO (LT)

10057 MAZZALAI,P., Dr - Engr
* Via 4 Novembre 11
38014 GARDOLO (TN)

10769 MAZZON, A., Dr - Engr
* A.N.A.S. - Compartimento della
Viabilita - Venezia Giulia - Friuli
34100 TRIESTE

10498 MAZZONCINI, F., Dr
* c/o EUROTECO
Via Galliari 11
20156 MILANO

10058 MAZZUCATO, A., Dr - Engr
* Centro Geotecnico Veneto
+ Via Ognissanti 39
35100 PADOVA

3610 MEARDI, G., Prof - Engr
* Via Battistotti Sassi 29
+ 20158 MILANO

3611 MEARDI, P., Dr - Engr
* Via Castelmorrone 4
+ 20129 MILANO

10770 MEGALE, P.G., Dr - Engr
* Via dei Carabinieri 28
57100 LIVORNO

3613 MEISINA, L., Dr
* Via Campari 12
27100 PAVIA

10059 MELGARA, R., Mr
* Via Duca d'Aosta 3
15040 BOZZOLE (AL)

3614 MELIDORO, G., Prof - Engr
* C.so V. Emanuele 35
+ 70072 PALO DEL COLLE (BA)

10771 MELLI, F., Mr
* Viale del Castani
00040 ARICCIA (Roma)

7047 MENNELLA, V., Dr - Engr
* Via Chiugiana 280
06074 ELLERA (PG)

4607 MERCOGLIANO, F., Dr - Engr
* P.za E. De Leva 18
80136 NAPOLI

10060 MERCURI, R., Mr
* Via Serlio 8/2
20139 MILANO

10501 MERIGGI, E., Dr
* Via Triulziana 34
20097 METANOPOLI
S. DONATO MILANESE (MI)

10441 MERIGGI, R., Dr - Engr
* Via Marco Volpe 12
30100 UDINE

10495 MERLI, G., Dr - Engr
* Via Archimede 80
00193 ROMA

3617 MESIRCA, R., Dr - Engr
* Via Segheria 9
39012 MERANO (BZ)

3618 MESSINA, G., Dr - Engr
* Via Santi Pietro e Paolo 34
00144 ROMA

10772 MICIELI, V., Mr
* Via Manfredi 10
97012 CHIARAMONTE GULFI (RG)

10061 MILANA, A., Mr
* c/o Dr Puleo Vincenzo
V.le Matteotti 64
18100 IMPERIA ONEGLIA

10062 MINNA, D., Mr
* Via P. Belon 135
00169 ROMA

7377 MIONE, M., Dr
* Via F. Meda 155
00157 ROMA

10773 MIZZAN, A., Dr
* Via Vira 7
21034 COCQUIO (VA)

10063 MOFFA, R., Mr
* V.le Martiri d'Ungheria 17
82100 BENEVENTO

10064 MOHAMED, A., Mr
* Via Carlo Tenca 5/9
00159 ROMA

10065 MOLINARO, P., Mr
* Via F. Gregorovius 3
00179 ROMA

10774 MONACO, A., Dr - Engr
* Via Fregene 67
00183 ROMA

3620 MONACO, I., Dr - Engr
* Via C.Ederle 38
37100 VERONA

4378 MONGIOVI', L., Dr
* Via Giachery 8
90142 PALERMO

6272 MONINI, S., Dr - Engr
* Via Terranova 28
44100 FERRARA

10442 MONTANI, S., Dr - Engr
* Via Panama 74
00198 ROMA

7379 MONTELEONE, V., Dr
* Lung.re Anguillara 11
00153 ROMA

7380 MONTUORI, C., Prof.- Engr
* Ist. Idraulica e Costruz. Idraul.
Via Claudio 21
80125 NAPOLI

10066 MORABITO, G., Dr
* Via 27 Aprile 8
\+ 54100 MASSA

3622 MORALDI, G., Prof - Engr
* L.go Messico 15
00198 ROMA

10775 MORAMARCO, G., Dr
* Studio Geotec. Geologico
Via Pasquale Giocoli
75017 SALANDRA (MT)

10067 MORELLI, V., Dr - Engr
* ENEL/DCO
Via G.B. Martini 3
00198 ROMA

3623 MORGANTE, S., Prof - Dr
Ist. Mineralogia e Petrogr.
P. le Europa 1
34127 TRIESTE

3624 MORI, M., Dr - Engr
* Via V. Emanuele 27
55043 LIDO DI CAMAIORE (LU)

10542 MORSELLI, G., Mr
* Via Risorgimento 4
93012 GELA

3626 MORTARI, R., Dr
* Istituto di Geologia-Cita
Universitaria
P.le delle Scienze
00185 ROMA

10068 MORTILLARO, F.P., Dr
* Via Nino Bixio 9
30030 OLMO DI MAERNE (VE)

3627 MOSCO, E., Dr - Engr
* Via L. Bodio 37
00191 ROMA

3628 MOSCONI, B., Dr
* Via Crescenzi 1
24100 BERGAMO

3629 MOTTA, A., Dr - Engr
* ENEL-CPCIE di Venezia
C.so del Popolo 111
30172 VENEZIA MESTRE

3630 MOTTA, S., Dr - Engr
* Via R.R. Garibaldi 42
00145 ROMA

3631 MOUTON, J., Dr - Engr
* Via Igea 49
00135 ROMA

10562 MUGELLI, M., Mr
* Via G. Marconi 82
50131 FIRENZE

10070 MUSSO, A., Dr - Engr
* Via Eugenio l'Emiro 11
90124 PALERMO

7381 MUZZI, F.P., Dr - Engr
* CNEN-Divisione Sicurezza Contr.
V.le Regina Margherita 125
00198 ROMA

6273 MYALLONNIER, M., Dr - Engr
* Via Pascoli 3
24100 BERGAMO

10570 NAPOLI, G., Mr
* Via Gliannuzzi 5
87030 CASTIGLIONE SCALO (CS)

10071 NAPOLITANO, L., Mr
* C.so V. Emanuele 118
80053 CASTELLAMMARE DI STABIA (NA)

4039 NAPPI, A., Dr - Engr
* Via Posillipo 281
80123 NAPOLI

10776 NASCA, T., Dr - Engr
* c/o ANAS
V. le Leopardi 126
33100 UDINE

10777 NASCE', M., Dr - Engr
* Via Mazzini 82
40122 BOLOGNA

10072 NATTA, M., Mr
* Via Alessandria 5
15040 VALLE S. BARTOLOMEO (AL)

3632 NEGRI, L., Dr - Engr
* C.so Regio Parco 1
10152 TORINO

7050 NENZI, A., Mr
* c/o Soc. NENZI
Via G. Galilei 16
20090 SEGRATE (MI)

3633 NERVO, V., Dr - Engr
* Via Roma 43/21
36061 BASSANO DEL GRAPPA (VI)

3634 NEWBURG, S., Dr - Engr
* Via G. Verdi 35
80133 NAPOLI

332 NICCOLAI, C., Dr - Engr
* c/o Montedil
Via Meraviglia 2
20123 MILANO

10073 NICOLA, A., Mr
* C.so Cavour 61
27036 MORTARA (PV)

626 NICOTERA, P., Prof.- Engr
* Ist. Geologia Applicata
\+ P.le Tecchio
80125 NAPOLI

10075 NISIO, P., Dr - Engr
* Via Dante De Blasi 31
00151 ROMA

10074 NOBILE, L., Mr
* Dipartimento Difesa del Suolo
Universita della Calabria
87100 COSENZA

7052 NOCCA, G., Dr - Engr
Via Belli 9
27100 PAVIA

10778 NOCILLA, N., Dr - Engr
* Via S. Cappello 16
90100 PALERMO

10503 NOLI, M., Dr - Engr
* Via Giotto 134
52100 AREZZO

10076 NOVA, R., Dr - Engr
* Istituto Scienza e Tecnica delle
Costruzioni
Politecnico
P.za L. da Vinci 32
20133 MILANO

10779 NOVELLINI, G., Dr
* Via Cav. di Vitt. Veneto 4
46010 REDONTESCO

10780 NUCERA, E., Mr
* Centro Residenziale
Universita degli Studi della
Calabria
Blocco 11 - App. 3 - Piano Terra
87036 ARCAVACATA DI RENDE (CS)

86 OBERTI, G., Prof. - Engr
* ISMES
Via S. Sandri 2
20121 MILANO

10443 OGLIOTTI, C., Dr - Engr
* Via Amipsia 61
00124 ROMA

10077 OLCESE, A., Dr
* Via C. Cabella 226/27
16122 GENOVA

10781 OLIVA, G., Dr - Engr
* Via Posillipo 276
80123 NAPOLI

7053 OLIVERO, S., Prof. - Engr
* Via Baglivi 5
\+ 00161 ROMA

10078 OLIVETI, P., Dr - Engr
* Coop. Enotria
87054 ROGLIANO (CS)

10558 ONOFRI, C., Mr - Geom.
* ESTRAMED
Via Campobello
00040 POMEZIA

7383 ORAZI, O., Mr
* V.le delle Province 114/B/16
00161 ROMA

6274 ORLANDINI, G., Dr
* Via C. Colombo 15
\+ 84019 VIETRISUL MARE (SA)

10782 ORSOGNO, L., Dr
* c/o Geo/Apulia
Via Sabotino 5
71100 FOGGIA

7054 OTTAVIANI, M., Dr - Engr
* Istit. di Arte Mineraria
Via Eudossiana 18
00184 ROMA

10081 PAGANO, D.
* Via Sciuti 112
90144 PALERMO

10082 PAGANO, M., Dr
* C.so Canalgrande 98
41100 MODENA

10084 PAGLIONE, G., Dr
* C.so Montegrappa 15
16137 GENOVA

7384 PAGLIANO, C., Dr - Engr
* Istituto Scienze delle Costruzioni
Via Monte d'Oro 28
00186 ROMA

10083 PAGLIANO, L.P., Dr - Engr
* Centro Sperimentale Stradale ANAS
00060 CESANO DI ROMA

7055 PAIOLA, A., Dr
* Via Marzanica 4/A
\+ 24100 BERGAMO

10559 PALAZZI, M., Dr - Engr
* V.le della Liberta 388
53042 CHIANCIANO TERME (SI)

10783 PALERMO, M., Mr
* Via di Niso 80
80124 NAPOLI

10784 PALLEGI, M., Dr
* Via Cimabue 2
52100 AREZZO

7385 PALMIERI, A., Mr
* Via Suvereto 10
00139 ROMA

10523 PALTRINIERI, C., Dr - Engr
* Via Provinciale 86
41036 MEDOLLA (MO)

3641 PANDOLFI, C., Prof. - Engr
* Via Genserico Fontana 5
00136 ROMA

7056 PANUNZI, G., Dr - Engr
* Circonvallaz. Appia 19
00179 ROMA

7386 PAOLANTONIO, P., Dr - Engr
* Via Nevio 76
80122 NAPOLI

3642 PAOLELLA, G., Dr - Engr
* Villa Eugenia
56042 CRESPINA (FI)

3643 PAOLINA, R., Dr - Engr
* ENEL
Via Bertola 40
10122 TORINO

7387 PAPARO FILOMARINO, M., Dr - Engr
* Via Tasso 91
80121 NAPOLI

10533 PAPPAS, C., Mr
* Via Tiberio 62B
80124 NAPOLI

10444 PARATORE, E., Dr - Engr
* Via Cassia 531
00189 ROMA

10538 PARBONI, F., Dr
* Via Rodriguez R. Pereira 27
00136 ROMA

10785 PARMIGIANI, B., Dr - Engr
* Via V. Bellini 23
20052 MILANO

10786 PASQUALINI, E., Dr - Engr
* Ist. Scienza delle Costruzioni
Politecnico di Torino
Fac. Ingegneria
Corso Duca degli Abruzzi 24
10129 TORINO

10514 PASQUESI, M., Dr - Engr
* FONDEDILE SpA
Plazza Buenos Aires 5
00198 ROMA

10086 PASQUINI, M., Mr - Geom.
* Via F. Redi 63
50144 FIRENZE

3645 PASSARO, A., Prof. - Engr
* Via Gen. G. Orsini 30
80132 NAPOLI

4040 PASTORE, R., Dr - Engr
* Via Nicola Fabrizi 133
65100 PESCARA

10087 PATANE', P.L., Mr
* Via Q.F. Pittore 20
00136 ROMA

10534 PATRICOLO, F., Dr - Engr
* Via Cagna 56
09100 CAGLIARI

10505 PAVAN, G., Dr - Engr
* C.so di Porta Romana 51
20122 MILANO

3646 PAZZAGLI, G., Dr
* Via Appia Nuova 52
00183 ROMA

6275 PEDOIA, R., Dr
* Direttore Tec. SCOT s.r.l.
V.le Etiopia 18
00199 ROMA

10787 PEDRONI, S., Dr - Engr
* c/o ENEL/CRIS
Via Ornato 90/14
20162 MILANO

3649 PELLEGRINO, A., Prof - Engr
* Via Andrea d'Isernia 28
80122 NAPOLI

7389 PELLEGRINO, G., Dr - Engr
* Soc. Fondedile
Via G. Verdi 35
80133 NAPOLI

10543 PERCIVALLI, L., Dr - Engr
* Via Martani 5
06049 SPOLETO

7390 PEROCCO DELLA MEDUNA, T., Dr - Engr
* Via G. Leopardi 29
20123 MILANO

3652 PEROSINO, A., Dr - Engr
* Via Albaro 8A/1
16145 GENOVA

10788 PERRONE, P., Dr - Engr
* Via Constantinopoli 8
74014 LATERZA (TA)

10445 PERTOSA, A., Dr - Engr
* Via Filzi 1/18
16166 GENOVA

10089 PERWANGER, W., Dr - Engr
* Studio d'Ingegneria
Ing. W. Perwanger
C.so Italia 32
39100 BOLZANO

6277 PESCE, V., Dr
* P.le Asia 21
00144 ROMA

10789 PETRUCCI, P., Mr
* Q.re S. Francesco St. 146,16
67100 L'AQUILA

3653 PETRUCCO, P., Dr - Engr
* Via Tiepolo 15
33100 UDINE

10508 PETTINARI, L., Dr - Engr
* Via Lepri 8
60013 CORINALDO (AN)

10090 PICARELLI, L., Dr - Engr
* Via V. Scala 12
00128 NAPOLI

7391 PICASSO, M., Dr
* ATLAS COPCO ITALIA
Via F.lli Gracchi 39
20092 CINISELLO BALSAMO (MI)

7392 PICCIO, A., Dr
* Via Digione 7
\+ 27100 PAVIA

3657 PICCIONE, M., Dr - Engr
* Via Sierra Nevada 37
00144 ROMA

3658 PICCIRILLI, V., Dr
* P.za Marconi 6
02013 ANTRODOCO (RI)

3659 PICIOCCHI, F., Dr - Engr
* Costruzioni U. Girola SpA
Via Nomentana 303
00162 ROMA

10091 PIERSANTI, A., Mr
* Strada di S. Filomena 12
05100 TERNI

3660 PIETRA, W., Dr - Engr
* Via Sebastiano Veniero 2
20148 MILANO

7393 PIFFER, R., Dr - Engr
* Via Liguria 3
27100 PAVIA

4041 PINO, C., Dr
* Soc. Lindquist & C.
V. le Europa 55
00144 ROMA

7395 PIOVANO, G., Dr - Engr
* Studio Ingegneria Mineraria
Via Villa della Regina 1
10131 TORINO

3661 PISTILLI, G., Prof.- Engr
* Calata Trinita Maggiore 4
80134 NAPOLI

10092 PIVETTA, A., Dr - Engr
* Via del Bosco 12
95100 S.AGATA LI BATTIATI (CT)

10790 PIZZICANELLA, S., Mr
* Via E. Duse 37
00197 ROMA

3662 POGGIO, S., Dr - Engr
* Via Paullo 7
20135 MILANO

10093 POLI, L., Dr
* Via Paitoni 13
25010 REMEDELLO (BS)

10094 POLI, S., Mr
* Via R. Fucini 91
00137 ROMA

10791 POLIGIONI, G., Dr - Engr
* Via Porto Fluviale 38
00154 ROMA

10792 POLIMANTI, A., Dr
* Via Mogadiscio 23
00199 ROMA

10793 POLISICCHIO, A., Dr - Engr
* Via Martiri di Gerace 8
88100 CATANZARO

3663 POLIZZI, B., Dr - Engr
* Via Fara 30
20124 MILANO

10485 POLLONI, G., Dr
* Via Mentana 8
22100 COMO

10095 PONE, E., Dr - Engr
* "Idrotecneco"
61047 S. LORENZO IN CAMPO (PS)

10794 PORRI, C.A., Dr - Engr
* Via Isole Eolie 7
00141 ROMA

3664 POZZI, P., Dr
* Via Germanico 203
00192 ROMA

10096 PREMICI, A., Dr
* Via M. Dodero 11/6
16146 GENOVA

10097 PRESTININZI, A., Dr
* Via Isabella d"Este 14/9
00164 ROMA

10098 PREVEDINI, C., Dr - Engr
* PONTEGGI EST Srl
C.so di Porta Nuova 3
20121 MILANO

10099 PREVIATELLO, P., Dr
* Centro Geotecnico Veneto
\+ Via Ognissanti 39
35100 PADOVA

7399 PUGLISI, S., Dr - Engr
* Via Monte Pollino 2
00141 ROMA

7398 PUGLIESE, G., Dr - Engr
* P.za Conciliazione 4
20123 MILANO

10446 PULCINI, S., Dr
* Via Brigiotti 14
64020 S.NICOLO'A TORDINO (TE)

3665 PUTTI, A., Dr - Engr
* Prato della Valle 60
35100 PADOVA

10100 QUATTRINI, G., Mr
* Via della Resistenza 7
01037 RONCIGLIONE (VT)

10795 RABBIOSI, G., Mr
* Piazza Mazzini 12
39100 BOLZANO

4042 RABBIOSI, T., Dr - Engr
* C.so Italia 10
39100 BOLZANO

10796 RADAELLI, E., Dr
* c/o Radaelli & Castellotti Snc.
Via XXV Aprile 10
20090 SEGRATE (MI)

3666 RADINA, B., Prof - Engr
* P.za Luigi di Savoia 24
\+ 70121 BARI

7400 RAGGI, C., Mr
* Via Curtatone 21
36100 VICENZA

10797 RAGNI, F., Mr
* Via Risorgimento 28
06024 GUBBIO (PG)

7401 RAGOSTA, G., Dr - Engr
* Impresa Astaldi
Via Po 9
00198 ROMA

10101 RANALLETTA, D., Mr
* Via Degli Equi 8
00185 ROMA

7402 RANTUCCI, G., Dr - Engr
* Via Segesta 25
00179 ROMA

428 REBAUDI, A., Dr - Engr
* C.so Re Umberto 138
10134 TORINO

7403 REGGIO, F., Mr
* Via Pitagora 5
03036 ISOLA LIRI (FR)

10448 RENDINA, M., Dr
* Via Rinaldo d'Aquino 19
90135 PALERMO

10102 RETACCHI, F., Dr - Engr
* Borgo Angelico 6
00193 ROMA

7408 RIBACCHI, R., Prof. - Engr
* Via Damiano Chiesa 32
+ 00136 ROMA

10103 RIBAS HERMELO, J.A., Dr - Engr
* DIGIAC-DAP - Secons
HABANA (CUBA)

7404 RICCA, G., Mr
* V.le della Republica 134
87100 COSENZA

10798 RICCARDELLI, N., Mr
* Via Benaco 5
00199 ROMA

10104 RICCARDI, A., Mr
* Via Pascoli 17
27058 VOGHERA (PV)

10497 RICCARDI, F., Dr - Engr
* c/o DI•PENTA SpA
Via Agrigento 5
00161 ROMA

10105 RICCARDO, L., Mr
* Via delle Antiche Mura 68
71043 MANFREDONIA (FG)

3669 RICCERI, G., Dr - Engr
* Via Configliachi 5
35100 PADOVA

7405 RICCI, B., Mr
* Via Montesilvano 18
00156 ROMA

10106 RICCI, M., Dr
* Via C.A. Nallino 21
00147 ROMA

3670 RICCI, R., Dr
* Via Flaminia 435
00196 ROMA

10799 RICCIARDI, C., Dr - Engr
* Via Michelangelo Testa 8
84100 SALERNO

7406 RICCIARDI, F., Dr - Engr
* ENEL/CPCIE
Viale Kennedy 5
80125 NAPOLI

7407 RICCIONI, R., Dr - Engr
* ISMES
V.le Giulio Cesare 29
24100 BERGAMO

3671 RICCIUTI, G., Dr
* Via Polibio 21
00136 ROMA

10449 RIDENTE, I., Mr
* Via Sottotenente Iannaccone 14
83100 AVELLINO

10800 RIGA, G., Dr
* Via Zamellati 11
88046 S. EUFEMIA LAMEZIA (CZ)

10107 RINAURO, A., Dr - Engr
* Via Petrarca 20
80123 NAPOLI

3673 RINOLDI, G., Dr - Engr
* P.zza G. Cesare 5
20145 MILANO

10108 RIPANUCCI, G., Dr - Engr
* Via dell'Acqua Bullicante 260
00177 ROMA

7409 RIVAZIO, F., Mr - Geom.
* Fondedile Soc.
Via G. Verdi 35
80133 NAPOLI

10109 RIZZO, A., Mr - Geom.
* V.le Virgilio 76
74100 TARANTO

10801 ROBERTI, P., Mr
* Via Bandeleo 10
10064 PINEROLO (TO)

10802 ROBOTTI, F., Dr - Engr
* c/o S.I.S. Geotechnica
Via Roma 15
20090 SEGRATE (MI)

10803 RONCONI, D., Dr - Engr
* Via T.G. Corrado 11
83100 AVELLINO

10111 RONZONI, E., Dr - Engr
* Via M. Di Cefalonia 5
24100 BERGAMO

7410 ROSSETTI, M., Dr - Engr
* ENEL/CPCIE
V.le Kennedy 5
80125 NAPOLI

3676 ROSSI, A., Dr - Engr
* Via B. da Quintavalle 13
06081 ASSISI

10112 ROSSI, F., Dr - Engr
* Via R. Pereira 201
00136 ROMA

10500 ROSSI, L., Dr
* c/o C.C.C. FONDISA SpA
Via Cicerone 35
00193 ROMA

10450 ROSSI, P., Mr
* V.le Quartara 7/4
16148 GENOVA

6278 ROSSI, P.P., Dr - Engr
* Via F. Baracca 1
+ 24100 BERGAMO

6279 ROSSI, S., Dr - Engr
* SACIF
\+ Via G. Verdi 35
80133 NAPOLI

10451 ROSSI, U., Dr - Engr
* Via Cimbali 64
90142 PALERMO

10113 ROSSO, M.M., Mr
* P.za Vittorio Veneto 10
10123 TORINO

10114 ROTATORI, A., Dr - Engr
* Via F. Bandiera 4
61040 S. VITO SUL CESANO (PS)

3679 ROTUNDI, L., Dr - Engr
* c/o Soc. DI-PENTA
Via Agrigento 5
00161 ROMA

3681 ROUL, M., Dr - Engr
* Via G. De Min 2
32100 BELLUNO

10804 ROVELLA, A., Dr - Engr
* Via XX Settembre 6
87036 RENDE (CS)

10115 RUBERL, E., Dr
* Via Gherardini 2
20145 MILANO

7412 RUBERTO, S., Dr - Engr
* Via Placida 6
98100 MESSINA

3680 RUBINO, D., Dr - Engr
* V.le della Vittoria 53
92100 AGRIGENTO

10805 RUSSO, D., Dr
* II Traversa Nicolardi (Parco Aviolo) 24
80131 NAPOLI

3682 RUSSO, F., Dr - Engr
* Via A. Baccarini 124
00179 ROMA

10806 RUSSO, F.P., Dr
* Via Venezia 2
82100 BENEVENTO

7413 RUSSO, R., Dr - Engr
* Via Passerini 2
20052 MONZA

10807 RUSTIA, V., Mr
* Via Ginestreto 67
00040 ARICCIA (Roma)

7414 SABATINELLI, R., Mr
* Via Castiglione d'Orcia 28
00149 ROMA

10119 SABINI, G., Dr - Engr
* Via Cilea 265
80127 NAPOLI

3683 SACCHI, R., Dr - Engr
* Via Pier Capponi 12
20145 MILANO

3684 SACCO, F., Dr - Engr
* Via Pignatelli Aragona 82
90141 PALERMO

10120 SAFFIOTI, G., Mr
* Via R. Pugliese 40
89015 PALMI (RC)

3685 SALLUSTO, G., Dr - Engr
* Via Gavinana 4
00192 ROMA

6281 SALMINCI, D., Dr - Engr
* P.za Civitella Paganico 12
00139 ROMA

10121 SALVINI, G., Dr
* Via Cappuccio 13
20123 MILANO

10122 SAMPAOLO, A., Dr
* ENEL-DSR/CRIS
Via Ornato 90/14
20162 MILANO

6282 SAMPERI, G., Dr - Engr
* Via Aquilea 5
90144 PALERMO

3687 SANSONI, R., Dr - Engr
* Via Gramsci 13
20090 OPERA (MI)

10123 SANTOMAURO, A., Mr
* Via E. Caterina 60
84100 SALERNO

10808 SANTORO, W.M., Mr
* Via Antonio Santoro 41
00155 ROMA

3690 SANTOVITO, L., Dr - Engr
* Via Felice Grossi Gondi 51
00162 ROMA

3691 SANTUCCI DE MAGISTRIS, E., Dr - Engr
* V.le Winspeare 19
80127 NAPOLI

3692 SAPIO, G., Prof - Engr
* Salita Arenella 25
80129 NAPOLI

10124 SARNATARO, G., Mr
* Tr. sa F.sco Provenzale 11
80141 NAPOLI

10502 SARTI, L., Dr
* Via dei Giroli 40
25085 GAVARDO (BS)

10125 SARTI, S., Dr
* P.za Vega 5
00056 LIDO DI OSTIA (Roma)

10452 SAVERI, E., Dr
*
ENEL-CTN
Servizio Civile, V.le Regina Margherita 137
00198 ROMA

10535 SAVERI, L., Dr - Engr
*
c/o ENEL
V.le Regina Margherita 137
00198 ROMA

10126 SCAFARELLI, F., Dr - Engr
*
FONDEDILE
Via G. Verdi 35
80133 NAPOLI

10809 SCALFATI, D., Mr
*
Via P. Aimone 4
00185 ROMA

10127 SCALORBI, R., Mr - Geom.
*
Via Bonaventura Somma 14
00100 ROMA

10453 SCANU, G., Dr
*
Via S. Rosalia 14
07010 BENETUTTI (SS)

10544 SCARPELLI, G., Mr
*
Via A. Fioravanti·6
00128 ROMA

7416 SCASSERRA, C., Dr
*
V.le Principe di Piemonte 68
86100 CAMPOBASSO

10810 SCAVIA, C., Mr
*
Via Colle delle Finestre A./bis
10059 SUSA (TO)

10536 SCHEMBRI, B., Mr
*
V.le A. Boito 50
00199 ROMA

3695 SCHLEIFER, A., Dr - Engr
*
Via Buonarroti 30
34141 TRIESTE

3696 SCIABA', A., Dr - Engr
*
Via Cesarea 2/12
16121 GENOVA

10454 SCIOTTI, M., Prof.
*
c/o Ist. di Geologia Applicata
\+
Via Eudossiana 18
00184 ROMA

10811 SCIUTTINI, I., Mr
*
06068 TAVERNELLE (PG)

10812 SCOPELLITI, S., Mr
*
Corso Vercelli 42/B
10155 TORINO

10129 SEGNANA, F., Dr - Engr
*
TITANIA, SpA
Via F. Denti 1
20133 MILANO

10128 SEGU', F., Mr
*
V.le Colgi 65
27100 PAVIA

10504 SELLER, S., Dr - Engr
*
Via Masaccio 4
57100 LIVORNO

10130 SELLERI, G., Dr - Engr
*
ENEL-SOIC
Via Carducci 1/3
20123 MILANO

3697 SEMBENELLI, P., Dr - Engr
*
Via Cimabue 20
20148 MILANO

3698 SEMENZA, M., Dr - Engr
*
Viale E. Dandolo 5
\+
30126 VENEZIA-LIDO

3699 SEMPI, A., Mr - Geom.
*
V.le Elvezia 4
20154 MILANO

10455 SEMPIO, G., Dr - Engr
*
Via Maestri Comacini 9
27100 PAVIA

10813 SESTILI, M., Dr - Engr
*
Via Famagosta 2
00100 ROMA

4043 SFONDRINI, G., Dr
*
Istituto di Geologia
\+
P. le Gorini 15
20133 MILANO

10131 SFRISO, M., Dr
*
Via Veneto 17
45017 LOREO (RO)

10818 SGOBINO, F., Dr
*
Comunita Montana del Gemonese
Via della Roggia 330
33013 GEMONA del FRIULI

10132 SICILIA, A., Dr
*
Pal. Gallo - De Marco
Commenda
87036 RENDE (CS)

10133 SIKORSKI, V., Dr
*
Via Boncompagni 30
\+
20139 MILANO

10515 SILVESTRI, L., Mr
*
Via A. Boito 15
00199 ROMA

7888 SILVESTRI, T., Dr - Engr
*
C.so V. Emanuele 167/III
80121 NAPOLI

10134 SIMONETTI, G., Dr - Engr
*
Via Roma 277
90040 SAN CIPIRELLO (PA)

10516 SIMONETTI, S., Dr - Engr
* c/o SOGENE
Cantiere Asse Viario
Via D. De Roberto
80143 NAPOLI

6284 SIOTTO, G., Dr - Engr
* Via Tola 21
09100 CAGLIARI

10511 SIRO, L., Dr - Engr
* Salita al Promontorio 23
34123 TRIESTE

4044 SOCCODATO, C., Dr - Engr
* Via Caio Canuleio 121
00174 ROMA

3704 SOLLAZZO, R., Dr - Engr
* V.le Giuseppe Di Vittorio 64
71100 FOGGIA

10135 SORANZO, M., Dr - Engr
* Centro Geotecnico Veneto
\+ Via Ognissanti 39
35100 PADOVA

10819 SPADAFORA, F., Mr
* Via Rupe 26
88050 PENTONE (CZ)

7417 SPALATRO, A., Dr - Engr
* Via Mascherpa 3/B
27100 PAVIA

7418 SPERANZA, V., Dr - Engr
* FONDEDILE
V.le De Filippis 228
88100 CATANZARO

7419 SPILOTRO, G., Dr - Engr
* Via De Amicis 2
70125 BARI

10137 SPORTELLI, G., Mr
* Via Germano Sommeiller 11
00185 ROMA

3705 STANINI, A., Mr
* Via Ghislandi 31
24100 BERGAMO

10138 STARACE, R., Dr - Engr
* Via Tommaseo 2
80121 NAPOLI

6285 STARNI, I., Dr
* Studio Tecnico Geologico
Via Citta Nuova 26
39049 VIPITENO (BZ)

10512 STAZZONE, G., Dr
* Via J. Kennedy 9
31015 CONEGLIANO VENETO (TV)

10139 STIRPE, C., Mr
* Via Umberto I 96
03020 TORRICE (FR)

539 STRAGIOTTI, L., Prof - Engr
* Ist. Arte Mineraria
\+ C.so Duca degli Abruzzi 24
10129 TORINO

10820 SUPERINA, R., Dr
* c/o Radaelli & Castellotti Snc.
Via XXV Aprile 10
20090 SEGRATE

10140 TAGNANI, C., Dr - Engr
* ISMES
V.le Giulio Cesare 29
24100 BERGAMO

10513 TAMAGNINI, B., Dr - Engr
* Via Cesare Baronio 145
00179 ROMA

10456 TANCREDI, G., Dr - Engr
* Via Cremuzio Cordo 11
00136 ROMA

3708 TARCHI, P., Dr - Engr
* Tecnimont/Montedison
P.za Turr 3
20149 MILANO

10141 TASSI, T., Mr
* V.le Garibaldi 94
30172 MESTRE (VE)

6286 TATA, G., Dr - Engr
* P.za di Villa Carpegna 43
00165 ROMA

3709 TEDESCHI, C., Dr - Engr
* Via Petrarca 79
80122 NAPOLI

3712 TERRACINA, F., Dr - Engr
* Via F. Dall'Ongaro 99
00152 ROMA

4046 TERRANOVA, R., Dr
* Via Ekengren 21/2
\+ 16033 LAVAGNA (GE)

10525 TESIO, G., Dr
* Via Norico 4
00183 ROMA

3713 TESTI, S., Dr - Engr
* Via Oberdan 106
10135 TORINO

3714 THOMOS, M., Dr - Engr
* EDOK SA - ETER SA
Via Akadimias 43
ATHENS 135 (GREECE)

6288 TOCCHIO, G., Dr - Engr
* Via Luigi Zoja 15
20153 MILANO

10457 TOFANI, A., Mr
* L. go Giuseppe Veratti
2º Gruppo
00146 ROMA

10821 TOGLIANI, G., Dr - Engr
* PALER-SA Lugano
Piazza Brocchi
CH - 6926 MONTAGNOLA

10142 TOMASI, G., Dr
* Museo Tridentino di Scienze Naturali
Via Calepina 14 C.P. 168
38100 TRENTO

10144 TOMBINI, B., Mr
* Via Meravigli 13
20123 MILANO

10545 TONTI, F., Dr - Engr
* Via d'Aquino 74
74100 TARANTO

10822 TONTODONATI, U., Mr
* Via Sestriere 13/bis
10024 B.S.P. MONCALIERI (TO)

10823 TORNABENE, P., Dr
* Viale Emporium (S.Leone) 94
92100 AGRIGENTO

3719 TORNAGHI, R., Dr - Engr
* Via G. D'Arezzo 8
\+ 20145 MILANO

10824 TORO, F., Dr - Engr
* Via S. Valentino 2
00197 ROMA

7420 TORRICE, E., Mr
* Via 24 Maggio 32
04100 LATINA

10146 TORRIGIANI, M., Dr - Engr
* Via Dassori 12
16131 GENOVA

10147 TORTI, G., Dr - Engr
* Via Miami 13
27100 PAVIA

6289 TOSI, A., Mr - Geom.
* Via Montanari 17
40138 BOLOGNA

10148 TOTTONE, A., Mr - Geom.
* Via Cacciano 12
67011 ARISCHIA (AQ)

4049 TOZZI, G., Dr - Engr
* Via Salaria 300
00199 ROMA

10149 TRIPICIANO, L.M., Dr - Engr
* ENEL/CPCT
Via Cardano 10
20124 MILANO

3721 TROIANO, V., Dr - Engr
* Via Agrigento 5
00161 ROMA

3722 TROISI, S., Dr - Engr
* 5 Trav. Re David 34
70125 BARI

10150 TROMBETTA, C., Dr - Engr
* Via Vincenzo Tizzani 64
00151 ROMA

7422 TULLIO CATALDO, F., Dr - Engr
* Via Petrarca 141/M
80122 NAPOLI

7423 TUSA, A., Dr - Engr
* Via Eduardo Carapelle 14
90100 PALERMO

4050 TUTINO, A., Dr - Engr
Suit-305-3911 Carrigan Court
BURNABY-B.C. V3N4J7
CANADA

3723 UMILTA', G., Dr - Engr
* Via La Masa 37
90139 PALERMO

10561 URSO, G., Dr
* Geotecneco SpA
61047 S.LORENZO IN CAMPO (PS)

10825 VACCA, R., Mr
* Via Carlo Gremo 33
10040 LEINI'(TO)

3724 VAIA, F., Dr
* Ist. Geologia e Paleontologia
Universita
P.le Europa
34127 TRIESTE

10152 VALENTE, G., Dr - Engr
* Via di Villa S. Maria 14
80122 NAPOLI

3725 VALENTINI, G., Dr - Engr
* L.go Sicilia 13
\+ 70125 BARI

6291 VALESINI, R., Dr
* Ingegneria e Geotecnica
Via Camozzi 95
24100 BERGAMO

3726 VALLETTA, N., Dr - Engr
* C.so del Popolo 111
30172 MESTRE (VE)

7424 VALORE, C., Dr - Engr
* Salita Partanna 3
90133 PALERMO

3727 VALOROSO, S., Dr - Engr
* Via F. Saporito 31
81031 AVERSA (CE)

3728 VANDONI, C., Dr - Engr
* V.le Tunisia 13
20124 MILANO

7425 VANDONI, G., Dr - Engr
* V.le Tunisia 13
20124 MILANO

10154 VANNUCCHI, G., Dr - Engr
* C.so Italia 29
50123 FIRENZE

7426 VAROSIO, G., Dr - Engr
* Societa Italiana Impianti
P.za Piccapietra 9
16121 GENOVA

10155 VARVERIS, M., Mr
* Via Volta 27
27100 PAVIA

10826 VASSALO, V., Dr - Engr
* Via dell'Aia 15
90014 CASTELDACCIA (PA)

10156 VECCHIO, C., Mr
* Kimates SpA
V. le Elvezia 18
20154 MILANO

3730 VECELIO, P., Engr
* Lung. re Flaminio 58
00196 ROMA

3731 VEDER, C., Dr - Engr
* Rechbauerstrasse 12
A-8010 GRAZ (AUSTRIA)

3732 VENTRIGLIA, U., Prof - Engr
* Via Eudossiana 18
\+ 00184 ROMA

10537 VENTURA BORDENCA, G., Dr
* Via A. Manzoni 13
92100 AGRIGENTO

6292 VENTURA, P., Dr - Engr
Via Veronica Gambara 16
00137 ROMA

10827 VERONESE, L., Dr
* Via Maggiore Toselli 32
90143 PALERMO

10157 VICENZETTO, V., Dr
* Geo-Studio
Via Municipio 18
35040 VILLA ESTENSE (PD)

10158 VIELMO, A., Dr - Engr
* Via General Giardino
35013 CITTADELLA (PD)

111 VIELMO, I., Dr - Engr
* Via F. Ferruccio 10
20145 MILANO

3736 VIGGIANI, C., Prof - Engr
* Via Posillipo 281
80123 NAPOLI

10458 VILLANI, A., Mr
* Via Trieste 1
71036 LUCERA (FG)

7427 VILLARINI, G., Dr - Engr
* Via Boncompagni 61
00187 ROMA

7428 VINALE, F., Dr - Engr
* Ist Tecnica delle Fondazioni e
Costruzioni in Terra
Via Claudio 21
80125 NAPOLI

10526 VIOLA, C., Dr
* Via Machiavelli 2
20145 MILANO

10159 VIOLA, T., Mr
* Via Madonna delle Grazie 13
65010 FARINDOLA (PE)

3737 VIPARELLI, C., Prof - Engr
* Rampe Brancaccio 8
80132 NAPOLI

3738 VIPARELLI, M., Prof - Engr
* Via Salvator Rosa 253
80136 NAPOLI

10527 VITA, M., Mr
* Via Casale Nuovo 8
85050 SATRIANO DI LUCANIA (PZ)

10160 VITALE, A., Mr - Geom.
* Fondedile
Via G. Verdi 35
80133 NAPOLI

10517 VITTURI, A., Dr
* Via Bellano 28
35100 PADOVA

7429 VITULLO, C., Dr - Engr
* Geosonda
Via Girolamo da Carpi 1
00196 ROMA

10548 VIVALDI, L., Mr. - Geom.
* V. le dei Bizantini 45
88100 CATANZARO

10459 VOLONNINO, N., Mr
* Via Epomea 481
80126 NAPOLI

7430 VOLPE, L., Dr
* c/o ENEL-DCO-SG
\+ V. le Regina Margherita 137
00198 ROMA

10162 VOLPINI, S., Dr
* Via Arco dei Cenci 7
00186 ROMA

10163 VOZZI, E., Dr - Engr
* V.le Virgilio 20
74100 TARANTO

10164 VUILLERMIN, E., Prof.
* COMITGEO
Via Perini 66
38100 TRENTO

10165 VULCANO, A., Dr - Engr
* C.so d'Italia 96/B
87100 COSENZA

3740 WETTER, G., Dr - Engr
* Via Baldissera 2
20129 MILANO

3741 WOLF, E., Dr - Engr
* c/o Autostrada Messina-Palermo
Contrada Praia
98060 PATTI

10166 ZACCAGNO, F., Mr - Geom.
* Via dell'Aquila 26
67011 ARISCHIA (AQ)

10167 ZACCHI, C., Dr
* Via Clitunno 2
00198 ROMA

10828 ZACCO, F., Dr
* Vialle della Libertá 163
90143 PALERMO

4051 ZAFARANA, F., Dr
* Via G. Vagliasindi 9
95126 CATANIA

10829 ZAMA, G., Mr
* Via XX Settembre 23/1
48018 FAENZA (RA)

7431 ZAMBELLONI, F., Dr - Engr
* Via Filippo Marchetti 19
00199 ROMA

10168 ZAMBRANO, A., Mr
* Via C. Africano 35
84100 SALERNO

3742 ZANELLA, G., Dr - Engr
* Via Rovereto 4
37100 VERONA

3743 ZANETTO, L., Dr - Engr
* Via V. Gussi 21
20059 VIMERCATE (MI)

10169 ZANINETTI, A., Dr - Engr
* Via C. Goldoni 34
21100 VARESE

10170 ZANNIER, G., Dr - Engr
* Via Zuccoli 19
20125 MILANO

10460 ZANON, A., Dr - Engr
* c/o Torno & C.SpA
Via Albricci 7
20122 MILANO

10171 ZANON, P., Dr - Engr
* Via Brentari 1
38100 TRENTO

3744 ZARETTI, L., Dr - Engr
* Via Ciro Menotti 21
20129 MILANO

10172 ZELGER, P., Dr - Engr
* Provincia Autonoma di
39100 BOLZANO

3745 ZIA, R., Dr
* Istituto di Geologia
Via S. Maria 53
56100 PISA

3746 ZILBERSTEIN, L., Mr
* Via G. Verdi 35
80133 NAPOLI

3747 ZILLER, M., Dr - Engr
* C.C.C. Fondisa
Via Cicerone 35
00193 ROMA

10173 ZOPPELLARO, R., Dr
* Via Carpini - Cond. Carpini
45026 LENDINARA (RO)

3748 ZOTTI, G., Dr
* c/o Radaelli & Castellotti Snc.
Via XXV Aprile 10
20090 SEGRATE (MI)

10546 ZOTTI, N., Mr
* P.za Riola 6
82030 PONTE (BN)

10461 ZUCCATO, P., Dr
* Via Morazzone 12
21100 VARESE

10174 ZUCCHELLI, P., Mr
* Via Prati 6
24020 ARDESIO (BG)

7432 ZULIAN, D., Dr - Engr
* Via Suvereto 10
00139 ROMA

SUPPORTING MEMBERS

7115 DIPENTA S.p.A.
Via Agrigento 5
00161 ROMA

7231 GEOTECNA - STUDIO APPLICAZIONI GEOTECHNICHE
Via Roncaglia 14
20146 MILANO

8148 ISMES - ISTITUTO SPERIMENTALE MODELLI E STRUTTURE
Viale Giulio Cesare 29
24100 BERGAMO

10557 NENZI S.p.A.
Via Galilei 16
20090 SEGRATE (MILANO)

7122 QUADRIO CURZIO S.p.A.
Via Locateli 6
MILANO

7550 SGI - STUDIO GEOTECNICO ITALIANO
Via Ripamonti 89
20139 MILANO

JAPAN

JAPANESE COMMITTEE FOR ISRM

Japan Society of Civil Engineers, Yotsuya 1-chome, Shinjuku-ku, TOKYO, 160

Chairman: Mr T. Misukoshi
Secretary: Mr T. Kawagoe

Total Number of Ordinary Members - 162
Total Number of Supporting Members - 62

ORDINARY MEMBERS

3861 ABE, M.
Instructor
Aramaki-Aza-Aoba, Sendai
MIYAGI, 980

3862 ADACHI, K.
Vice Manager of Research Dept
3154, Ohbuchi, Fuji
SHIZUOKA, 417

13284 * ADACHI, T.
Associate Professor
Gokasho
UJI-SHI, 611

13285 * AKAGI, T.
Associate Professor
Eisei-cho 2-1, Aichi Prefecture
TOYOTA CITY, 471

3863 AKAI, K., Prof.
Professor
Yoshida-Hommachi, Sakyo-ku
KYOTO, 606

3816 AKIMOTO, M.
Lecturer
223, Toroku, Ohemachi
KUMAMOTO, 862

7853 AOKI, M.
5-2-16, Turumaki, Setagaya-ku
TOKYO, 154

13286 DOSHIDA, T.
Masukoso 2F, 33 Wakaba-cho
Kashiwa-shi
CHIBA, 277

10965 ENDO, G.
Lecturer
4-170, Nishi-Okubo, Shinjuku-ku
TOKYO, 160

3865 * FUJII, K.
Lecturer
2-1, Minami-Josanjima-cho
TOKUSHIMA, 770

2563 FUJII, T.
7-22-11, Miyamoto, Funabashi,
CHIBA, 273

10966 FURUMI, K.
Chief Researcher
2597, Shinomiya, Hiratsuka
KANAGAWA, 254

6736 + FURUTA, M.
Geological Engineering
Tsukisamu Higashi 4-Jo, 9-chome
Toyohira-ku, Sapporo
HOKKAIDO, 062

3866 FURUZUMI, M.
Assistant
3-5, Ueda 4-chome, Morioka
IWATE, 020

3867 HADA, M.
Assistant
Taiko-heights 101, 6-13-16,
Higashi-Narashino, Narashino
CHIBA, 275

10967 HAGIMORI, K.
3-5-8, Hamaguchi-Nishi, Suminoe-ku
OSAKA, 559

2564 HATANAKA, S.
Vice-Chief (Civil Engineering) of
Construction Department
5, Nakanoshima 3-chome, Kita-ku
OSAKA, 530

2565 HATANO, T.
Director
1-6-1, Ohtemachi, Chiyoda-ku
TOKYO, 100

10968 HAYAMIZU, H.
Chief, Mining Machine Section
7-17-17, Todoroki, Setagaya-ku
TOKYO, 158

2566 HAYASHI, K., Prof.
Professor
533, Onaga-cho
HIROSHIMA, 730

2567 HAYASHI, M.
Director
131-7, Wakamatsu, Abiko
CHIBA, 270-11

7854 HAYASHI, Y.
Chief of Bridge Construction
Mukaijima-cho, Mitsugi-gun
HIROSHIMA, 722

10969 HIBINO, S.
Senior Research Engineer
4543-203, Terada, Toride
IBARAKI, 302

2568 HIKINO, S.
Manager, Dept. of Hydro Power
1, Toshin-cho, Higashi-ku
NAGOYA, 461

3869 HIRAMATSU, Y., Prof.
Professor
45, Yoshida-Shimoohji-cho
KYOTO, 606

10970 HISATAKE, M.
Research Assistant
Yamadakami, Suita
OSAKA, 565

2569 HORIBE, T., Prof.
Professor
22.1, Aza-Izumizakiura, Tomizawa
SENDAI, 982

3871 HOSHINO, K.
Chief Geologist
1136-355, Imaizumi, Kamakura
KANAGAWA, 247

6738 * HOSOKAWA, M.
Executive Director
322, Serikyaku, Urasoe
OKINAWA, 901-21

2570 IIDA, K., Prof.
Professor
1247, Yachikusa, Yakusa-cho, Toyota
AICHI, 470-03

2571 IIDA, R.
Director, Planning and Research
Administration Dept
1, Asahi, Toyosato-cho, Tsukuba
IBARAKI, 300-26

3872 IIJIMA, H.
Manager
5-14-11, Tamadaira, Hino
TOKYO, 191

3874 INOUE, M., Prof.
Professor
2-39-1, Kurokami
KUMAMOTO, 860

2573 IROBE, M., Prof.
Professor
4-20-1, Nakamachi, Machida
TOKYO, 194

3875 ISHIHARA, T.
Director, General Manager
2-1-1, Nihonbashi-Muromachi, Chuo-ku
TOKYO, 103

7855 ISHIJIMA, Y.
Assistant Professor
Kita 12-jo, Nishi 8-chome, Kita-ku
SAPPORO, 060

13287 + ISHIKAWA, K.
Assistant General Manager
2-5-19, Nishi-Hommachi
Nishi-ku, Osaka-shi
OSAKA, 550

7856 ISOBE, T., Prof.
Professor
Kita 13-jo, Nishi 8-chome
SAPPORO, 060

10971 ITO, H., Prof.
Professor
550-31, Akishino-cho
NARA, 631

2572 ITO, I., Prof.
Professor
270, Daimon-cho, Muromachi-Marutamachi-
Agaru, Kamikyo-ku,
KYOTO, 602

13288 ITO, T.
Professor
Yamadakami, Suita-shi
OSAKA, 565

2574 IWABUCHI, S.
Chief of a Technical Section
6-5, 1-chome, Shinjuku, Shinjuku-ku,
TOKYO, 160

13289 IWAMOTO, J.
President
1-6-3, Taito, Taito-ku
TOKYO, 110

13290 KANDA, S.
6-20-5,Hinosato, Munakata,
FUKUOKA, 811-34

10996 KANEKO, K.
Research Assistant
2-39-1, Kurokami
KUMAMOTO, 860

3877 KAWAHARA, M.
Associate Professor
1-13-27, Kasuga, Bunkyo-ku
TOKYO, 112

10972 KAWAI, H.
Chief of Design Section,
Higashi Itaya, Neo-vill, Motosu-D.,
GIFU, 501-15

3878 KAWAI, T.
1-23, Ankoji-cho 3-chome, Takatsuki,
OSAKA, 569

13291 KAWAI, T.
Professor of 2nd Division
7-22-1, Roppongi, Minato-ku
TOKYO, 106

2575 KAWAMOTO, T., Prof.
* Professor
Furo-cho, Chikusa-ku, Nagoya
AICHI, 464

3921 KIKUCHI, K.
\+ Senior Research Engineer
14-17, Kaga 1-chome, Kashiwa
CHIBA, 277

7857 KINOSHITA, S., Prof.
Professor
Kita 13-jo, Nishi 8 chome, Kita-ku
SAPPORO, 060

2577 KITAHARA, Y.
Chief
1646, Abiko, Abiko
CHIBA, 270-11

3880 KITAMURA, N., Prof.
Professor
Aobayama, Sendai
MIYAGI, 980

10973 KITAOKA, M.
Graduate Student
201, Kopo-Izeki,
2-273, Nakaarai, Tokorozawa
SAITAMA, 359

3881 KIYAMA, H.
Associate Professor
101, Minami 4-chome
KOYAMA-CHO, 680

10974 KOBAYASHI, H.
Research Worker
Onogawa 16-3, Yatabe-cho, Tsukuba-gun
IBARAKI, 305

2578 KOBAYASHI, R., Prof.
Professor
Aramaki-aza-Aoba, Sendai
MIYAGI, 980

10975 KOBAYASHI, S., Prof.
Professor
Yoshida-Hommachi, Sakyo-ku
KYOTO, 606

2579 KOBAYASHI, Y.
\+ Associate Professor
Gokanosho, Uji
KYOTO, 611

13292 KOBAYASHI, Y.
Vice Chief of Foreign Business Division
Harashima Building, 6, San-ei-cho
Shinjuku-ku
TOKYO, 160

3882 KOIDE, H.
Senior Scientist
8, Kawada-cho, Shinjuku-ku
TOKYO, 162

11000 KOJIMA, K.
\+ Associate Professor
7-3-1, Hongo, Bunkyo-ku
TOKYO, 113

7858 KOJIMA, K.
Adviser
6-3, Sakuragawa-cho, Nishikubo
Shiba
Minato-ku,
TOKYO, 105

10976 KOMADA, H.
Research Engineer
Abiko 1646, Abiko
CHIBA, 270-11

2580 KOMAI, I.
President
Miyoshi Building, 2-2-1
Koenji-Kita, Suginami-ku
TOKYO, 166

534 KONDO, N.
Manager of Construction Dept
3-3-22, Nakanoshima, Kita-ku
OSAKA,530

10997 KOSUGI, M.
Senior Engineer
Onogawa 16-3, Yatabe-cho,
Tsukuba-gun
IBARAKI, 305

10977 KURIYAGAWA, M.
Senior Researcher
Onogawa 16-3, Yatabe-cho
Tsukuba-gun
IBARAKI, 305

13293 KUROKAWA, T.
Chief Technical Division
14-14, Ginza 8-chome, Chuo-ku
TOKYO, 104

2581 KUWABARA, K.
\+ Senior Researcher
1-504, Nedo 427-5, Kashiwa,
CHIBA, 113

3884 MAE, I.
Senior Research Engineering
Construction Division
5-487-79, Hanakoganei, Kodaira
TOKYO, 113

13294 MASUDA, T.
Assistant
7-3-1, Hongo, Bunkyo-ku
TOKYO, 113

13310 MATSUKI, K.
Instructor
Aramaki Aza-Aoba
SENDAI

7859 MATSUYOSHI, K.
* Chief of Technical Institute
\+ 1233, Yono, Yono,
SAITAMA, 338

3886 MIKI, G., Prof.
* Professor
22-1, Roppongi 7-chome, Minato-ku
TOKYO, 106

10998 MIKI, K.
Chief Director
Sankyo Bldg., 3-172, Motomachi, Naniwa-ku
OSAKA, 556

7860 MIMAKI, Y.
Deputy Manager of Construction Department
1-3, 1-chome, Uchisaiwai-cho,
Chiyoda-ku
TOKYO, 100

10978 MISAWA, S.
Senior Researcher
Onogawa 16-3, Yatabe-cho, Tsukuba-gun
IBARAKI, 305

3887 * MITANI, T.
Chief Director
91, Ishizaka, Fuji
SHIZUOKA, 417

3888 MITSUI, H.
Geotechnical Engineer
299-7, Shimo-Makuri, Koshigaya
SAITAMA, 343

2583 MIZUKOSHI, T.
Highest Adviser
1-3, 1-chome, Uchisaiwai-cho
Chiyoda-ku
TOKYO, 100

10979 MIZUTA, Y.
Assistant Professor
Tokiwadai, Ube
YAMAGUCHI, 755

2584 MOGI, K., Prof.
3-29-4-501, Yatsu, Narashino
CHIBA, 275

3892 * MORI, H.
President
1-11-5, Kudan-Kita, Chiyoda-ku
TOKYO, 102

13309 * MORI, Y
President
912-23, Seta-cho,Ise-shi
MIE-KEN, 516

10980 MORIE, T.
Chief Engineer
2-16-1, Kyobashi, Chuo-ku
TOKYO, 104

3893 MURAKAMI, T.
Instructor
Aramaki-Aza-Aoba, Sendai
MIYAGI, 980

10981 * MURATA, H.
Assistant Professor
1460-1, Shinko, Hayato-cho, Aira-gun
KAGOSHIMA, 899-51

3894 * MURAYAMA, S., Prof.
Professor Emeritus
26, Shichiku-Kamiumenoki-cho,Kita-ku
KYOTO, 603

13295 MUROI, I.
Chief Research Worker
4-chome, Karita-cho, Sumiyoshi-ku
OSAKA, 558

3895 * NAKABORI, K.
Head
4-11, Esaka 3-chome, Suita
OSAKA, 564

3896 NAKAGAWA, M.
Engineer
1-8-5, Fujimino, Hiratsuka
KANAGAWA, 259-12

3897 NAKAJIMA, S.
1120-5, Suzuya, Yono
SAITAMA, 338

10982 NAKAMURA, J.
Researcher
2594, Shinomiya, Hiratsuka
KANAGAWA, 254

7861 * NAKANO, R.
Chief of 1st Laboratory
2-1-2, Kan-nondai, Yatabe-cho
Tsukuba-gun
IBARAKI, 300-22

3898 NAKAO, K.
Chief Research Engineer
344-1, Nasemachi, Tozuka-ku
Yokohama,
KANAGAWA

10983 NISHIDA, K.
2-2-19, Daitakubo, Urawa
SAITAMA, 336

6004 * NISHIDA, Y., Prof.
Professor
2-4-20, Kodachino, Kanazawa
ISHIKAWA, 920

13296 * NISHIGAKI, Y.
324-18, Jio-cho, Kashihara
NARA, 634

3899 NISHIMATSU, Y., Prof.
Professor
3,7-chome, Hongo, Bunkyo-ku
TOKYO, 113

10984 NISHIMURA, M.
3-5-8, Hamaguchi-Nishi, Sumioe-ku
OSAKA, 559

13297 NOJIMA, T.
Vice President
4-22-26, Chofugaoka, Chofu City
TOKYO, 182

127 NOSE, M.
Vice President
8-2, Marunouchi 1-chome
Chiyoda-ku
TOKYO, 100

3902 OGATA, Y.
Researcher
Onogawa 16-3, Yatabe-cho, Tsukuba-gun
IBARAKI, 305

3904 OHMI, M.
Assistant Professor
Kurokami 2-chome
KUMAMOTO, 860

10985 OHNISHI, Y.
Assistant Professor
Yoshida-Hommachi, Sakyo-ku
KYOTO, 606

13298 OKA, T.
Ginza 8-14-14, Chuo-ku
TOKYO, 104

10986 OKABAYASHI, T.
Assistant
Shinko 1460-1, Hayato-cho, Aira-gun
KAGOSHIMA, 899-51

6003 + OKAMOTO, R.
Chief of Geology Section
8-5-503, Kashima, Hachioji
TOKYO, 192-03

2587 * OKAMOTO, S.
President
1-4-18 Anagawa
CHIBA, 280

10987 OKAMURA, H.
Associate Professor
2-31-9, Kurokami
KUMAMOTO, 860

13299 OKUBO, S.
Associate Professor
7-3-1, Hongo, Bunkyo-ku
TOKYO, 113

10988 OKUMURA, T., Prof.
Professor
Hakusan 4-3-3, Bunkyo-ku
TOKYO, 112

3400 * OKUSA, S., Prof.
Professor
1000, Orido, Shimizu
SHIZUOKA, 424

2588 ONODERA, T., Prof.
Technical Adviser
2-2-12, Nakamagone, Ota-ku
TOKYO, 143

6741 OTSUKA, Y.
Geotechnical Engineer
3-4-17, Etchujima, Koto-ku
TOKYO, 135

6742 SAEKI, M.
Civil Engineer
3rd Masuda Bldg., 10, Nishi-Okubo
3-chome, Shinjuku-ku
TOKYO, 160

3906 * SAITO, M.
Director and Chief Engineer
2-25-12, Ikego, Zushi
KANAGAWA, 249

3907 SAITO, T.
Graduate Student
4-3-5, Ueda, Morioka
IWATE, 020

6743 SAITO, T.
Assistant
Yoshida-Hommachi, Sakyo-ku
KYOTO, 606

3909 SAKURAI, S., Prof.
Professor
Rokkodaimachi, Nada-ku
KOBE, 657

7862 * SASSA, K.
Associate Prof
Yoshida - Hommachi, Sakyo-ku
KYOTO, 606

10999 SATO, M.
General Manager
Matsukawa Bldg
2-33, Shikitsu-cho, Naniwa-ku
OSAKA, 556

3910 SATO, T.
Assistant Professor
27-1, Mizumoto-cho, Muroran
HOKKAIDO, 050

3911 SHIDOMOTO, Y.
Civil Engineer
6-7, Tenjin 1-chome
Chuo-ku
FUKUOKA, 810

13300 * SHIJUUKU, Y.
Tohkoh 1-4, Asahikawa-shi
HOKKAIDO, 078-11

13301 SHIMOTANI, T.
Lecturer
3-1, 7-chome, Hongo, Bunkyo-ku
TOKYO, 113

3912 SHIRAKAWA, O.
Sectional Manager for Technic
8-2, Marunouchi 1-chome, Chiyoda-ku
TOKYO, 100

10989 SUGAWARA, K.
Assistant Professor
2-31-9, Kurokami
KUMAMOTO, 860

13302 SUJAKU, T.
President
Kohbunsha Bldg., 1631, Ikebukuro 3
Toshima-ku
TOKYO, 171

2590 SUZUKI, K.
Technical Adviser
1-4-28, Mita, Minato-ku
TOKYO, 108

10990 SUZUKI, T.
Senior Engineer of Rock Mechanics Division
2-2-19, Daitakubo, Urawa
SAITAMA, 336

2592 TAKAHASHI, H.
Vice-Chief of Hydro-Power Division
19-2, Minami-Otsubo, Ueda
Tenpaku-cho, Tenpaku-ku, Nagoya
AICHI, 468

13303 * TAKAHASHI, K.
4-16-2, Uzumoridai, Higashinada-ku
Kobe-shi
HYOGO, 658

50 TAKANO, M.
President
Tominkogyo-Shinjuku-Bldg
2-8-7, Shinjuku, Shinjuku-ku
TOKYO, 160

3914 TAKATA, A.
Senior Engineer
Onogawa 16-3, Yatabe-cho, Tsukuba-gun,
IBARAKI, 305

13304 TAKEUCHI, N.
Research Assistant of 2nd Division
7-22-1, Roppongi, Minato-ku
TOKYO, 106

10991 * TAKEUCHI, T.
Chief of Rock Mechanics Division
2-2-19, Daitakubo, Urawa
SAITAMA, 336

110 + TANAKA, H.
Managing Director
2-9-20, Noda, Fukushima-ku
OSAKA, 553

10992 TANAKA, S.
2-19, Daitakubo 2-chome, Urawa
SAITAMA, 336

6745 TANAKA, T.
Chief Geologist
Tsukisamu Higashi 4-jo, 9-chome
Sapporo
HOKKAIDO, 062

10993 TANIMOTO, C.
Research Associate
Yoshida-Hommachi, Sakyo-ku
KYOTO, 606

13305 * TANO, H.
U. Teacher
Tamura-machi, Koriyama-shi
FUKUSHIMA, 963

10994 TERADA, M.
Assistant Professor
Yoshida-Hommachi, Sakyo-ku
KYOTO, 606

6746 * TOHNO, I.
Assistant
2-12-1, O-Okayama, Meguro-ku
TOKYO, 152

10995 TOMINAGA, Y.
Associate Professor
165, Koen-cho, Kitami
HOKKAIDO, 090

13306 TSUGAKI, A.
Manager of Business Section
1-2-7, Moto-Akasaka, Minato-ku
TOKYO, 107

3916 UEMURA, T., Prof.
Professor
8050, Igarashi-Nino-machi
NIIGATA, 950-21

3917 USHIDA, M.
Instructor
Aza-Aoba, Aramaki, Sendai
MIYAGI, 980

3918 YAIRI, K.
Associate Professor
1466, Nagara-Jonouchi
GIFU, 502

6641 * YAMAGAMI, T.
Lecturer
2-1, Minami-Josanjima-cho
TOKUSHIMA, 770

7864 YAMAGATA, M.
2-2.1, Tenpozan-cho, Imabari
EHIME, 794

2594 YAMAGUCHI, U., Prof.
Professor
3-1, Hongo 7-chome, Bunkyo-ku
TOKYO, 113

13307 YAMAKAWA, S.
Manager, Foreign Department
Room 5, Seizan-so House, 1-2-22
Nishi-Ochiai, Shinjuku-ku
TOKYO, 161

6749 + YAMANOUCHI, T., Prof.
Professor
10-1, Hakozaki 6-chome, Higashi-ku
FUKUOKA, 812

2596 YASUE, T.
Chief of Slope Failure Section
1, Oaza-Asahi, Toyosato-cho
Tsukuba
IBARAKI, 300-26

235 YOSHIDA, M.
Senior Managing Director
3-3-22, Nakanoshima, Kita-ku
OSAKA, 530

2599 YOSHIKOSHI, M..
Executive Director
1-21-10, Minamidaira, Hino
TOKYO, 191

13308 YOSHIMURA, H.
Director
SHINKANSEN Construction Dept.
Aobadai 2-11-14, 22-1, Meguro-ku
TOKYO, 153

3920 YOSHIMURA, K.
5-15-312, Matsuhama-cho, Ashiya
HYOGO, 659

2600 YOSHINAKA, R.
* Assistant Professor
+ 255, Shimo-Okubo, Urawa
SAITAMA, 338

SUPPORTING MEMBERS

13222 AOKI CONSTRUCTION CO. LTD
4-15, Ohyodominami 1-chome
Ohyodo-ku
OSAKA, 531

13223 CHEMICAL GROUTING CO., LTD
2-1-1, Nishi-Shinjuku, Shinjuku-ku
TOKYO, 160

13224 CHIZAKI KOGYO CO., LTD
23-1, 2-chome, Nishi-shinbashi, Minato-ku
TOKYO

13225 CHUDEN ENGINEERING CONSULTANTS INC.
2-3-30, Deshio
HIROSHIMA

13232 CHUGOKU ELECTRIC POWER CO., INC.
4-33, Ko-machi
HIROSHIMA

13233 CHUO KAIHATSU CORPORATION
3-13-5, Nishiwaseda, Shinjuku-ku
TOKYO, 160

13234 CTI ENGINEERING CO., LTD
Ninth Chuo Bldg., 2, Nihonbashi-Honcho
4-chome, Chuo-ku
TOKYO, 103

13235 DAIHO CONSTRUCTION CO., LTD
1-24-4, Shinkawa, Chuo-ku
TOKYO, 104

13236 DIA CONSULTANT CO., LTD
Kohbunsha Bldg., 1631, Ikebukuro 3
Toshima-ku
TOKYO

13275 ELECTRIC POWER DEVELOPMENT CO. LTD
8-2, Marunouchi 1-chome, Chiyoda-ku
TOKYO, 100

13237 FUJITA CORPORATION
6-15, Sendagaya 4-chome, Shibuya-ku
TOKYO

13226 FUKKEN CHOSA SEKKEI CO., LTD
2-10-11, Hikari-machi
HIROSHIMA

13227 HAZAMA-GUMI, LTD
2-5-8, Kita-Aoyama, Minato-ku
TOKYO

13276 HOKKAIDO ENGINEERING CONSULTANT CO., LTD
Tsukisamu Higashi 4-9-5
Toyohira-ku, Sapporo,
HOKKAIDO

13228 HOKUDEN INDUSTRY CO., LTD
3-1, Sakurabashidori
TOYAMA

13229 HOKUDEN KOGYO CO., INC.
1,3-chome, Higashi 1-jo, Kita,
Chuo-ku
Sapporo
HOKKAIDO

13230 IDOWR ENGINEERING CO., LTD
2-1, Koenji Kita 2-chome,
Suginami-ku
TOKYO

13231 I.N.A. CIVIL ENGINEERING CONSULTANTS CO., LTD
Suido-cho Bldg., Suido-cho, 22-1
Shinjuku-ku
TOKYO

13238 INOUE-GUMI CO., LTD
Nishikubo, Handa-cho, Mima-gun
TOKUSHIMA

13239 JAPAN DEVELOPMENT & CONSTRUCTION CO., LTD
9-9, 4-chome, Minato-ku,
TOKYO

13240 KAIHATSU DOBOKU CONSULTANTS CO., LTD
6-5, 1-chome, Nishi-shinjuku,
Shinjuku-ku
TOKYO

13241 KAJIMA CORPORATION
2-7, 1-chome, Moto-Akasaka,
Minato-ku
TOKYO

13282 KAWASAKI GEOLOGICAL ENGINEERING
Omorikita 1-11-1, Ota-ku
TOKYO

13242 KISO-JIBAN CONSULTANTS CO., LTD
1-11-5, Kudan-Kita, Chiyoda-ku
TOKYO

13243 KOKUSAI AERIAL SURVEYS CO., LTD
2, Rokuban-cho, Chiyoda-ku
TOKYO

13244 KONOIKE CONSTRUCTION CO., LTD
27, 4-chome, Kitakyuhojimachi
Higashi-ku
OSAKA

13254 KOWA CHIKAKENSETSU CO., LTD
17, 3-chome, Kawagishi-cho
NIIGATA

13246 KUMAGAI-GUMI CO., LTD
17-1, Tsukudo-cho, Shinjuko-ku
TOKYO

13283 KYOWA CHIKAKAIHATSU CO., LTD
2-9-11, Iwamoto-cho, Chiyoda-ku
TOKYO

13247 KYUSHU ELECTRIC POWER CO., INC.
1-82, Watanabedori 2-chome, Chuo-ku
FUKUOKA

13248 MAEDA CONSTRUCTION CO., LTD
10-26, Fujimi 2-chome, Chiyoda-ku
TOKYO

13249 MEIJI CONSULTANTS CO., LTD
24-1, Nishishinbashi 3-chome
Minato-ku
TOKYO, 105

13250 MITSUI CONSTRUCTION CO., LTD
Iwamoto-cho Mitsui Building, 3-10-1
Iwamoto-cho, Chiyoda-ku
TOKYO

13251 NIHON PUBLIC ENGINEERING CO., LTD
3-24-9, Nishi-Shinbashi, Minato-ku
TOKYO

13281 NIPPON GEOPHYSICAL PROSPECTING CO.LTD
2-12, 2-chome, Nakamagome, Ohta-ku
TOKYO

13280 NIPPON KOEI CO., LTD
5-4, Kojimachi, Chiyoda-ku
TOKYO

13252 NISHIMATSU CONSTRUCTION CO., LTD
20-10, 1-chome, Toranomon, Minato-ku
TOKYO

13279 NITTOC CONSTRUCTION CO., LTD
14-14, Ginza 8-chome, Chuo-ku
TOKYO, 104

13253 OHBAYASHI-GUMI, LTD
3, 2-chome, Kanda Tsukasa-cho,
Chiyoda-ku
TOKYO

13254 OKUMURA CORPORATION
1-2-3, Matsuzaki-cho, Abeno-ku
OSAKA, 545

13255 OYO CORPORATION
2-1, Otsuka 3-chome, Bunkyo-ku
TOKYO, 112

13256 PENTA-OCEAN CONSTRUCTION CO., LTD
2-2-8, Koraku, Bunkyo-ku
TOKYO

13257 SATO KOGYO CO., LTD
8, Honcho 4-chome, Nihonbashi,
Chuo-ku
TOKYO

13258 SHIKOKU ELECTRIC POWER CO., INC.
2-5, Marunouchi, Takamatsu City
KAGAWA

13259 SHIMIZU CONSTRUCTION CO., LTD
16-1, Kyobashi 2-chome, Chuo-ku
TOKYO, 104

13260 TAISEI CORPORATION
25-1, Nishi-Shinjuku 1-chome,
Shinjuku-ku
TOKYO

13261 THE CHUBU ELECTRIC POWER CO., INC.
1, Toshin-cho, Higashi-ku
NAGOYA, 461-91

13262 THE HOKKAIDO ELECTRIC POWER CO., INC.
Higashi 1-chome, O-dori
SAPPORO

13263 THE HOKURIKU ELECTRIC POWER CO., INC.
3-1, Sakurabashi-dori
TOYAMA, 930

13264 THE JAPAN ATOMIC POWER CO.
6-1, 1-chome, Ote-machi, Chiyoda-ku
TOKYO, 100

13265 THE KANSAI ELECTRIC POWER CO., INC.
3-22, 3-chome, Nakanoshima, Kita-ku
OSAKA

13277 THE NEW JAPAN ENGINEERING
CONSULTANTS, INC.
Maruzen Oil Bldg., 3, 1-come,
Nagahoribashi-suji
Minato-ku,
OSAKA

13266 THE TOKYO ELECTRIC POWER CO., INC.
1-1-3, Uchisaiwai-cho, Chiyoda-ku
TOKYO

13267 THE ZENITAKA CORPORATION
2-11, Nishihoncho 2-chome,
Nishi-ku
OSAKA, 550

13268 TOBISHIMA CORPORATION
3-28, 2-chome, Kudan Minami,
Chiyoda-ku
TOKYO

13269 TODA CONSTRUCTION CO., LTD
7-1, Kyobashi 1-chome, Chuo-ku
TOKYO

13270 TOHOKU ELECTRIC POWER CO., INC.
7-1, 3-chome, 1-bancho
SENDAI

13271 TOKAI ELECTRICAL INSTALLATION CO., LTI
20-31, Sakae 1-chome, Naka-ku
NAGOYA

13272 TOKYO ELECTRIC POWER SERVICES CO., LTD
Hibiya-Chunichi Bldg., 1-4, Uchisaiwai-cho
2-chome, Chiyoda-ku
TOKYO

13273 WEST JAPAN ENGINEERING CONSULTANTS, INC.
1-82, 2-chome, Watanabe-dori, Chuo-ku
FUKUOKA

13278 YACHIYO ENGINEERING CO., LTD
10-21, 1-chome, Nakameguro, Meguro-ku
TOKYO

13274 YONDEN ENGINEERING CO., INC.
1-4, 3-chome, Kamino-cho, Takamatsu
KAGAWA

MEXICO

SOCIEDAD MEXICANA DE MECANICA DE ROCAS, AC

Camino Santa Teresa 187, Vila Olimpica, MEXICO 22, DF

President: Mr J. Gamboa Flores

Total Number of Ordinary Members - 110

MOROCCO

COMITE MAROCAIN DE MECANIQUE DES SOLS ET DES ROCHES

25, Rue d'Azilal, CASABLANCA

President: Mr M. Moussaoui
Secretary: Mr M. Mariotti

Total Number of Ordinary Members - 99

ORDINARY MEMBERS

6313 ABDELHOUHAB, A., Mr - Engr
* 4 Rue Zagora (Ex rue de Nice)
RABAT

6317 ALLALI, M., Mr - Engr
* Chef d'Arrondissement des Travaux
Publics
d'EL JADIDA

172 AMSELLEM, L., Mr - Engr
* Subdivision des Travaux Publics
(Maritimes)
B.P. 23
KENITRA

6324 AUBERT
* Directeur de la Societe
Les Grands Travaux Routiers
5 Bd Ben Yacine
CASABLANCA

1722 AZZAOUI, A., Mr - Engr
* Subdivision des Travaux Publics
SETTAT

6328 BEL HADJ, M., Mr - Engr
* Secretaire General du Ministere des
Travaux Publics et des Communications
RABAT

1727 BEKKALI, A., Mr - Engr
* Direction de l'Hydraulique
B.P. 525
RABAT-CHELLAH

6343 BELLEFKIH, M.
* Directeur de la Societe Maghrebine de
Construction et d'Entretien des Routes
196b Bd Ibn Tachfine
CASABLANCA

13012 BEN AMAR
* Service Regional de Casablanca
de la Division de l'Hydraulique
61 Bd Camille des Moulins
CASABLANCA

6384 BENCHEIKH, L., Mr - Engr
* Office National des Chemins de Fer
19 Avenue Allal Ben Abdallah
RABAT

1725 BENJELLOUN
* Directeur General du Bureau des
Recherches et de Participations
Minieres
27 Avenue Moulay Hassan
RABAT

6383 BENLAHBIB, Mr - Engr
* Chef de l'Amenagement du Barrage
de Timin'outine
Direction de l'Hydraulique
MARRAKECH

6386 BENZEKRI, E.M., Mr - Engr
* Chef de l'Amenagement du Barrage
d'El Massira
Casier Postal
SETTAT

6387 BERGERET, P.,
* Directeur General des Tuileries et
Briqueteries du Maroc
MOHAMMEDIA

13013 BURTE, Mr - Engr
* Laboratoire Public d'Essais et d'
Etudes
25 Rue d'Azilal
CASABLANCA

2119 BZIOUI, M., Mr - Engr
* Direction de l'Hydraulique Service
"Etudes"
B.P. 525
RABAT-CHELLAH

1730 CARANCHINI, J., Mr - Engr
* Service des Ports Secondaires
Ministere des Travaux Publics et
des Communications
RABAT

1733 CHAOUI, A., Mr - Engr
* Directeur-Adjoint de l'Hydraulique
B.P. 525
RABAT-CHELLAH

1735 CHENTOUF, A., Mr - Engr
* Subdivision des Travaux Publics
d'AL HOCEIMA

13014 CHOUKAILI, Mr - Engr
* Chef du Departement Geotechnique
au Laboratoire Public d'Essais et
d'Etudes
25 Rue d''Azilal
CASABLANCA

1740 DAOUDI, M., Mr - Engr
* Directeur du Bureau de Controle
Technique de la Construction
Rue de Figuig
RABAT

1743 DE SAMBUCY, E., Mr - Engr
* Arrondissement des Travaux Publics
d'AGADIR

1744 DINIA, N., Mr - Engr
* Directeur de l'Hydraulique
B.P. 525
RABAT-CHELLAH

1746 DIOURI, M., Mr - Engr
* Direction du Bureau de Recherches et
de Participations Minieres
27 Avenue Moulay Hassan
RABAT

13015 DUHON, E., Mr - Architect
* 44 Rue Omar Slaoui
CASABLANCA

13016 ESPRESATI, A., Mr - Engr
* Chef du Departement Physique-Chimie
au Laboratoire Public d'Essais et d'
Etudes
25 Rue d'Azilal
CASABLANCA

1753 GHISSASSI, A.
* Ministre des Finances
RABAT

1754 GLAOUI, O., Mr - Engr
* Subdivision des Travaux Publics
FES-NORD

1758 HAINAUT, J.P.
* Directeur de la Societe des
Basaltes Marocains
17 Rue Berthelot
MEKNES

1759 HAKIMI, A.
* Directeur du Laboratoire Public
d'Essais et d'Etudes
25 Rue d'Azilal
CASABLANCA

1760 HARDI, Mr - Engr
* Service des Bases Aeriennes
Arrondissement du Sud
CASABLANCA/ANFA

13017 ISLAH, M., Mr - Engr
* Chef de l'Amenagement du Barrage de
Tamzaourt sur l'Oued Issen
B.P. 432
AGADIR

13018 JAOUI, A., Mr - Engr
* Chef de la Division de l''Exploitation
Direction de l'Hydraulique
B.P. 525
RABAT-CHELLAH

6396 JBILOU, A., Mr - Engr
* Ministere des Travaux Publics et des
Communications
Division des Transports et des
Equipements
RABAT

6397 KABBAJ, M.
* Directeur des Routes
Ministere des Travaux Publics et
des Communications
RABAT

6398 KACHFI, L., Mr - Engr
* Subdivision des Travaux Publics
ESSAOUIRA

13019 KACIMI, Dr - Engng Geologist
* Laboratoire Public d'Essais et
d'Etudes
25 Rue d'Azilal
CASABLANCA

13020 KADIRI, A., Mr - Engr
* 7 Rue Benghazi
RABAT

6399 KARMOUNI, M., Mr - Engr
* Arrondissement des Travaux Publics
SAFI

6400 LAHLOU, A.
* Directeur de l'O.R.M.V.A. du Loukkos
KSAR EL KEBIR

1763 LAHRICHI, O.
* Directeur de l'Omnium Technique
Marocain
11 Rue LLapebie
CASABLANCA

2125 LAKHDARI, Mr - Engr
* Office National des Chemins de Fer
19 Avenue Allal Ben Abdallah
RABAT

7467 LAMRINI, H., Mr - Engr
* Arrondissement des Travaux Publics
FES

13021 LAMZIBRI, M., Mr - Engr
* Service Regional de la Direction
de l'Hydraulique
Place El Maghrib El Arabi
B.P. 203
KENITRA

1766 LAYACHI, S.M., Mr - Engr
* Directeur des Ports Secondaires
Ministere des Travaux Publics
RABAT

1767 LUNARDI, M., Mr - Engr
* 49 Cite du C.I.L.
RABAT/AVIATION

13026 MAAROUFI, Mr - Engr
Directeur du Port de
CASABLANCA

7471 MALTI, D., Mr - Engr
* Prefecture de
CASABLANCA

1770 MARCHAND, R., Mr - Engr
* Laboratoire Public d'Essais et
d'Etudes
Parc des Travaux Publics,
Quartier Industriel
AGADIR

1771 MARIOTTI, M.,
* Directeur-Adjoint du Laboratoire
Public d'Essais et d'Etudes
25 Rue d'Azilal
CASABLANCA

1773 MAZURAT, J.C.,
* Directeur de l"Entreprise Larbi
Ben Ghali
11 Rue des Orchidees
CASABLANCA

6403 MENNI, M., Mr - Engr
* Societe Maghrebine de Construction
et d'Entretien des Routes
196 Bd Ibn Tachfine
CASABLANCA

2126 MEZIANE, A., Mr - Engr
* Inspecteur General
Ministere de l'Equipement et de la
Promotion Nationale
RABAT

1775 M'KINSI, A.
* Rue des Princesses
RABAT/SOUISSI

1776 MOUGERY, R., Mr - Engr
* Laboratoire Public d'Essais et
d'Etudes
Arrondissement des Travaux Publics
OUJDA

1778 MOUSSAOUI, M.
* Directeur de l'Office National des
Chemins de Fer
19 Avenue Allal Ben Abdallah
RABAT

13022 NACIRI, H., Mr - Engr
* Chef du Departement de l'Hydraulique
au Laboratoire Public d'Essais et
d'Etudes
25 Rue d'Azilal
CASABLANCA

1780 NATAF, M., Dr - Engng Geologist
* Service d'Etudes des Gites Mineraux
RABAT

3210 NOUARI, L., Mr - Engr
* Chef du Service Materiaux au
Laboratoire Public d'Essais et
d'Etudes
25 Rue d'Azilal
CASABLANCA

13023 NOUIJ, Mr - Engr
* Amenagement du Port de Jorf El Asfar
PAR EL JADIDA

1782 ORLIAC, M., Mr - Engr
* Directeur de Geobema
63 Bd Abdellatif Ben Kaddour
CASABLANCA

1783 PAGANO
* Directeur de Maghreb - Etudes
34 Bd Mohammed V
CASABLANCA

13024 RENIE, Mr - Engr
* Service Fondations
Laboratoire Public d'Essais et
d'Etudes
25 Rue d'Azilal
CASABLANCA

1790 ROBERT, F.
* Directeur au Maroc du Bureau d'Etudes
Coyne et Bellier
Villa Najathe - La Pinede
RABAT

1791 RUHARD, J.P., Dr - Geologist
* Direction de l'Hydraulique,
Service des Ressources en Eau
B.P. 525
RABAT/CHELLAH

13025 SAADI-MOUSSA
* Ministre de l'Energie et Recherches
Minieres
RABAT

1792 SAMIR, A., Mr - Engr
* Direction des Routes
Ministere des Travaux Publics
RABAT

1784 SIBERCHICOT, M., Mr - Engr
* Place Ibn Tachfine
RABAT

1795 SIBONI, A.,
* Directeur de la Regie d'Aconage du Port de CASABLANCA

6414 SMIRES-BENNANI, A., Mr - Engr
* Secretaire General du Ministere du Commerce, de l'Industrie et des Mines RABAT

6415 SOUFI, M., Mr - Engr
* Office National des Chemins de Fer 19 Avenue Allal Ben Abdallah RABAT

13027 TAZI
Directeur General de l'O.N.E. 65 Rue Aspirant Lafuente CASABLANCA

6416 TAHIRI - JOUTEI, M.L., Mr - Engr
* Arrondissement des Travaux Publics de CASABLANCA

13028 VERDUGO, C.L., Mr - Architect
* 12 Zankat Al Yamama RABAT

1802 ZAZ, A., Mr - Engr
* Directeur-Adjoint de la Regie d'Aconage du Port de CASABLANCA

13057 Representative of:
A.T.B.A. 25, Rue d'Azilal CASABLANCA

13058 Representative of:
ALPINCONSULT - MOTOR - COLUM BUS - Ingenieurs-Conseils S.A. 24 Rue de Meknes RABAT

13059 Representative of:
BUREAU MEGHREB - ETUDES 34 Bd Mohammed V CASABLANCA

13060 Representative of
INGEMA (Coyne et Bellier) 21 Rue Idriss El Akbar RABAT

13061 Representative of:
CHAUFFOUR-DUMEZ 5 Rue Enseigne de Vaisseau Yves Gay CASABLANCA

13062 Representative of:
CHERIFIENNE DES TRAVAUX AFRICAINS (C.T.R.A.) 1 Bd du Fourat CASABLANCA

13063 Representative of:
COMPAGNIE MAROCAINE DES GOURONDS ET BITUMES 105 Rue Emir Abdelkader CASABLANCA/AIN BORDJA

13064 Representative of:
ELECTRO - WATT - Ingenieurs-Conseils/ S.A. 13 Rue de Demnate RABAT

13065 Representative of:
E N A T R A Rue de Sergent Rene Montanon CASABLANCA

13066 Representative of:
LABORATOIRE PUBLIC D'ESSAIS ET d'ETUDES 25 Rue d'Azilal CASABLANCA

13067 Representative of:
L'AMENAGEMENT DU BARRAGE DE TAMZAOURT SUR L'OUED ISSEN B.P. 432 AGADIR

13068 Representative of:
LA ROUTE MAROCAINNE 291 Bd Mohammed V CASABLANCA

13069 Representative of:
LES GRANDS TRAVAUX ROUTIERS 5 Bd Abdellah Ben Yacine CASABLANCA

13070 Representative of:
OFFICE NATIONAL DES CHEMINS DE FER 19 Avenue Allal Ben Abdellah RABAT

13071 Representative of:
OFFICE NATIONAL DE L'ELECTRICITE 65 Rue Aspirant Lafuente CASABLANCA

13080 Representative of:
OFFICE REGIONAL DE LA MISE EN VALEUR AGRICOLE DU LOUKKOS (ORMVAL) KSAR EL KEBIR

13072 Representative of:
OMNIUM TECHNIQUE MAROCAIN 11 Rue Lapebie CASABLANCA

13073 Representative of:
REGIE D'ACONAGE DU PORT DE CASABLANCA

13074 Representative of:
S E P R O B 7 Rue Mohammed Diouri CASABLANCA

13075 Representative of:
Services d'Etudes des Gites
Mineraux
Directions des Mines et de la Geologie
RABAT

13076 Representative of:
S.M.E.R.
196 Bd Ibn Tachfine
CASABLANCA

13077 Representative of:
SOCOTEC
37 Rue des Ait Ba Amrane
CASABLANCA

13078 Representative of:
SOLETANCHE
B.P. 215
RABAT

13079 Representative of:
SONDAGES INJECTIONS-FORAGES
B.P. 2023
CASABLANCA

NETHERLANDS

Membership: Total Number of Corresponding Members - 3
Total Number of Supporting Members - 2

CORRESPONDING MEMBERS

1852 MANDL, G., Dr
Koninklijke/Shell Exploratie en
Produktie Laboratorium
Vormerlaan 6,
Postbus 4560
RIJSWIJK (Z.H.)

4443 RENGERS, N., Dr - Geologist
International Institute for Aerial
Survey and Earth Sciences
144 Boulevard 1945
P.O. Box 6
ENSCHEDE

346 VOUTE, C., Prof - Consulting
Geologist
* Postal Code 7500 BB
P.O. Box 1079
ENSCHEDE

SUPPORTING MEMBERS

10583 A. A., BALKEMA Publishers
P.O. Box 1675
NL 3000 BR ROTTERDAM

13029 SOIL MECHANICS LABORATORY
Stieltijesweg 2
P.O. Box 69
DELFT

NEW ZEALAND

NEW ZEALAND GEOMECHANICS SOCIETY

New Zealand Institution of Engineers, P.O. Box 12-241, WELLINGTON NORTH

Vice President: Mr M. J. Pender

Total Number of Ordinary Members - 29
Total Number of Corresponding Members - 1

ORDINARY MEMBERS

7541 + BELL, D. H.
Geology Dept
University of Canterbury
Private Bag
CHRISTCHURCH

12034 + BORRIE, G. W.
P.O. Box 30-897
LOWER HUTT

7974 BRYANT, J. M.
M.W.D., Central Laboratories
P.O. Box 30-845
LOWER HUTT

12035 + BUNTING, D. G.
19 Kenwyn Ave
CHRISTCHURCH, 5

12045 DOIG, K. D.
859 Heaphy Terrace
HAMILTON

12036 FAMA, M. E., Dr
14 Malcolm St
HAMILTON

7542 FOOKES, R. W. J.
P. O. Box 12-004
WELLINGTON

7544 GULLIVER, C. P.
45 Park Rise
Cambells Bay
AUCKLAND, 10

7545 JONES, L. S.
Mines Dept
P.O. Box 6342, Te Aro
WELLINGTON

7168 * + KEMP, B. J.
P.O. Box 5101
Lambton Quay
WELLINGTON

8176 * KENNEDY, H. J. F.
56 Miles Cresc.
Newlands
WELLINGTON, 4

8177 + LEWANDOWKSI, R.J.
C/- Coffey & Ptners
12 Waterloo Road
North Ryde
N.S.W. 2113 AUSTRALIA

12037 + MACFARLANE, D. F.
C/- M.W.D., Private Bag
TWIZEL

12038 MCINTOSH, I. S.
P.O. Box 6342
WELLINGTON

8178 * + MCLEAN, J. D.
C/- Cuthbertson, Megget Reservoir
St. Mary's Loch
Selkirkshire
SCOTLAND

7976 MILLER, P. J.
Rangipo North Residency
Tongariro Power Residency
TURANGI

12039 + PATERSON, B. R.
C/- University of Canterbury
Private Bag
CHRISTCHURCH

12040 * + PENDER, M. J., Dr
Civil Engineering Dept
University of Auckland
Private Bag
AUCKLAND

12041 + PERRIN, N. D., Dr
N.Z. Geological Survey
P.O. Box 30-368
LOWER HUTT

7169 + READ, S. A. L.
P. O. Box 30-368
LOWER HUTT

7170 + RIDDOLLS, B. W., Dr
N.Z. Geological Survey
P.O. Box 30368
LOWER HUTT

7547 RILEY, P. B.
P.O. Box 6345
AUCKLAND

12042 ST. GEORGE, J. D.
Dept Mineral Technology
University of Otago
P.O. Box 56
DUNEDIN

12043 SPARROW, W. R.
+ 90A Waiatarua Road
Remuera
AUCKLAND

8181 SUGGATE, R. P., Dr
* P.O. Box 30-368
+ LOWER HUTT

7549 TAYLOR, D. K.
* P.O. Box 5271, Wellesley Street
+ AUCKLAND

8182 TAYLOR, R. G.
* 49 Wai-iti Crescent
+ LOWER HUTT

12044 TOAN, D. V., Dr
* 43 Murdock Road, Grey Lynn
AUCKLAND, 2

8183 WELLS-GREEN, P. S.
* 50 Connaught Street
+ Blockhouse Bay
AUCKLAND

CORRESPONDING MEMBERS

591 DODD, T. A. H., Mr - Civil Engr
Senior Lecturer in Civil Engng
Canterbury University
Private Bag
CHRISTCHURCH 1

NORWAY

NORSK BERGMEKANIKKGRUPPE

The University of Trondheim, Department of Mining Engineering, 7034 TRONDHEIM NTH

Secretary: Mr Tore Lasse By

Total Number of Ordinary Members - 144
Total Number of Supporting Members - 1

ORDINARY MEMBERS

7650 ANDERSEN, U. H., Mr - Civil Engr
Consulting Engr
c/o Norsk Olje A/S
Postboks 457
N-9401 HARSTAD

6367 ASKILDSEN, K., Mr - Civil Engr
Manager
Bastadveien 98
N-1370 ASKER

6419 ASPHAUG, S., Mr - Engr
Sales
Atlas Copco A/S
Lade Alle 69
N-7000 TRONDHEIM

13134 ASTING, G., Mr - Civil Engr
Consulting Engr
c/o Elliot Strømme A/S
Drammensveien 126
N-OSLO 2

13135 BABIC, M. M., Mr - Mining Engr
Consulting Engr
Rogneveien 24
N-1480 SKATTUM

13136 BAKKEN, R. E., Mr - Engr
Research Engr
Norwegian Institute of Rock Blasting Techniques
Forskningsveien 3b
N-OSLO 3

3347 * BARBO, T. F., Mr - Civil Engr
Chief Engr
NOTEBY - Norsk Teknisk Byggekontroll A/S
Thv. Meyersgt. 9
N-OSLO 5

538 BARKEY, H., Mr - Geologist
Research Engr
Geological Survey of Norway
Geophysical Department
Boks 3006
N-7000 TRONDHEIM

7651 BARLINDHAUG, S., Mr - Engr Geologist
Consulting Engr
NOTEBY - Norsk Teknisk Byggekontroll A/S
Postboks 757
N-9001 TROMSØ

13137 * BARTON, N., B.Sc., Ph.D
Rock Mechanics Engr
Norwegian Geotechnical Institute
Postboks 40 - Tasen
N-OSLO 8

3349 BAUMANN, L., Mr - Civil Engr
Assistant Director
Dyno Industrier A/S
Tollbugt. 22
N-OSLO 1

13138 BENDIKSEN, S., Mr - Civil Engr
Manager
NVE - State Power System
N-4230 SAND

6420 * BERDAL, B. A., Mr - Civil Engr
Consulting Engr
Maries vei 20
N-1322 HØVIK

275 BERGE, I., Prof.- Mining Engr
U. Prof.
The University of Trondheim
Department of Mining Engineering
N-7034 TRONDHEIM-NTH

3351 BERGH, C. J., Mr - Civil Engr
Consulting Engr
Wdm. Thranesgt. 98
N-OSLO 1

7652 BJERKAN, G., Mr - Engr
Technical Engr
The University of Trondheim
Department of Mining Engineering
N-7034 TRONDHEIM-NTH

441 BJERKAN, M., Mr - Engng Geologist
Manager
Løkenasfaret 1
N-1473 SKARER

3354 BJERKAN, R. K., Mr - Civil Engr
Manager
Løkenasfaret 1
N-1473 SKARER

6421 BJØRGE, A., Mr - Civil Engr
Consulting Engr
Dyno Industrier A/S
Tollbugt. 22
N-OSLO 1

6422 BJØRLYKKE, S., Mr - Civil Engr
Consulting Engr
VIAK A/S
Strandgaten 221
N-5000 BERGEN

6423 BLINDHEIM, T.O., Mr - Civil Engr
University Teacher
The University of Trondheim
Department of Geology
N-7034 TRONDHEIM-NTH

7653 BOLLINGMO, P., Mr - Geological Engr
Chief Engr
NOTEBY - Norsk Teknisk Byggekontroll A/S
Thv. Meyersgt. 9
N-OSLO 5

258 BRANDTZAEG, A., Dr techn.
Prof. em.
The University of Trondheim
N-7034 TRONDHEIM - NTH

6424 BREDELI, T., Mr - Mining Engr
Chief Engr
Titania A/S
N-4380 HAUGE I DALANE

3357 BROCH, E., Dr - Geological Engr
University Teacher
The University of Trondheim
Department of Geology
N-7034 TRONDHEIM-NTH

7654 BRØNDBO, K., Mr - Mining Engr
Managing Director
Orkla Industrier A/S
N-7341 LØKKEN VERK

13139 BRAATEN, E. M., Mr - Engng Geologist
Geological Engr
Cementstøperiet
Sluppenvn. 12
N-7000 TRONDHEIM

7830 BRATEN, T., Mr - Engng Geologist
Contracting
A/S Veidekke
Postboks 70
N-1344 HASLUM

7655 BUEN, B., Dr - Geological Engr
Sr. Consulting Engr
Ing. A.B. Berdal A/S
Maries vei 20
N-1322 HØVIK

13140 BY, T. L., Mr - Mining Engr
Research Engr
The University of Trondheim
Department of Mining Engineering
N-7034 TRONDHEIM-NTH

13141 CARLSEN, K. B., Mr - Mining Engr
Research Engr
The University of Trondheim
Department of Mining Engineering
N-7034 TRONDHEIM-NTH

13142 DANIELSEN, A., Mr
Contractor
Brokvik Maskinstasjon A/S
N-9400 HARSTAD

7656 DANIELSEN, S. W., Mr - Civil Engr
Consulting Engr
NOTEBY - Norsk Teknisk Byggekontr-oll A/S
Thv. Meyersgt. 9
N-OSLO 5

6429 DANNEVIG, N. Th., Mr - Engng Geologist
Contracting
F. Selmer A/S. Boks 25
N-1701 SARPSBORG

7657 DRAKE, J., Mr - Civil Engr
Consulting Engr
Fagersta Haak A/S
Ole Deviks vei 26
N-OSLO 6

13143 ENGEBAK, G., Mr - Engr
Contractor
Engebaks Brønnboring
Trondheimsvn. 42
N-OSLO 5

7658 ERIKSEN, F. R., Mr - Mining Engr
Chief Engineer
The University of Trondheim
Mining Research Department
N-7034 TRONDHEIM-NTH

13144 FISKUM, K., Mr - Civil Engr
Consulting Engr
c/o Taugbøl & Øverland
Plogveien 3b
N-OSLO 5

13145 FOSSHEIM, A., Mr - Civil Engr
Division Engr
Narvik Komm. E-verk
Postboks 55
N-8501 NARVIK

13146 FOSSUM, K., Mr - Mining Engr
Research Engr
Norwegian Institute of Rock
Blasting Techniques
Forskningsveien 3b
N-OSLO 3

7661 FREDRIKSEN, U., Mr - Geological Engr
Division Engr
Municipality of Oslo
Geotechnical Department
Kingos gt. 22
N-OSLO 4

7662 GARSHOL, K., Mr - Geological Engr
Contracting
c/o Ing. Thor Furuholmen A/S
Karl Johansgt. 39
N-OSLO 1

13147 GEIS, H.-P., Mr - Geologist
Chief Geologist
Elkem-Spigerverket A/S
Postboks 4002
N-OSLO 4

13148 GUSTAVSEN, T. K., Mr - Geological Engr
Consulting Engr
Ing. A.B. Berdal A/S
Maries vei 20
1322 HØVIK

13149 HALVORSEN, A., Mr - Engng Geologist
Consulting Engr
A/S Geotam
Wdm. Thranesgt. 98
N-OSLO 1

13150 HANSEN, A., Mr - Mining Engr
Manager
Kastanjevn. 14e
N-3000 DRAMMEN

13151 HANSEN, O. J., Mr - Contractor
Brødrene Hansen A/S
N-4600 KRISTIANSAND S

6430 HATLE, J., Mr - Mining Engr
Chief Engr
NOTEBY - Norsk Teknisk Byggekontroll A/S
Thv. Meyersgt. 9
N-OSLO 5

6431 HATLE, U., Mr - Geological Engr
Nedre Ringvoll loc
N-1313 VØYENENGA

7663 HEIMLI, P., Mr - Engng Geologist
Consulting Engr
NOTEBY - Norsk Teknisk Byggekontroll A/S
Thv. Meyersgt. 9
N-OSLO 5

276 HELTZEN, A. M., Mr - Mining Engr
Head of Department
Norwegian Institute of
Rock Blasting Techniques
Forskningsveien 3b
N-OSLO 3

3361 * HOFFMANN, W., Mr - Civil Engr
Chief Consulting Engr
A/S Geoteam
Wm. Thranesgt. 98
N-OSLO 1

6433 HOLMSEN, T. W., Mr - Mining Engr
Manager
Dahleflate
N-2670 OTTA

13152 HOTVEDT, O., Mr - Civil Engr
Research Engr
NVE - State Power System
Ulla-Førre
N-4230 SAND

7665 HOV, A., Mr - Engr
The University of Trondheim
Department of Geology
N-7034 TRONDHEIM-NTH

7666 HUGSTED, T., Prof - Civil Engr
U. Prof.
The University of Trondheim
Division of Construction Engr
N-7034 TRONDHEIM-NTH

13153 HØYGAARD, B., Mr - Civil Engr
Consulting Engr
NOTEBY - Norsk Teknisk Byggekontroll A/S
Thv. Meyersgt. 9
N-OSLO 5

13154 HALAND, S., Mr - Engr
Consulting Engr
ARGO A/S, Jaktlia 1, boks 23
N-1345 ØSTERAS

6436 JOHANSEN, J., Mr - Civil Engr
Manager
Dyno Konsulent A/S
Postboks 779 - Sentrum
N-OSLO 1

13155 JOHANSEN, P.M., Mr - Civil Engr
Research Engr
Norwegian Geotechnical Institute
Postboks 40, Tasen
N-OSLO 1

7667 Johansen, T., Mr - Civil Engr
Chief Engr
c/o Ing. Thor Furuholmen A/S
Karl Johansgt. 39
N-OSLO 1

13156 Johansson, K.W.,
Dept Man.
Progas A/S
Ryensvingen 1
N-OSLO 6

7668 Johansson, S., Mr - Engr
Consulting Engr
Göteborgs Byggmästartjänst AB
Föreningsgt. 2
S-411 27 GÖTEBORG

3362 * JOHNSEN, M. G., Mr - Civil Engr
Technical Director
Astrup & Aubert A/S
Postboks 6056, Etterstad
N-OSLO 6

3363 JØSANG, O., Mr - Geologist
Consulting Engr
Norwegian Road Research Laboratory
Gaustadalleen 25
N-OSLO 3

13157 KARLSEN, K., Mr - Civil Engr
Research Engr
Norwegian Institute of Rock Blasting Techniques
Forskningsveien 3b
N-OSLO 3

7669 KIEL, L.G., Mr - Civil Engr
Chief Engr
Gml. Drammensvn. 187c
N-OSLO 2

13158 KIELLAND, B., Mr - Civil Engr
Manager
Kolsvik Kraftverk
N-8940 TERRAK

3365 KJØLBERG, R., Mr - Engng Geologist
Consulting Engr
A/S Geoteam
Wdm. Thranesgt. 98
N-OSLO 1

3366 KLETTE, E. V., Mr - Mining Engr
Head of Division
The Labour Inspection
Postboks 8103 Dep.
N-OSLO 1

3840 KLÜVER, B. H., Mr - Geological Engr
Chief Engr
Norsk Hydro a.s.
Rafnes
N-3912 HERRE

6428 KNUTSON, F., Mr - Civil Engr
Chief Engr
Dyno Industrier A/S
Tollbugt. 22
N-OSLO 1

13159 KOLLSTRØM, J. I., Mr - Geological Engr
Consulting Engr
Ing. A.B. Berdal A/S
Maries vei 20
N-1322 HØVIK

13160 LAMBACH, G. J., Mr - Civil Engr
Research Engr
Norwegian Institute of Rock Blasting Techniques
Forskningsveien 3b
N-OSLO 3

13161 LARSEN, J. O., Mr - Civil Engr
Research Engr
Norwegian Geotechnical Institute
Postboks 40, Tasen
N-OSLO 8

13162 LARSEN, T. M., Mr - Mining Engr
Research Engr
Norwegian Institute of Rock Blasting Techniques
Forskningsveien 3b
N-OSLO 3

6438 LARSSON, T., Mr - Civil Engr
Consulting Engr
Dyno Industrier A/S
Tollbugt. 22
N-OSLO 1

3368 LEISTAD, E., Mr - Mining Engr
Consulting Engr
Dyno Industrier A/S
Tollbugt. 22
N-OSLO 1

3370 LI, B., Dr - Mining Engr
University Teacher
The University of Trondheim
Department of Mining Engineering
N-7034 TRONDHEIM-NTH

13163 LIEN, S., Mr - Geological Engr
Research Engr
The University of Trondheim
Department of Geology
N-7034 TRONDHEIM-NTH

7670 LILE, O. B., Dr - Mining Engr
University Teacher
The University of Trondheim
Department of Petroleum Technology and Applied Geophysics
N-7034 TRONDHEIM-NTH

6441 LIE, R. K., Mr - Civil Engr
Chief Engr
Industrial Civil Defence
Drammensveien 40
N-OSLO 2

3371 LIEN, R., Dr - Geological Engr
Senior Engr
Norwegian Geotechnical Institute
Postboks 40, Tasen
N-OSLO 8

13164 LUDVIGSEN, E. S., Mr - Mining Engr
Research Engr
The University of Trondheim
Department of Mining Engineering
N-7034 TRONDHEIM-NTH

13165 LUNDE, B., Mr - Mining Engr
Consulting Engr
Motordrift A/S
Brynsengvn. 1
N-OSLO 6

3373 * LUNDE, J. A., Mr - Civil Engr
Senior Engr
Norwegian Geotechnical Institute
Postboks 40, Tasen
N-OSLO 8

3374 LUNESTAD, S., Mr - Engr
Senior Technician
Norwegian Institute of Rock
Blasting Techniques
Forskningsveien 3b
N-OSLO 3

3375 * LØKEN, T., Mr - Geologist
Research Engr
Norwegian Geotechnical Institute
Postboks 40, Tasen
N-OSLO 8

13166 LØVDAL, K., Mr - Engr
Consulting Engr
Progas A/S
Ryensvingen 1
N-OSLO 6

6443 MARKUSSEN, O., Mr - Mining Engr
Managing Director
Franzefoss Bruk A/S
Postboks 1
N-1349 RYKKINN

13167 MARSTEIN, B., Mr
Øverbergvn. 22
N-1315 NESØYA

13168 MEDBØ, F., Mr - Civil Engr
Consulting Engr
Vestfjorden Avløpsselskap
Deriksveien 1, boks 150
N-1360 NESBRU

13169 MOHN, K.-F., Mr - Engng Geologist
Contractor
c/o Chr. Grøner A/S
Maries vei 20
N-1322 HØVIK

3377 MYRVANG, A., Dr - Mining Engr
University Teacher
The University of Trondheim
Department of Mining Engineering
N-7034 TRONDHEIM-NTH

7672 NEEB, P. R., Mr - Engng Geologist
Research Engr
Geological Survey of Norway
Leiv Eirikssonsv. 39
N-7000 TRONDHEIM

13170 NILSEN, B., Mr - Geological Engr
Research Engr
The University of Trondheim
Department of Geology
N-7034 TRONDHEIM-NTH

13171 NYGAARD, T. H., Mr - Engr
Troms Kraftforsyning
N-8048 SKIBOTN

6444 ORMASEN, J., Mr
Chief Driller
A/S Hardmetall
N-1380 HEGGEDAL

13172 ORNAES, J., Mr - Mining Engr
Chief Engr
Grong Gruber A/S
N-7894 LIMINGEN

6445 * PALMSTRØM, A., Mr - Engng Geologist
Consulting Engr
Ing. A.B. Berdal A/S
Maries vei 20
N-1322 HØVIK

7675 RISE, L., Mr - Geological Engr
University Teacher
The University of Trondheim
Department of Geology
N-7034 TRONDHEIM-NTH

13173 ROHDE, J. K. G., Mr - Civil Engr
Research Engr
Norwegian Geotechnical Institute
Postboks 40, Tasen
N-OSLO 8

7676 ROKOENGEN, K., Dr - Geological Engr
Research Engr
Continental Shelf Institute
Hakon Magnussonsgt. 1b
N-7000 TRONDHEIM

13174 ROMSLO, N., Mr - Civil Engr
Consulting Engr
I/S Øvre Otra Anl.kontor
N-4694 BYKLE

13175 ROSKA, S., Mr
Terje Woldsgt. 8
N-9800 VADSØ

7677 RUSSNES, B. F., Mr - M.Sc.
Consultant Engr
Utbygn.avd.Sogn og Fjordane
Fylkeskommune
N-5840 HERMANSVERK

3379 RAAEN, K., Mr - Geological Engr
Consulting Engr
NOTEBY - Norsk Teknisk
Byggekontroll A/S
Thv. Meyersgt. 9
N-OSLO 5

7678 SANDBERG, J. C., Mr - Civil Engr
Consulting Engr
Hedemark Kraftverk
N-2300 HAMAR

6447 SANDSTRØM, H., Mr - Engr
Managing Director
Innovation & Co
Adler Salvius väg 47
S-15600 TULLINGE

7679 SANDVIK, K. O., Mr - Engng Geologist
Research Engr
Geological Survey of Norway
Leiv Eirikssonsv. 39
N-7000 TRONDHEIM

3380 SCHACH, R., Mr - Civil Engr
Research Engr
Norwegian Institute of
Rock Blasting Techniques
Forskningsveien 3b
N-OSLO 3

249 SELMER-OLSEN, R., Prof - M.Sc.
* U. Prof
The University of Norway
Department of Geology
N-7034 TRONDHEIM-NTH

6448 SIGURDSSON, E., Mr - Civil Engr
ISTAK, Islenzkt Verktak h.f.
Sudurlandsbraut 6
REYKJAVIK (ISLAND)

7680 SKOGVANG, A., Mr - Engr
Consulting Engr
Norsk Hydro a.s.
Bygdø Alle 2
N-OSLO 2

7681 SLOREBY, E., Mr - Civil Engr
Manager
Steinvegen 10
N-2600 LILLEHAMMER

7682 SMEMO, Ø., Mr - Mining Engr
Manager
A/S Norcem, avd. Dalen
N-3850 BREVIK

13176 SOLEM, O., Mr - Civil Engr
Consulting Engr
NOTEBY - Norsk Teknisk Byggekontroll A/S
Damsgardsgaten 167-169
N-5031 LAKSEVAG

7683 SOMMERFELDT, W., Mr - Engr
Mine Superintendent
A/S Norcem, Kalksteinsbruddet
Langøen
N-3081 HOLMESTRAND

13177 STANGELAND, G., Mr
Contractor
Postboks 3087
N-4301 SANDNES

6449 STEFFENSEN, K., Mr - Civil Engr
Consulting Engr
c/o Ing. Chr. Grøner A/S
Postboks 14
N-1322 HØVIK

13178 STEPHANSSON, O., Prof.
U. Prof.
Högskolan i Lulea
LULEA

7684 STOKKELAND, H., Mr - Engr
Consulting Engr
Østlandskonsulent A/S
Ferjestadveien 24
N-1600 FREDRIKSTAD

13179 STRØM, E., Mr - Civil Engr
* Consulting Engr
NOTEBY - Norsk Teknisk Byggenkontroll A/S
Thv. Meyersgt. 9
N-OSLO 5

13180 SUND, R., Mr - Engr
Föreningsgatan 2
S-41127 GÖTEBORG

537 SVINNDAL, S., Mr - Geologist
Research Engr
Ministry of Industry, Mining Dept
Boks 3006
N-7001 TRONDHEIM

7686 SYSTAD, H., Mr - Civil Engr
Consulting Engr
NOTEBY - Norsk Teknisk Byggekontroll A/S
Stakkevollveien 69
N-9000 TROMSØ

13181 SAETERBØ, E., Mr - Civil Engr
Manager
NVE - State Power System
Eidfjordanleggene
N-5783 EIDFJORD I HARDANGER

13182 SAETHER, L.Kr., Mr - Civil Engr
Consulting Engr
c/o Elliot Strømme A/S
Drammensveien 126
N-OSLO 2

13183 SAETRE, K., Mr - Civil Engr
Manager
Norwegian State Railways
Geotechnical Department
Storgt. 33
N-OSLO 1

7688 TAUGBØL, T., Mr - Civil Engr
Consulting Engr
Taugbøl & Øverland A/S
Plogveien 3b
N-OSLO 6

13184 THIDEMANN, A., Mr - Civil Engr
Chief Engr
Nord-Trøndelag Elektrisitetsverk
Kongensgt. 42
N-7700 STEINKJER

13185 TINGVOLDLI, A., Mr - Civil Engr
Chief Engr
Sør-Trøndelags Kraftselskap
Ingv. Ystgaards vei 1
N-7000 TRONDHEIM

3381 TEPPER, A., Mr - Engr
* Manager
Grunnboring A/S
Pilestredet 17
N-OSLO 1

6451 VAGSTEIN, A. B., Mr - Civil Engr
Consulting Engr
NOTEBY - Norsk Teknisk Byggekontroll A/S
Thv. Meyersgt. 9
N-OSLO 5

6452 VALSTAD, T., Mr - Civil Engr
* Consulting Engr
Norwegian Geotechnical Institute
Postboks 40, Tasen
N-OSLO 8

7689 VASSHAUG, B., Mr - Civil Engr
Contracting
F. Selmer A/S
Innspurten 5
P.b. 6035, Etterstad
N-OSLO 6

6454 VEDNER, E., Mr - Civil Engr
Manager
Maskin A/S Pay & Brinck
Brobekkveien 62b, Risløkken
N-OSLO 5

13186 VELDE, H., Mr
Contractor
H. Velde Maskin A/S
N-4056 TANANGER

3382 WABAKKEN, A., Mr - Civil Engr
* Contractor
Entreprenørservice A/S
Sandviksveien 12
N-1322 HØVIK

13187 ØISETH, T., Mr
* Contractor
Entreprenørservice A/S
P.B. 48
N-1322 HØVIK

13188 ØVSTEDAL, E., Mr - Engng Geologist
Research Engr
Norwegian Road Research
Laboratory
Gaustadalleen 25
N-OSLO 3

13189 AAGAARD, B., Mr - Engng Geologist
Consulting Engr
Oslo Lysverker
N-5745 AURLAND

13190 AALDE, O., Mr - Civil Engr
Chief Engr
Jernbeton A/S
Mellomila 24
N-7000 TRONDHEIM

7832 AASGAARD, R., Mr - Civil Engr
Managing Director
c/o Ing. Thor Furuholmen A/S
Karl Johansgt. 39
N-OSLO 1

SUPPORTING MEMBERS

7269 NOTEBY
Norsk Teknisk Byggekontroll A/S
Thv. Meyersgt. 9
OSLO 5

PARAGUAY

ASSOCIACION PARAGUAYA DE GEOTECNIA

Avda. Espana 959, C.C. 336, ASUNCION

President: Dr G. Vera Morinigo
Secretary: Mr L Codas

Total Number of Ordinary Members - 23

ORDINARY MEMBERS

13575 ALVARENGA F., D.A.

13576 ANDRADA, M.

13577 ARCE OTANO

13578 BANKS M., L.

13579 BELLASAI, C.

13580 BORDON DEL PUERTO, M.

13595 BOSIO, J.J.

13581 CHASE, M.A.

13582 CODAS E., L.E.

13583 DE SALVO, O.

13584 FRANCO, V.H.

13585 GOLBERG. E.A.

13586 LARAN, C.

13587 LINARES S., R.A.

13588 LLANES V., A.

13589 LOPEZ Bosio, A.

13590 MORALES, S.

13591 PALMIERI A., J.H.

13592 PARADERA, J.

13593 SOTOMAYOR S., F.

13594 SPINZI M., A.M.

13596 VERA MORINIGO, G.

13597 VILLALBA, J.

PORTUGAL

SOCIEDADE PORTUGUESA DE GEOTECNIA (SPG)

Laboratorio Nacional de Engenharia Civil, Av. do Brasil, P-1799 LISBOA CODEX

President: Mr J. Moura Esteves
Secretary: Mr A. M. S. Pinelo

Total Number of Ordinary Members - 70
Total Number of Supporting Members - 2

ORDINARY MEMBERS

1654 AIRES-BARROS, L. A., Prof. - Mining Engr
* U. Prof.
\+ Instituto Superior Tecnico
Av. Rovisco Pais
P-1000 LISBOA

1658 ALVES, A.O.F., Mr - Civil Engr
* R. Dr. Nunes da Ponte 76
P-4100 PORTO

7487 AMARAL, P.A., Dr - Geologist
* R. Mangabeira 546, Apt 201
\+ Bairro Santo Antonio
30 000 BELO HORIZONTE (BRAZIL)

1695 ARAUJO, F.A., Dr - Geologist
* R. Sa Ferreira 83, Apt 702
\+ Copacabana
RIO DE JANEIRO (BRAZIL)

10603 ARMELIN, J.L., Dr - Engng Geologist
\+ Laboratorio Central de Engenharia Civil
15378 ILHA SOLTEIRA, SP (BRAZIL)

10604 BAIRAO, C.A., Dr - Geologist
* Av. Santo Amaro 3558
\+ 04556 BROOKLYN SP (BRAZIL)

1692 BARROSO, M.J.G., Mr - Mining Engr
\+ Research Officer
Laboratorio Nacional de Engenharia Civil
Av. do Brasil
P-1799 LISBOA CODEX

7485 CADMAN, J.D., Dr - Civil Engr
* Q.I. 21 - Conjunto 7, Casa 8 - SHI/SUL
\+ 70 000 BRASILIA, DF (BRAZIL)

7447 CALLIARI, N.A., Mr - Civil Engr
* R. Teixeira Coelho 280, Apt 5
CURITIBA, Parana (BRAZIL)

1665 CARLOS, A.P., Mr - Civil Engr
* R. Diogo Fernandes 18,5º Dir.
\+ P-1700 LISBOA

10605 CARVALHO, F.F., Dr - Geologist
* Urbanizacao da Portela, Lote 180,6ºEsq
\+ P-2685 SACAVEM

10606 CARVALHO, J.A.R., Dr - Geologist
* Av. Rainha D. Amelia 18, 6º Dir.
\+ P-1600 LISBOA

7185 CARVALHO, J.M., Dr - Geologist
* Urbanizacao da Portela, Lote 47, 3º Dir.
\+ P-2685 SACAVEM

7496 COELHO, F.B., Dr - Geologist
\+ R. do Arsenal 146, 2º
P-1100 LISBOA

7186 CORTEZ, J.A.S., Mr - Mining Engr
* R. Martim Moniz 707, 2º Dir.
\+ P-4100 PORTO

10607 CUNHA, A.J.V.P., Mr - Civil Engr
Trainee Research Officer
Laboratorio Nacional de Engenharia Civil
Av. do Brasil
P-1799 LISBOA CODEX

6217 ESTEVES, J.M., Mr - Civil Engr
* Research Officer
\+ Laboratorio Nacional de Engenharia Civil
Av. do Brasil
P-1799 LISBOA CODEX

10608 ESPINOZA, M.T., Mr - Civil Engr
* Pasaje Diego Puerta 221
QUITO (ECUADOR)

10609 FERRAZ, J.L., Mr - Civil Engr
Laboratorio Central de Engenharia Civil
15378 ILHA SOLTEIRA, SP (BRAZIL)

12032 FERREIRA, A.M.G.dos S., Mr - Civil Engr
* Estrada Militar, lote G, 3ºDto.
2780 OEIRAS

7146 FERREIRA, C.C., Mr - Civil Engr
* Av. Estados Unidos da America 82,
\+ 5º Esq.
P-1700 LISBOA

1651 FERREIRA, H.N., Mr - Civil Engr
* Principal Research Officer
\+ Laboratorio Nacional de Engenharia Civil
Av. do Brasil
P-1799 LISBOA CODEX

7189 FIGUEIRA, J.M.G., Mr - Civil Engr
* Calc. de Palma de Baixo 4
\+ P-1600 LISBOA

540 FORJAZ, V.H.L., Dr - Geologist
* Instituto de Geociencias dos Acores
\+ Junta Geral
P-9500 PONTA DELGADA (ACORES)

7200 GAMA, S.C., Dr - Geologist
* Av. Antonio Augusto de Aguiar 84, 2º Dir.
\+ P-1000 LISBOA

7190 GANDOLFI, N., Dr - Geologist
R. Adolfo Catani 1356
13560 SAO CARLOS, SP (BRAZIL)

10610 GERALDES, D.P.A.A., Mr - Civil Engr
Laboratorio Nacional de Engenharia Civil
Av. do Brasil
P-1799 LISBOA CODEX

10611 GOMES, L.M.S., Mr - Mining Engr
R. da Cruz Vermelha 13, 6º Dir.
P-1600 LISBOA

7147 GRACA, J.G.C., Mr - Civil Engr
Research Officer
Laboratorio Nacional de Engenharia Civil
Av. do Brasil
P-1799 LISBOA CODEX

1693 GUERRA, J.R., Mr - Civil Engr
* R. Pedro Nunes 45, 5º Dir.
P-1000 LISBOA

7149 LEITE, A.G.M., Dr - Geologist
* R. Manuel Correia Gomes 4, r/c Esq.
\+ P-1500 LISBOA

7150 LIMA, R.S., Mr - Civil Engr
* R. Pinheiro Machado 22, Bloco C,5º
Laranjeiras - RIO DE JANEIRO, RJ (BRAZIL)

321 LOPES, J.J.A.B., Mr - Civil Engr
* Research Officer
Laboratorio Nacional de Engenharia Civil
Av. do Brasil
P-1799 LISBOA CODEX

10612 MAGALHAES, J.G.S., Dr - Geologist
* Gabinete da Area de Sines
\+ P-7520 SINES

7151 MARANHAO, N., Dr - Geologist
* R.S. Francisco Xavier 278, Apt 709
\+ RIO DE JANEIRO, RJ (BRAZIL)

1386 MARTINS, J.B., Mr - Civil Engr
* R. Dr. Joao Antunes Guimaraes 67, 2º Esq.
P-4700 BRAGA

7093 MELLO, F.M., Mr - Civil Engr
* R. Real Grandeza 219
RIO DE JANEIRO, RJ (BRAZIL)

313 MENDES, F.M., Prof. - Mining Engr
* U. Prof.
R. Cidade de Cabinda 22, 2º Dir.
P-1800 LISBOA

10613 MIRANDA, A.J.C.M., Mr - Mining Engr
* Av. Alvares Cabral 65, 1º
\+ P-1200 LISBOA

7202 MOREIRA, J.E., Mr - Civil Engr
* S.Q.S. 308, Bloco A, Apt 210
\+ 70 000 BRASILIA, DF (BRAZIL)

10614 NEVES, A.M.P.P., Mr - Civil Engr
* Av. Oscar Monteiro Torres 57, 1º
\+ P-1000 LISBOA

10615 NEVES, E.J.L.M., Mr - Civil Engr
* Research Officer
Laboratorio Nacional de Engenharia Civil
Av. do Brasil
P-1799 LISBOA CODEX

7489 OLIVEIRA, M.F., Dr - Geologist
* R. Luis de Camoes 105, 3º Esq.
\+ Alges
P-1495 LISBOA

1653 OLIVEIRA, R.A.M., Dr - Geologist
* Research Officer Division Head
\+ Laboratorio Nacional de Engenharia Civil
Av. do Brasil
P-1799 LISBOA CODEX

7154 PEREIRA, A.S.C., Dr - Geologist
* R. Curry Cabral, lote 8, r/c Esq.
\+ Paco de Arcos
P-2780 OEIRAS

7230 PIMENTEL, V.M.J., Mr - Civil Engr
* Av. Marginal 58 (Chale Madalena), 2º
\+ P-2765 ESTORIL

7156 PINELO, A.M.S., Mr - Civil Engr
* Trainee Research Officer
Laboratorio Nacional de Engenharia Civil
Av. do Brasil
P-1799 LISBOA CODEX

10616 PINHARANDA, J.F.S., Mr - Mining Engr
* R. Pedro Cristo 1, 7º Dir.
LISBOA -5

7157 PINHO, M.A.M., Mr - Civil Engr
* R. do Paraiso 6
P-4000 PORTO

10684 PINTO, J.L., Mr - Civil Engr
* Research Officer
Laboratorio Nacional de Engenharia Civil
Av. do Brasil
P-1799 LISBOA CODEX

10617 PINTO, P.S.S., Mr - Civil Engr
* Trainee Research Officer
Laboratorio Nacional de Engenharia Civil
Av. do Brasil
P-1799 LISBOA CODEX

7481 RAMALHAL, F.J.S., Dr - Geologist
* Caixa Postal 6870
\+ LUANDA (R.P. ANGOLA)

10618 RAQUEL, J.A.C., Dr - Geologist
* Av. dos Bombeiros Voluntarios, lote 118, 8º Dir.
\+ Miraflores, Alges
P-1495 LISBOA

7447 RAYMUNDO, G.C., Dr - Geologist
* R. Alberto de Campos 40, Apt 1905
Ipanema - RIO DE JANEIRO, RJ (BRAZIL)

10619 RIBEIRO, S.W.M., Mr - Civil Engr
* R. Montevideo 1436
\+ ASUNCION (PARAGUAY)

372 ROCHA, M.C.M., Prof. - Civil Engr
* U. Prof., Research Engr
\+ Av. Estados Unidos da America 95, 3º Dir.
P-1700 LISBOA

10620 RODRIGUES, F.M.C.P., Mr - Civil Engr
* Principal Research Officer,
\+ Division Head
Laboratorio Nacional de Engenharia Civil
Av. do Brasil
P-1799 LISBOA CODEX

1659 RODRIGUES, J.D., Dr - Geologist
* Research Officer
\+ Laboratorio Nacional de Engenharia Civil
Av. do Brasil
P-1799 LISBOA CODEX

7199 RODRIGUES, L.F., Dr - Geologist
* Trainee Research Officer
\+ Laboratorio Nacional de Engenharia Civil
Av. do Brasil
P-1799 LISBOA CODEX

1655 SANTOS, J.A., Mr - Civil Engr
* R. Carlos Mardel 110, 1º
P-1900 LISBOA

7192 SILVA, J.E.C.M., Mr - Civil Engr
* Av. da Republica 20, 2º
P-1000 LISBOA

3851 SILVEIRA, A.F., Mr - Civil Engr
Principal Research Officer,
Dept Head
Laboratorio Nacional de Engenharia Civil
Av. do Brasil
P-1799 LISBOA CODEX

1655 SILVERIO, A., Dr - Physicist
* Research Officer, Division Head
\+ Laboratorio Nacional de Engenharia Civil
Av. do Brasil
P-1799 LISBOA CODEX

7488 SOARES, J.E.S., Mr - Civil Engr
* R. Real Grandeza 86, Apt 706
\+ RIO DE JANEIRO, RJ (BRAZIL)

7193 SOEIRO, F.A., Mr - Civil Engr
* R. Tenente Espanca 40, 1º Esq.
\+ P-1000 LISBOA

10621 SOUSA, H.R., Mr - Civil Engr
* Pr. de Alvalade 6
\+ P-1700 LISBOA

7507 TEIXEIRA, H.A.S., Dr - Geologist
* S.H.I/Sul, Q.I. 7/14 - Casa 6
\+ 70 000 BRASILIA, DF (BRAZIL)

6220 VARGENS, J.R.C., Mr - Civil Engr
* Av. Oceanica 511
SALVADOR, Bahia (BRAZIL)

10622 VELOSO, E.J.A.A.S., Dr - Geologist
\+ R. da Veronica 126, 3º Dir.
P-1100 LISBOA

6218 VIEIRA, C.P.M., Dr - Geologist
* Centro Commercial, Bloco G, 32-F
\+ Nova Oeiras
P-2780 OEIRAS

SUPPORTING MEMBERS

370 ELECTRICIDADE DE PORTUGAL
Orgao Central, Logistica
Av. Infante Santo 15-40
1300 LISBOA

439 LABORATORIO NACIONAL DE ENGENHARIA CIVIL
101 Av. do Brasil
1799 LISBOA CODEX

ROMANIA

Total Number of Corresponding Members - 3

CORRESPONDING MEMBERS

598 CIOCARDEL, R., Dr - Geologist
Strada Dumbrava Rosie 24
Etaj 2; Apart. 4
Sectorul II
BUCHAREST

7255 HARSULESCU, A.I., Dr - Geological Engr
Str. Grigore Moisil, 4 Bloc 6
Scara 2, Etaj 5, Apart. 65, Sector 2
BUCHAREST

6703 SIMA, N.,
Alee Iuliu Valaori, 8
Sector 4
BUCHAREST

SAUDI ARABIA

Total Number of Corresponding Members - 2

CORRESPONDING MEMBERS

13195 ASLAM, M., Dr - Engng Geologist
P.O. Box 728
AL-KHOBAR

13194 SHEHATA, W.M., Dr - Engng Geologist
Faculty of Earth Sciences
P.O. Box 1744
JEDDAH

SOUTH AFRICA, REPUBLIC OF

SOUTH AFRICAN NATIONAL GROUP ON ROCK MECHANICS

P.O. Box 61809, Marshalltown 2107

Chairman: Dr H. Wagner
Secretary: Mr A.J. McDonald

Total Number of Ordinary Members - 192
Total Number of Supporting Members - 3

ORDINARY MEMBERS

10179 ABBOTT, P., Mr - Geologist
+ Site Geologist
c/o P.O. Box 321
BARBERTON 1300

11073 ADAMS, D.J., Mr - Geologist
Assistant Strata Control Officer
Rock Mechanics Department
Anglo American Corporation - Technical Dev. Services
P.O. Box 3
WELKOM 9460

8188 ANDERSON, L., Mr - Mechanical & Mining Engr
Consulting Engineer
The Robbins Co. Africa (Pty) Ltd
P.O. Box 995
ROODEPOORT 7125

8148 ARNOLD, D.A., Mr
Rock Mechanics Officer
17 Karee Crescent
Flamingo Park
WELKOM 9460

10573 AULD, G.J., Mr - Mining Engr
Professional Assistant
Mining Operations Laboratories
P.O. Box 61809
MARSHALLTOWN 2107

10574 BELOVAY, W.E.F., Mr - Engng Geologist
Assistant Research Officer
20, 10th Avenue
ORANGE GROVE 2192

3825 BENZINGER, E.W., Mr
Laboratory Manager
Geomechanics Division
N.M.E.R.I. - C.S.I.R.
P.O. Box 395
PRETORIA 0001

13209 BHANA, J.H., Mr
Asst. Strata Control Officer
8 Christine Street
Northview
KIMBERLEY 8301

6853 BIONDI, P., Mr - Geologist
Engineering Geologist
S.A. Railways
Room 511
Vrystaat Building
BRAAMFONTEIN 2001

6794 BONIFACE, A., Mr - Civil Engr
Civil Engr
P.O. Box 316
GILLITTS 3603

10180 BOUCHER, G.G., Mr - Engng Geologist
30 Battery Reef
Randfontein Estates
RANDFONTEIN

10175 BRADSHAW, R.A., Mr - Geologist
Geologist
11 Sussex Street
CLAREMONT 7700

10686 BRASLER, A.J., Mr - Engr
C.S.I.R.
P.O. Box 395
PRETORIA 0001

8613 BRINCH, G.H., Mr
Rock Mechanics Officer
7, 5th Avenue
BLYVOORUITSIG 2504

6070 BUDAVARI, S., Prof. - Mining Engr
Professor of Rock Mechanics
Department of Mining Engineering
University of the Witwatersrand
Jan Smuts Avenue
JOHANNESBURG 2001

6836 BURGERS, A., Mr - Civil Engr
Civil Engr
Ninham Shand & Partners
P.O. Box 6101
ROGGEBAAI 8012

8150 BURKE, T.J., Mr - Civil Engr
* Director
P.O. Box 48144
ROOSEVELT PARK 2129

3140 CAHNBLEY, H.K.R., Dr - Mining Engr
Group Rock Mechanics Engineer
18 Jacqueline Avenue
NORTHCLIFF 2195

2559 CARTER, C.A., Mr - Civil Engr
Ninham Shand & Partners
P.O. Box 6101
ROGGEBAAI 8012

6838 CLANAHAN, C.R.H., Mr - Civil Engr
Civil Engineer
230 Julius Jeppe Street
Waterkloof
PRETORIA 0181

3139 CLUR, E.N., Mr
Manager
Paddy McDonnel (Pty.) Ltd
P.O. Box 1666
JOHANNESBURG 2000

7229 * COCKCROFT, T.N., Mr - Civil Engr
Senior Engineer
Van Niekerk, Kleyn & Edwards
ICL House
Hertzog Boulevard
CAPE TOWN 8001

10558 COETZER, F.W.C., Mr - Mining Engr
Senior Scientist
P.O. Box 608
BRITS 0250

8152 COETZER, S.J., Mr
Technical Officer
Geomechanics Division
N.M.E.R.I. - C.S.I.R.
P.O. Box 395
PRETORIA 0001

7117 COLE, K.L., Mr
Engineers Representative
Gibb, Hawkins & Partners
Private Bag 306
JAGERSRUST 3354

10623 COOPER, P.H., Mr - Mining Geologist
Managing Director
Spedel (Pty.) Ltd
P.O. Box 56380
PINEGOWRIE 2123

3404 CROUS, N.G., Mr
Rock Mechanics Officer
35, 8th Street
FOCHVILLE 2515

3221 CSATARY, C.J., Mr - Mining Engr
Mining Engineer
Johannesburg Con. Investment Ltd
P.O. Box 231
JOHANNESBURG 2000

10687 DACE, T.W., Mr - Geologist
Rock Mechanics Engineer
Rock Mechanics Department
West Driefontein G M Co. Ltd
Private Bag
CARLETONVILLE 2500

7121 DAVIES, B., Mr
Rock Mechanics Officer
Rock Mechanics Dept
Western Holdings Limited
P.O. Box 3
WELKOM 9460

7446 DAVIES, P.H., Mr - Civil Engr
Partner
Jeffares and Green
P.O. Box 220
PIETERMARITZBURG 3200

13210 DAWE, S.G., Mr
Rock Mechanics Officer
20 C R Swart Street
De Bruin Park
ERMELO 2350

7230 * + DE BEER, J.H., Mr - Engng Geologist
Engineering Geologist
Ove Arup & Partners
P.O. Box 52285
SAXONWOLD 2132

3232 DE JONGH, C.L., Mr
Rock Mechanics Engineer
Goldfields of S.A. Ltd
P.O. Box 1167
JOHANNESBURG 2000

10683 DE KLERK, R., Mr
Rock Mechanics Officer
18, Daffodil Street
Jim Fouche Park
WELKOM 9460

13211 DELPORT, J.E., Mr - Civil Engr
Assistant Engineer
P.O. Box 9644
NEWCASTLE 2940

11074 DE MAAR, W.J., Mr - Geologist
Rock Mechanics Engineer
Planning Department
Westonaria Gold Mining Co. Ltd
P.O. Box 66
WESTONARIA 1780

10584 DE SAINT SIMON, P., Mr - Mining Engr
Branch Manager
P.O. Box 11129
JOHANNESBURG 2000

13570 DE WET, M., Mr - Civil Engr
Department of Civil Engineering
University of Stellenbosch
STELLENBOSCH 7600

7961 DE WIT, B., Mr - Civil Engr
Civil Engineer
D.L. Webb & Associates
P.O. Box 1856
DURBAN 4000

10194 + DICK, A., Mr - Engng Geologist
Senior Engineering Geologist
24 Clifford Avenue
Highlands Estate
CAPE TOWN 8001

8185 DIX, M.N., Mr
P.O. Box 1282
WELKOM 9460

3222 * DU PLESSIS, J.G., Mr
Chief Engineer
Department of Water Affairs
Private Bag X313
PRETORIA 0001

7176 DU TOIT, C.T., Mr - Civil Engr
District Engineer
P.O. Box 440
WORCESTER 6850

3212 DU TOIT, M.C., Mr
Strata Control Officer
P.O. Box 846
WELKOM 9460

10176 * EAST, D.R., Mr - Civil Engr
Associate Partner
P.O. Box 31176
BRAAMFONTEIN 2017

6653 EKAT, F.W.G.D., Mr
Rock Mechanics Engineer
S.W.A. Lithium Mines
P.O. Box 49
KARIBIB 9160

6236 EMERE, G.T.G., Mr - Geologist
Assist. Div. Rock Mechanics Engineer
Harmony G.M. Co. Ltd
P.O. Box 1
Glen Harmony
VIRGINIA 9435

10177 * EVERITT, P.R., Mr - Civil Engineer
Engineer (Materials Section)
78 Forsdick Road
Carrington Heights
DURBAN 4001

8154 FERGUSON, G.A., Mr - Mining Geologist
Rock Mechanics Engineer
Dale House
Belmont Vale
MAIDENHEAD SL6 6BH
UK

3833 FORBES, A.C., Mr - Engng Geologist
Engineering Geologist
P.O. Box 572
BELLVILLE 7530

10559 FOURIE, H.J. de W.L., Mr
Section Head
c/o Technical Services Dept
P.O. Box 1
GLEN HARMONY 9435

7980 FOX, M., Mr - Mechanical Engr
Mechanical Engineer
Atomic Energy Board
Private Bag X256
PRETORIA 0001

7955 + GALLIERS, R.M., Mr - Engng Geologist
Engineering Geologist
c/o M.J. Mountain
Vineyard Centre
Vineyard Road
CLAREMONT 7700

7177 GAY, N.C., Dr - Geologist
Principal Professional Assistant
Mining Operations Laboratories
P.O. Box 61809
MARSHALLTOWN 2107

8167 GENIS, L.M., Mr - Geologist
Chief Mine Geologist
c/o O'Kiep Copper Co. Ltd
P.O. Box 17
NABABEEP 8265

6839 + GEORGE, D.L., Mr - Geologist
Geologist
P.O. Box 559
PIETERMARITZBURG 3200

10563 * GIDDINGS, T.R., Mr - Civil Engr
Geotechnical Engineer
P.O. Box 3132
CAPE TOWN 8001

7981 * GOODMAN, J.A., Mr - Civil Engr
Deputy Project Director
P.O. Box 31021
BRAAMFONTEIN 2017

7418 GRAHAM, P.C., Mr - Mechanical Engr
Consulting Engr
P.O. Box 995
ROODEPOORT 1725

8614 GREAVES, R.A., Mr - Mining Engr
Contracts Manager - Civil Engineering
Grecon Mining Contracting Co.
P.O. Box 91035
AUCKLAND PARK 2006

488 GROBBELAAR, C., Mr - Civil Engr
Civil Engineer
461 Queens Crescent
LYNNWOOD 0081

6657 GRÜTZNER, W.C.G., Mr
Rock Mechanics Engineer
Western Areas G.M. Co. Ltd
Rock Mechanics Department
P.O. Box 57
WESTONARIA 1780

10560 GUEST, A.R., Mr - Geologist
P.O. Box 1160
HONEYDEW 2040

8155 HAINSWORTH, W.E., Mr - Civil Engr
Director
J.C. Theron, Burke & Isaac
P.O. Box 48144
ROOSEVELT PARK 2129

3142 HALLBAUER, D.K., Dr - Geologist
Geologist
Chamber of Mines of S.A.
Mining Laboratory
P.O. Box 61809
MARSHALLTOWN 2107

10195 HARRIS, R.R.W., Mr - Geologist
Project Engineer
P.O. Box 524
ISANDO 1600

6854 + HARTOPP, P.G., Mr - Engng Geologist
Engineering Geologist
P.O. Box 391152
BRAMLEY 2018

3812 HATTINGH, K.F., Mr
Strata Control Officer
11 van Riebeeck Drive
STILFONTEIN 2550

10637 HEPWORTH, N., Mr - Geologist
Student
58, 14th Avenue
EDENVALE 1610

7232 HILLYARD, E.G., Mr - Civil Engr
Resident Partner
P.O. Box 134
GORDONS BAY 7150

6814 * HOJEM, P.A., Mr - Civil Engr
Civil Engineer
Keeve Steyn & Partners
P.O. Box 31021
BRAAMFONTEIN 2017

8189 HOUGHTON, D.A., Mr - Engng Geologist
Engineering Geologist
Steffen Robertson and Kirsten Inc.
P.O. Box 8856
JOHANNESBURG 2000

8168 HUGO, C.G., Mr
Senior Rock Mechanics Technician
P.O. Box 17
NABAPEEP 8265

8190 IRVING, K.E., Mr - Geologist
Rock Mechanics Officer
P.O. Box 219
ST. MICHAELS-ON-SEA 4265

10688 JACOBS, F.M.J., Mr - Geologist
Rock Mechanics Engineer
Libanon G M Co. Ltd
P.O. Box 22
WESTONARIA 1780

11075 JAMES, J.V., Mr - Geologist
Strata Control Officer
Rock Mechanics Department
Anglo American Corporation
Technical Dev. Services
P.O. Box 3
WELKOM 9460

10185 JONES, D.C., Mr - Engng Geologist
Rock Mechanics Engineer
Rock Mechanics Dept
Libanon G.M. Co. Ltd
P.O. Box 22
WESTONARIA 1780

6815 KERSTEN, R.W.O., Mr - Geologist
Geologist
Anglo Transvaal Con. Investment Co. (Pty.) Ltd
P.O. Box 62379
MARSHALLTOWN 2107

7982 KIRSTEIN, A.C., Mr
Strata Control Officer
P.O. Box 17
KLERKSDORP 2570

8615 + KIRSTEN, H.A.D., Mr - Mining & Civil Engr
Engineer
P.O. Box 8856
JOHANNESBURG 2000

6658 KLOKOW, J.W., Mr - Geologist
Rock Mechanics Officer
Rock Mechanics Dept
West Driefontein G.M. Co. Ltd
Private Bag
CARLETONVILLE 2500

3383 KOTZE, T.J., Mr
Rock Mechanics Engineer
Group Rock Mechanics Dept
Winkelhaak Mines Ltd
P.O. Box 1
EVANDER 2280

3826 KOVACS, I.K.A., Mr
Senior Technical Officer
Geomechanics Division
N.M.E.R.I.- C.S.I.R.
P.O. Box 395
PRETORIA 0001

6300 KUBISCH, F., Mr
Senior Resident Engineer
P.O. Box 9916
General Post Office
(HONG KONG)

13213 LATEGAN, J.C., Mr
Regional Rock Mechanics Officer
P.O. Box 5161
EERSTE MYN 9466

8156 LAUBSCHER, D.H., Dr - Mining Engr
Geomechanics & Geological Consultant
P.O. Box 1100
BULAWAYO (ZIMBABWE)

13214 LEAR, C.D., Mr - Mining Engr
Strata Control Officer
Group Rock Mechanics Department
Bracken Mines Limited
P.O. Box 73
EVANDER 2280

3127 LEGGE, G.G.H., Mr
Partner
Watermeyer, Legge, Piesold & Uhlmann
P.O. Box 31176
BRAAMFONTEIN 2017

3213 LEIDING, F.Q.P., Prof.- Mining Engr
Prof. of Mining Engineering
Department of Mining Engineering
University of Pretoria
PRETORIA 0002

6241 LEWIS, B.P., Mr
Assist. Rock Mechanics Officer
P.O. Box 822
WELKOM 9460

10638 LILLY, P.A., Dr - Geologist
Lecturer
Department of Geology
University of the Witwatersrand
1, Jan Smuts Avenue
JOHANNESBURG 2001

8191 LINSELL, C.P., Mr - Mining Engr
Partner
c/o J.B. Mudd & Partners
3rd Floor, Schlesinger Centre
222 Smit Street
BRAAMFONTEIN 2001

10181 LOMBARD, S.B., Mr
Technical Assist - Rock Mechanics
9 Clubview Road
STILFONTEIN 2550

6842 * LOUDON, P.A., Dr - Civil Engr
Civil Engineer
c/o Van Niekerk, Kleyn & Edwards
ICL House
Hertzog Boulevard
CAPE TOWN 8001

8138 * LOUW, J.M., Dr - Civil Engr
Director
P.O. Box 905
PRETORIA 0001

7267 LUWES, W.H., Mr - Civil Engr
Director
P.O. Box 718
SILVERTON 0127

2557 * MACKELLAR, D.C.R., Mr - Civil Engr
Civil Engineer
Ninham Shand & Partners
P.O. Box 6101
ROGGEBAAI 8001

7233 MASCHEK, R.K.A., Mr - Engng Geologist
Research Engineer
Geomechanics Division
N.M.E.R.I. - C.S.I.R.
P.O. Box 395
PRETORIA 0001

628 MAUD, R.R., Dr - Civil Engr
Civil Engineer
Drennan, Maud & Partners
1103 Escoval House
437 Smith Street
DURBAN 4001

10570 McDONALD, A.J., Mr - Geophysicist
Professional Assistant
Mining Operations Laboratories
P.O. Box 61809
MARSHALLTOWN 2107

8157 McMURRAY, D.T., Mr - Mining Engr
Mining Consultant
40 Willow Road
NEWLANDS 7700

6243 MEIRING, H.A., Mr
Rock Mechanics Officer
P.O. Box 1328
KLERKSDORP 2570

8158 + METZGER, G.R., Mr - Engng Geologist
ROCK Mechanics Geologist
Drakon
Private Bag 653
BERGVILLE 3350

10196 MINNEY, D.S., Mr - Geologist
Strata Control Engineer
P.O. Box 512
WELKOM 9460

3396 MORE O'FERRALL, R.C., Mr - Mining Engr
Strata Control Engineer
P.O. Box 206
STILFONTEIN 2550

3215 MORTON, W.A.T., Mr
Asst. Managing Director
Hudemco (Pty.) Ltd
P.O. Box 1159
JOHANNESBURG 2000

10197 MOULD, P.A., Ms
Assistant Technician
Sunnyside House
University of the Witwatersrand
1 Jan Smuts Avenue
JOHANNESBURG 2001

6795 MOUNTAIN, M.J., Mr - Engng Geologist
* Engineering Geologist
1st Floor
Vineyard Centre
Vineyard Road
CLAREMONT 7700

7238 MULLER, E.F., Mr - Civil Engr
Civil Engineer
c/o Campbell, Bernstein & Irving
P.O. Box 37002
OVERPORT 4067

10182 MURIE, A., Mr - Engng Geologist
Rock Mechanics Officer
313 Long Road
Bedelia
WELKOM 9460

3388 MYBURGH, R.I.D.M., Mr
Director
P.O. Box 1179
WINDHOEK 9100

11076 NAISMITH, W.A., Mr - Engng Geologist
Collieries Rock Mechanics Officer
Douglas Colliery
PO VAN DYK'S DRIFT 2245

8169 NANGLE, J.B., Mr - Mining Engr
Assistant Manager
P.O. Box 17
NABABEEP 8265

6054 NETTERBURG, F., Dr - Engng Geologist
+ Engineering Geologist
Soil Engineering Division
N.I.R.R.-C.S.I.R.
P.O. Box 395
PRETORIA 0001

6855 NICOLAYSEN, L., Prof - Geologist
Director
Bernard Price Institute for
Geophysical Research
University of the Witwatersrand
1 Jan Smuts Avenue
JOHANNESBURG 2001

3849 NICOLL, A., Mr - Mining Engr
28 Greenhills Avenue
Greenhills
RANDFONTEIN 1760

10183 NIEUWOUDT, A.P.C., Mr - Geologist
Mine Geologist
P.O. Box 47
OKIEP 8270

6238 NINHAM, C.M., Mr
Strata Control Officer
Rock Mechanics Dept
Western Holdings Ltd
P.O. Box 3
WELKOM 9460

8159 NOBLE, K.R., Mr - Geologist
Rock Mechanics Officer
65 Keats Avenue
ORKNEY 2620

6582 NOCKLER, E., Mr
Rock Mechanics Officer
Blyvooruitzicht G.M. Co. Ltd
Rock Mechanics Dept
P.O. Box 7001
BLYVOORUITSIG 2504

6299 OLIVIER, H.J., Dr - Engng Geologist
+ Engineering Geologist
c/o Keeve Steyn & Partners
P.O. Box 31021
BRAAMFONTEIN 2017

3032 ORAWECZ, K.I., Dr
Department of Petroleum & Mining
Engineering
New Mexico Tech.
Socono
NEW MEXICO 87801 (U.S.A.)

7215 ORR, C.M., Mr - Engng Geologist
+ Engineering Geologist
Geological Survey
Private Bag X112
PRETORIA 0001

494 ORTLEPP, W.D., Mr - Mining Engr
Rock Mechanics Engineer
Rand Mines
P.O. Box 62370
MARSHALLTOWN 2107

13215 PAIGE-GREEN, P., Mr - Engng Geologist
Engineering Geologist
NITRR-CSIR
P.O. Box 395
PRETORIA 0001

8617 PATCHET, S.J., Dr - Mining Engr
Rock Mechanics Engineer
Anglo American Corp of S.A. Ltd
Technical Development Services
P.O. Box 20
WELKOM 9460

7956 PATTISON-BACON, C.J., Mr
Asst Rock Mechanics Officer
Rock Mechanics Dept
Vaal Reefs Exploration & Mining Co
P.O. Box 5010
VAAL REEF 2621

10561 POTHAS, W.J.C., Mr - Mining Engr
Rock Mechanics Engineer
3 Aster Drive
CARLETONVILLE 2500

2558 POWRIE, W.E., Mr
Consulting Engineer
Ninham Shand & Partners
P.O. Box 6101
ROGGEBAAI 8012

8170 PRETORIUS, J.F., Mr - Mechanical Engr
Director
P.O. Box 9222
JOHANNESBURG 2000

10585 RAS, D.J.R.M., Mr - Geologist
Rock Mechanics Officer
P.O. Box 232
COPPERTON 8944

8618 REEVES, P.F., Mr - Geologist
Assist. Rock Mechanics Engineer - Mining
Rock Mechanics Dept
Western Holdings Ltd
P.O. Box 3
WELKOM 9460

8160 REGNERY, F.J., Mr - Mining Engr
Mining Engineer
79 Van Beek Street
Doornfontein
JOHANNESBURG 2000

10684 RETIEF, J.A., Mr - Geologist
Geologist/Rock Mechanics Officer
P.O. Box 10
ALLANRIDGE 9490

10571 RETIEF, J.V., Mr - Civil Engr
Principal Scientist
Atomic Energy Board
Private Bag X256
PRETORIA 0001

3389 + ROBERTSON, A. McG., Mr
Engineer
Steffen, Robertson & Kirsten Inc.
415 Europa House
32 Plein Street
JOHANNESBURG 2001

6856 ROBSON, J.F., Mr - Engng Geologist
Engineering Geologist
c/o D.L. Webb & Associates
P.O. Box 31824
BRAAMFONTEIN 2017

10562 RUST, J.J.H., Mr
Director
P.O. Box 27678
SUNNYSIDE 0132

10639 RYDER, J.A., Dr - Electrical Engr
Head, Computer Applications Division
Mining Operations Laboratory
P.O. Box 61809
MARSHALLTOWN 2107

3033 SALAMON, M.D.G., Dr - Mining Engr
Research Adviser
Research Organization
Chamber of Mines of S.A.
P.O. Box 61809
MARSHALLTOWN 2107

2193 SCHMIDT, N.F.B., Mr - Civil Engr
Technical Manager
13, 10th Avenue
PARKTOWN NORTH 2193

11077 SCHULTZ, E.A., Mr - Civil Engr
41, Old Main Road
HILLCREST 3600

10689 SCHUMANN, E.H.R., Mr
Senior Professional Assistant
35, 8th Avenue
MELVILLE 2092

6581 SIMMS, J.C., Mr
Rock Mechanics Officer
Harmony G.M. Co. Ltd
P.O. Box 1
GLEN HARMONY 9435

13216 SPEARING, A.J.S., Mr - Mining Engr
Mining Graduate
Tennent House
P.O. Box 78
ORKNEY 2620

10198 SPENCER, D.A., Mr - Mining Engr
Mining Engineer
19 Franschhoek Street
Culemborg Park
RANDFONTEIN 1760

6240 SPENGLER, M.G., Mr
Rock Mechanics Officer
29 Palm Ave
Plantation
BOKSBURG 1460

10690 SPOTTISWOODE, S.M., Mr - Geophysicist
Seismologist
Blyvooruitzicht G M Co. Ltd
P.O. Box 7001
BLYVOORUITSIG 2504

13217 STACEY, B.T., Mr - Computer Scientist
Research Officer
Mining Operations Laboratory
P.O. Box 61809
MARSHALLTOWN 2107

1137 STACEY, T.R., Dr - Civil Engr
Civil Engineer
Steffen Robertson & Kirsten Inc.
P.O. Box 8856
JOHANNESBURG 2000

6844 STARFIELD, A.M., Prof
Prof. of Applied Mathematics
Dept of Applied Mathematics
University of the Witwatersrand
Jan Smuts Avenue
JOHANNESBURG 2001

6659 * STEFFEN, O.K.H., Mr - Civil Engr
Civil Engineer
Steffen Robertson & Kirsten Inc.
P.O. Box 8856
JOHANNESBURG 2000

8161 STEGER, P.H., Mr - Civil Engr
Assistant Engineer
10 Rodney Road
MALVERN 4093

10572 STEIJN, J.J., Mr
Professional Assistant
P.O. Box 188
RIVONIA 2128

10640 STEPHANSEN, S.J., Mr - Applied Mathematician
Research Officer
14 Altson Road
Moffat View
JOHANNESBURG 2197

7958 STILWELL, A.W., Mr
Asst Rock Mechanics Officer
159 Chaucer Avenue
ORKNEY 2620

10575 SUMMERS, J.W., Mr - Geologist
Strata Control Officer
Rock Mechanics Dept
Western Holdings Ltd
P.O. Box 3
WELKOM 9460

3144 * TAIT, R.J., Mr
Resident Senior Partner
P.O. Box 68
SWARTKOPS 6210

3231 TAUSSIG, S.G., Mr
Group Rock Mechanics Engineer
53 Rothesay Avenue
Craighall Park
JOHANNESBURG 2196

9015 TAYLOR, H.W., Mr - Geologist
Resident Geologist
3 Chinda Heights
SHABANI (RHODESIA)

7448 * THERON, J.C., Mr - Civil Engr
Director
J.C. Theron, Burke & Isaac
P.O. Box 3251
PRETORIA 0001

6791 THOMAS, J.H., Mr - Geologist
Geologist
P/A Laboratorium voor Grond Mechanica
DELFT (NETHERLANDS)

10586 TUCKER, E.G.R., Mr - Engr
Rock Mechanics Engineer
Harmony G.M. Ltd
P.O. Box 1
GLEN HARMONY 9430

8162 * TYRRELL, A.P., Dr - Civil Engr
Consulting Geotechnical Engineer
P.O. Box 399
RIVONIA 2128

10199 * VAN DER MERWE, D.H. Dr - Engng - Geologist
Director
P.O. Box 3173
PRETORIA 0001

8619 VAN DER MERWE, J.N., Mr - Engr
Chief Rock Mechanics Engineer
Doornfontein G.M. Co. Ltd
Private Bag
WELVERDIEND 2700

10200 + VAN DER MERWE, W.J., Mr - Engng - Geologist
Partner
P.O. Box 3132
CAPE TOWN 8000

7963 * + VAN DER VLUGT, R., Mr - Civil Engr
Civil Engineer
4 Garton Road
RONDEBOSCH 7700

8620 VAN DEVENTER, G.J., Mr
Strata Control Officer
173 Jan van Riebeeck Street
STILFONTEIN 2550

1136 VAN HEERDEN, W.L., Mr
Chief Research Officer
Geomechanics Division
N.M.E.R.I.-C.S.I.R.
P.O. Box 395
PRETORIA 0001

3852 VAN JAARSVELD, A.P., Mr - Engr
Engineer
34 Egret Street
Flamingo Park
WELKOM 9460

1968 + VAN SCHALKWYK, A., Mr - Engng - Geologist
Engineering Geologist
Geological Survey of S.A.
Private Bag X112
PRETORIA 0001

3141 * VAN TONDER, C.P.G., Mr - Civil Engr
Civil Engineer
Postbus 112
PIETERSBURG 0700

7179 VAN VUUREN, P.J.J., Mr - Mining Engr
Project Engineer
Technical Services
Stilfontein G.M. Co. Ltd
P.O. Box 1
STILFONTEIN 2550

7449 VAN WYK, J.J., Mr
Rock Mechanics Officer
29 Addison Road
ORKNEY 2620

8193 VEALE, V.F., Mr
Technical Assistant
84 Stilfontein Road
STILFONTEIN 2550

7108 VENTER, P.P., Dr - Geologist
Asst Consulting Geologist
41 Milner Street
WATERKLOOF 0181

8163 VIEWING, K.A., Dr - Geologist
Director
Institute of Mining Research
University of Rhodesia
P.O. Box MP 167
Mount Pleasand
SALISBURY (RHODESIA)

11078 VLOK, W., Mr - Engr
District Engineer
South African Railways & Harbours
P.O. Box 9727
NEWCASTLE 2940

1138 VOGLER, V.W.O.L., Mr
Senior Technical Officer
Geomechanics Division
N.M.E.R.I.-C.S.I.R.
P.O. Box 395
PRETORIA 0001

10691 VREEDE, F.A., Mr - Civil Engr
Head, Geomechanics Division
N.M.E.R.I.-C.S.I.R.
P.O. Box 395
PRETORIA 0001

506 WAGNER, H., Dr - Mining Engr
Director
23 Cecilia Avenue
RISIDALE 2195

7959 WALDECK, H.G., Mr - Mining Engr
Rock Mechanics Engineer
Rock Mechanics Dept
Libanon G.M. Co. Ltd
P.O. Box 22
WESTONARIA 1780

7180 + WEAVER, J.M., Mr - Engng Geologist
Engineering Geologist
364 Main Street
WATERKLOOF 0181

504 * WEBB, D.L., Dr
Consulting Engineer
D.L. Webb & Associates
P.O. Box 1856
DURBAN 4000

13571 WELTHAGEN, M., Mr - Mining Engr
Post Graduate - Mining
P.O. Box 51793
RANDBURG 2125

7818 WIGGETT, C.H., Mr - Mining Geologist
Rock Mechanics Engineer
Rock Mechanics Dept
Western Holdings Ltd
P.O. Box 3
WELKOM 9460

3143 WIID, B.L., Dr - Engng Geologist
Engineering Geologist
P.O. Box 17060
GROENKLOOF 0027

6858 * WILLIAMS, A.A.B., Dr - Civil Engr
Chief Research Engineer
Geotechnical Division
N.B.R.I.-C.S.I.R.
P.O. Box 395
PRETORIA 0001

3138 WILSON, A.A.T., Mr
Partner
26th Floor
Durban Bay House
333 Smith Street
DURBAN 4001

6792 * WILSON, L.C., Mr - Civil Engr
Partner
Ninham Shand & Partners
P.O. Box 1347
CAPE TOWN 8000

SUPPORTING MEMBERS

457 CHAMBER OF MINES OF SOUTH AFRICA
P.O. Box 61809
MARSHALLTOWN 2107

6992 GOLD FIELDS MINING AND DEVELOPMENT LTD
75 Fox Street
JOHANNESBURG 2001

10220 STEFFEN ROBERTSON AND KIRSTEN INC.
Geotechnical Engineers
Andersonstraat 20
Postbus 8856
JOHANNESBURG 2000

SOUTHEAST ASIA

SOUTHEAST ASIAN SOCIETY OF SOIL ENGINEERING (SEASSE)

Asian Institute of Technology, P.O. Box 2754, BANGKOK (THAILAND)

President: Dr T. S. Beng
Secretary: Dr A. S. Balasubramaniam

Total Number of Ordinary Members - 35
Total Number of Corresponding Members - 5

CORRESPONDING MEMBERS

10184 ADHAMI, N.u.Z., Dr - Geologist
+ Chief Geotechnical Division
National Engineering Services (Pakistan) Limited
402 Wapda House, Shahrah-e-Quaid-e-Azam
P.O. Box 1351
LAHORE (PAKISTAN)

10577 AHMED, M., Dr - Physicist, Geologist
Principal Geologist
National Engineering Services (Pakistan) Limited
405 Wapda House
LAHORE (PAKISTAN)

8197 HALIM, A. Dr - Geologist
36/B, Lawrence Road
LAHORE - 3 (PAKISTAN)

10578 HAMID, A., Dr - Geologist
Deputy Chief Geotechnical Division
403, Wapda House
LAHORE (PAKISTAN)

10214 LEMMENS, E., Dr - Geologist
12 Jalan Tera
BANDUNG (INDONESIA)

SPAIN

SOCIEDAD ESPANOLA DE MECANICA DE LAS ROCAS

Paseo Bajo de la Virgen del Puerto 3, MADRID 5

President: Prof. S. Uriel R.
Secretary: Prof. M. Romana R.

Total Number of Ordinary Members - 191
Total Number of Supporting Members - 2

ORDINARY MEMBERS

1519 * ADALID E., J.L., Mr - Road Engr
M.O.P. Direccion General de
Obras Hidraulicas
MADRID

1488 AGUILA G., A. del, Mr - Road Engr
Av. de Portugal 81
MADRID

1445 * ALCAIDE P., A., Mr - Road Engr
Section Head
Direccion Head
Direccion General de Carreteras M.O.P.
Nuevos Ministerios
MADRID

1606 ALLENDE G.B., J.L., Mr - Road Engr
Hidroelectrica Espanola, S.A.
Hermosilla 3
MADRID

11004 ALONSO D., J.J., Mr - Road Engr
INTECSA
Lazaro Galdiano 6
MADRID 16

11005 ALONSO P. de A., E., Prof - Road Engr
U. Prof
Jose Girona Salgado 31
BARCELONA

1521 * ALVAREZ M., A., Mr - Road Engr
M.O.P.
Divina Pastora 5
GRANADA

1489 ALVAREZ M., A., Mr - Road Engr
AUXINI
P.L. de Salamanca 8
MADRID

1410 + ANGOLOTI, A., J. M., Mr - Mining Engr
Dragados y Construcciones, S.A.
Agustin de Foxa 29
MADRID 16

1598 * ARTEAGA C., R.
Department Head
Guillen de Castro 92 - 25º
VALENCIA

11006 + AYALA C., F.J., Mr - Mining Engr
Instituto Geologico y Minero de
Espana
Rio Rosas 23
MADRID

1490 BARREDO De V., D., Mr - Builder
Procedimientos Barredo
Raimundo Fernandez Villaverde 45
MADRID 3

1908 BENET G., J., Mr - Road Engr
CUBIERTAS y M.Z.O.V., S.A.
Ayala 42
MADRID

1523 BENITO Q., L., Mr - Road Engr
SALTOS DEL SIL
Breton de los Herreros 43
MADRID

1588 BOADA E., M. F., Mr - Mining Engr
Exploration Head
Ensidesa
Fray Valentin Moran 68 - 6ºB
AVILES (Oviedo)

1570 BOBO M., J., Mr - Road Engr
6a. Jefatura Rgional de
Transportes Terrestres
Av. De Baron de Carcer 48
VALENCIA

1910 BOMPREZZI, C., Mr - Engr
Cimentaciones Especiales, S.A.
Av. del Generalisimo 20
MADRID 16

1911 BRAVO G., G., Mr - Road Engr
Confederacion Hidrografica del
Guadalquivir
Recogidas 48
GRANADA

11007 * BRIALES J., J., Mr - Road Engr
Pº de la Farola, 23
MALAGA

1901 BUTRAGUENO B., J.A., Mr - Mining Engr
+
IBERDUERO, S.A.
Av. de America 32
MADRID

7461 CABANAS C., R., Dr - Geologist
6a. Jefatura de Proyectos de Valencia
P°. del Mar s/n
VALENCIA

1599 CAMPO A., A. del, Mr - Road Engr
Saltos del Sil, S.A.
Narvaez, 28
MADRID

11009 CANALDA C., A., Mr - Road Engr
Bonetero 6
MADRID 16

1492 CANIZO P., L. del, Mr - Road Engr
* Comandante Zorita, 2-4°
MADRID 20

7462 CASTILLO R., M., Mr - Road Engr
Hidroelectrica Espanola, S.A.
MADRID

1494 COLL A., M., Mr - Road Engr
Geotecnia y Cimientos, S.A.
Agustin de Foxa 29
MADRID 16

1573 COMPTE G., J., Mr - Road Engr
Autopistas Concesionaria Espanola, S.A.
Plaza de Gala Placidia 1 y 3
BARCELONA

1526 CORRAL Y P., S., Mr - Road Engr
Nunez de Balboa 41
MADRID

11010 CORRAL G., J., Mr - Road Engr
GEOEXPERTS, S.A.
Av. del General Peron 22
MADRID

11019 CORTACANS T., J. A., Mr - Road Engr
* Entrecanales y Tavora, S.A.
Juan de Mena 8
MADRID

11011 CRIADO, B., F. J., Mr - Road Engr
9a. Jefatura Regional de Carreteras
Carretera de Madrid
VALLADOLID

7464 CUADRA I., L. de la, Mr - Mining Engr
Escuela Tecnica Superior de Ingenieros de Minas
Rios Rosas 23
MADRID

11012 DAPENA G., J. E., Mr - Road Engr
* Laboratorio del Transporte
+ Alfonso XII 3
MADRID

1452 DIAZ G., A., Mr - Road Engr
Cia. Sevillana de Electricidad, S.A.
Orense 11
MADRID 20

1575 DIAZ del R. y J., M., Mr - Road Engineer
RENFE - Estacion del Principe Pio
Juan XXIII 24
MADRID 3

1620 DUELO T., C., Mr - Road Engr
Dragados y Construcciones, S.A.
P°. de la Alameda de Osuna s/n
MADRID 32

1497 ECHEVARRIA C., M. R., Mr - Mining Engr
+ Instituto Geologico y Minero de Espana
Rios Rosas 23
MADRID

1453 EIGUREN B., G., Mr - Road Engr
Augusto Gonzalez Besada 2-10C
MADRID 18

7467 ESBERT A., R. M., Dr - Geologist
Universidad de Oviedo
Facultad de Ciencias
Departamento de Petrologia
OVIEDO

1576 ESCARIO U., V., Mr - Road Engr
* Laboratorio del Transporte y Mecanica del Suelo
Alfonso XII 3
MADRID

11013 ESCUDERO G., J., Mr - Road Engr
Hidroelectrica Espanola, S.A.
Hermosilla 3
MADRID

1412 FARINA L., P., Mr - Road Engr
* Dames and Moore Iberia, S.A.
Pedro Muguruza 8
MADRID 16

1499 FAYAS J., J. A., Mr - Road Engr
+ Servicio Geologico de Obras Publicas
Av. del Generalisimo 630
BARCELONA

1529 FEIJOO, M., J., Mr - Road Engr
Rua Petin
ORENSE

1392 FERNANDEZ B., M., Mr - Road Engr
* Sociedad de Reconocimientos Geofisicos, S.A.
J. y F. Serrano Suner 6
MADRID 13

7470 FERNANDEZ B. M., M., Mr - Road Engr
Sociedad de Reconocimientos Geofisicos, S.A.
J. y F. Serrano Suner 6
MADRID 13

7471 FERNANDEZ C., C., Mr - Road Engr
* Grijalba 9
MADRID 6

11014 FERNANDEZ C., F., Mr - Road Engr
* KRONSA
Comandante Zarita 2
MADRID 20

1531 FERNANDEZ M., J., Mr - Road Engr
Confederacion Hidrografica del Ebro
ZARAGOZA

1501 FERNANDEZ R., L. F., Mr - Road Engr
* General Yague, 8
MADRID

7472 FLOREZ A. J., Mr - Road Engr
* Dragados y Construcciones, S.A.
Orense 11
MADRID 20

11016 FORA B., J., Mr - Road Engr
IBERDUERO, S.A.
Gardoqui 8
BILBAO

11017 FORT L.-T., L., Mr - Road Engr
Entrecanales y Tavora, S.A.
Juan de Mena 6
MADRID

1623 FRANCOS R., L., Mr - Road Engr
Goya 44
MADRID

11018 GALINDO M., G., Mr - Road Research
* KRONSA
Commandante Zorita 2
MADRID 20

1624 GARCIA-ARENAL R., P., Mr - Road Engr
Cubiertas y MZOV, S.A.
Ayala 42
MADRID

1457 GARCIA M., F., Mr - Road Engr
Oficina Regional de Proyectos
Cervantes 23
OVIEDO

11020 GARCIA R., R., Mr - Road Engr
7a. Jefatura Regional de Carreteras
P°. de la Farola 23
MALAGA

1625 GARCIA R., J., Mr - Road Engr
* Claudio Coello 32
MADRID 1

1626 GARCIA DE S.yM.,F., Mr - Civil Engr
* INI
Pl. de Salamanca 8
MADRID

7474 GASPAR T., D. B., Mr - Mining Engr
Dragados y Construcciones, S.A.
Orense 11
MADRID 20

1895 GETE-ALONSO DE Y., A., Mr - Road Engr
* ENHER - P°. de Gracia 132
BARCELONA

1458 GIL E., L. E., Mr - Road Engr
Diputacion Provincial de Almeria
ALMERIA

7476 GIL M., F., Mr - Engr
* Managing Director
Industrial de Sondeos, S.A.
Mostoles
MADRID

1627 GODED V., J. M., Mr - Road Engr
Vallehermoso, 30
MADRID 15

1393 GOICOLEA Z., F. J., Mr - Road Engr
Escuela Tecnica Superior de
Ingenieros de Caminos
Ciudad Universitaria
MADRID

1535 GOMEZ C., C., Mr - Road Engr
Berlin 3
MADRID 2

1536 GOMEZ DE LA F., A., Mr - Road Engr
Confederacion Hidrografica del Ebro
General Mola 23
ZARAGOZA

4159 GOMEZ DE LA F., M.E., Mr - Road Engr
* Rios Rosas 36
MADRID

1537 GOMEZ L., G., Mr - Road Engr
* Direccion General de Obras
Hidraulicas
MADRID

1628 GOMEZ O., A., Mr - Road Engr
* AGOSA
Carlos III 4
SANTANDER

1503 GOMEZ DE P., M., Mr - Road Engr
Caracas 6
MADRID

11023 GONZALEZ DE V., L., Dr - Engng Geologist
* P°. de la Habana 33, 3°.
+ MADRID 16

11024 GONZALEZ S., A., Mr - Road Engr
Entrecanales y Tavora, S.A.
Juan de Mena 6
MADRID

1629 GRACIA B., V., Mr - Road Engr
Duque de Sesto 38
MADRID

1631 GUEDAN G., G., Mr - Road Engr
* Entrecanales y Tavora S.A.
Matias Montero 3
MADRID

4160 GUERRERO M.,R.R., Mr - Road Engr
Seccion de Construccion de Presas del MOP
Nuevos Ministerios
MADRID

1632 GUINEA Y O. De U., P.M., Mr - Road Engr
IBERDUERO, S.A.
Gardoqui 8
BILBAO

7477 GUITART Y DE G., J.L., Mr - Road Engr
Velasquez 26
MADRID

1578 * HACAR B., M.A., Mr - Road Engr
San Francisco de Sales 2
MADRID 3

1542 HERAS R., R., Mr - Road Engr
Instituto de Hidrologia
P.º Bajo de la Virgen del Puerto 3
MADRID 5

1460 HERNANDEZ R., C., Mr - Road Engr
Miguel Angel 24
MADRID

1504 * HERNANDO R., J., Mr - Road Engr
Cerdena 502
BARCELONA

1904 HERRERAS E., J.A., Mr - Road Engr
Av. de San Luis 95
MADRID 33

11027 HERRERO P., E., Mr - Road Engr
ENHER
Mallorca 274
BARCELONA

11028 HIDALGO B.,E., Mr - Mining Engr
Instituto Geologico y Minero
Rios Rosas 23
MADRID

11029 HOYO F.G., R. del, Mr - Road Engr
FENOSA
F. Macias 2
LA CORUNA

1505 IBARROLA Z., A., Mr - Mining Engr
Entrecanales y Tavora, S.A.
Juan de Mena 6
MADRID

11030 IZQUIERDO DEL F., J., Mr - Road Engr
Jefatura Provincial de Carreteras
Juan de Herrera 16
SANTANDER

1397 * + JIMENEZ S., J.A., Mr - Road Engr
Centro de Estudios y Experimentacion de Obras Publicas
Alfonso XII 3
MADRID

11031 LABRADA R., M., Dr - Geologist
RENFE
Direccion de Obras e Instalaciones
Av. Ciudad de Barcelona 2
MADRID

1544 LANDABASO A., J.A., Mr - Road Engr
IBERDUERO, S.A.
Gardoqui 8
BILBAO

1914 LARRIQUETA G., R., Mr - Mining Tech Engr
Cuesta de las Descargas 6
MADRID 5

1636 LEAL E., G., Mr - Road Engr
Leal Elizaran, S.A.
Alfonso Rodriguez Santamaria 8
MADRID 2

11033 LLAMAS G., P., Mr - Road Engr
Iberduero, S.A.
Gardoqui 8
BILBAO

11034 LLORENS A., M., Mr - Road Engr
Entrecanales y Tavora, S.A.
Juan de Mena 6
MADRID

1507 LORENTE DE NO C., C., Mr - Road Engr
Entrecanales y Tavora, S.A.
Diego de Leon 58, 3.ºA
MADRID 6

1591 MARIN L., J., Mr - Mining Engr
Via Layetana 5
BARCELONA

1466 MARRERO O'S., P.M., Mr - Road Engr
Dr Apolinario Macias 44
LAS PALMAS DE GRAN CANARIA

1508 * MARSAL M., R., Mr - Road Engr
Guzman el Bueno, 2
MADRID 15

1423 * + MARTIN B., P., Mr - Mining Engr
EPTISA
Arapiles 14
MADRID

1509 MARTIN DE LA S., F., Mr - Road Engr
IBERDUERO, S.A.
Gardoqui 8
BILBAO

1469 MILLA R., A., Mr - Road Engr
Espanoleto 20
MADRID 4

1425 MONTOTO S., M., Dr - Geologist
Universidad de Barcelona
Muntaner, 20
BARCELONA

7482 MORAN N., J.A., Mr - Ind Engr
Cimentaciones Especiales, S.A.
Fernan Gonzalez 39
MADRID 9

11037 MORRAS O.J.L., Mr
Entrecanales y Tavora
MADRID

11036 MOSCOSO DEL P. Y DEL A., F., Mr
Road Engr
Ferrovial, S.A.
P°. de Colon, s/n
BILBAO

7484 MUNOZ G., R., Mr - Road Engr
Confederacion Hidrografica
del Guadalquivir
Plaza de Espana
SEVILLA

7485 MUZAS L., F., Mr - Road Engr
* Pedro Teixeira, 16
MADRID

1602 NAVACERRADA F., G., Mr - Road Engr
Capitan Haya, 32
MADRID

1471 NAVALON G., N., Mr - Road Engr
Hidroelectrica Espanola, S.A.
Hermosilla, 3
MADRID

1946 NUNEZ O., J., Mr - Road Engr
* KRONSA
Comandante Zorita 2
MADRID 20

1426 OJEDA A., F., Mr - Road Engr
\+ Pl. Santa Ana, 9
MADRID

11038 OLAGUIBEL L., L., Mr - Road Engr
IBERDUERO, S.A.
Gardoqui, 8
BILBAO

1582 OLIVEROS R., F., Mr - Road Engr
RENFE
Estacion Principe Pio
MADRID

7486 ORDAZ G., J., Dr - Geologist
Universidad de Oviedo
Facultad de Ciencias
OVIEDO

11039 OROVIOGOICOECHEA,A., Mr - Ind. Engr
* Paulino Mendivil 3 - 4°.
LAS ARENAS (VIZCAYA)

1510 OSENDE C., J., Mr - Road Engr
* Cimentaciones Especiales, S.A.
Av. del Generalisimo 20
MADRID 16

11040 OTEO M., C., Mr - Road Engr
* Laboratorio del Transporte y
Mecanica del Suelo
Alfonso XII 3
MADRID

11041 PARRA V., R., Mr - Road Engr
* HICEOSA
Av. de Jose Antonio 62
MADRID

4166 PEDRERO G., A., Mr - Road Engr
* Euroestudios, S.A.
Castello 23
MADRID

1474 PEIRONCELLY A., J.M., Mr - Road Engr
* Serrano 32
MADRID

1926 PENA P., J.L., Mr - Mining Engr
* Instituto Geologico y Minero de
Espana
Rios Rosas 23
MADRID

7487 PEREZ S., A., Mr - Road Engr
* OFITECO
Amador de los Rios 8
MADRID 4

11043 PIRIS P., A.
Euroestudios
P. de Valdivia
MADRID 18

7488 PLAZA M., A., Mr - Road Engr
* Geotecnia y Cimientos, S.A.
Agustin de Foxa 29
MADRID

1593 PORTUGAL A., F., Mr - Mining Engr
Victor de la Serna 32
MADRID 16

1556 PRESMANES DE LA V.H., A., Mr - Road Engr
Dr. Maranon 1
SANTANDER

11044 PRIETO A., J., Mr
Entrecanales y Tavora, S.A.
Juan de Mena 8
MADRID

7489 PUERTAS P., F., Mr - Road Engr
Padilla 37
MADRID

1595 RAMIREZ O., P., Mr - Mining Engr
\+ Apartado 21
PAMPLONA

1477 REBOLLO A., A., Mr - Road Engr
* Raset 25 y 27
BARCELONA

1557 RIERA R., J., Mr - Road Engr
Confederacion Hidrografica del
Guadalquivir
Carcel Alta 5
GRANADA

1641 RIPA M., T., Mr - Road Engr
Cubiertas y MZOV
Ayala 42
MADRID

7490 RODRIGUEZ A., J.I., Mr - Mining Engr
Euroestudios, S.A.
Sancho el Sabio 18
SAN SEBASTIAN

1479 RODRIGUEZ-MIRANDA G. A., Mr - Road Engr
Nunez de Balboa 10
MADRID

1642 ROMANA R., M., Mr - Road Engr
* Alonso Cano 77
\+ MADRID

1929 ROMERO R., E., Mr - Road Engr
General Ampudia 14
MADRID 3

1512 ROYO S., J.L., Mr - Road Engr
* Amadeo Nerbo 3
MADRID

7492 RUBIO R., J.M., Mr - Road Engr
Avda. de Roma, 82
BARCELONA

1597 RUIZ DE V. Z., V., Mr - Mining Engr
ASINEL
Nunez de Balboa, 46
MADRID

7493 SANCHEZ DEL C. Y DEL R., J.M., Mr
Road Engr
Confederacion Hidrografica del Duero
Carcer Alta 5
GRANADA

1643 SANCHEZ DE M., M., Mr - Road Engr
Brasilia 27
MADRID 28

11045 SANCHEZ M., M., Mr - Road Engr
* Dracena 30
MADRID 16

1934 SANCHEZ R., J., Mr - Road Engr
* ESTEYCO, S.A.
Menendez Pidal 15
MADRID 16

1404 SAN MIGUEL A., A., Prof.
Diputacion 185
BARCELONA

1481 SANTOS G., F., Mr - Road Engr
\+ H.D.R. Torre de Madrid
Planta 7a - 14
MADRID

11046 SANTOS M., A., Mr - Road Engr
GEOEXPERTS, S.A.
Av. del General Peron 22
MADRID 20

1436 SANZ H.-S., T., Mr - Mining Engr
\+ Geotecnia y Cimientos, S.A.
Conde de Foxa 29
MADRID

11047 SANZ-PASTOR Y F. DE P., C., Mr -
Road Engr
Alfonso XII, 14
MADRID

1515 SANZ S., J.M., Mr - Road Engr
* Geotecnia y Cimientos, S.A.
Agustin de Foxa 29
MADRID 16

1516 SERRANO G., A., Mr - Road Engr
* Laboratorio del Transporte y
\+ Mecanica del Suelo
Alfonso XII 3
MADRID

7495 SERRANO N., J., Mr - Engr
IBERDUERO, S.A.
Gardoqui 8
BILBAO

1645 SERRANO S., E.F., Mr - Road Engr
Santa Ana 14
GUECHO (VIZCAYA)

11049 SOLER L., C., Mr - Ind. Engr
* Ganduxer 113
BARCELONA

7497 SORIGUE Z., J.M., Mr - Road Engr
* Confederacion Hidrografica del Ebro
Jose Antonio 29
LERIDA

1940 STUMP, H.
Carreteras S. Ma. Magdalena s/n
MADRID 16

13571 SUAREZ del R., L.
Universidad de Oviedo
OVIEDO

1483 TIRADO C., J.J., Mr - Road Engr
Hilarion Eslava 65
MADRID

1941 TORAL C., T., Dr - Geologist,Chemist
Novell, 79-81
BARCELONA

4168 TORAL S., J.R., Mr - Road Engr
Padre Damian 29
MADRID 16

11050 TORNER P., J.A., Mr - Road Engr
Confederacion Hidrografica
del Pirineo Oriental
Aragon 383
BARCELONA

1845 TORROJA C., J.A., Mr - Road Engr
Sanchez Pacheco, 61
MADRID

1564 TRINCADO S., J., Mr - Road Engr
General Mola, 84
MADRID

4169 UCEDA J., J.L., Mr - Road Engr
Santiago, 11
MADRID

7498 URENA F., R. de, Mr - Road Engr
Velasquez 86
MADRID 6

1517 URIEL O., A., Mr - Road Engr
Entrecanales y Tavora, S.A.
Juan de Mena 6
MADRID

1518 URIEL R., S., Mr - Road Engr
* Laboratorio del Transporte y
\+ Mecanica del Suelo
Alfonso XII 3
MADRID 7

1567 VALERO B., G., Mr - Road Engr
Cia. Sevillana de Electricidad, S.A.
Orense 11
MADRID 20

1406 VALLARINO C. DEL C., E., Mr - Road Engr
Av. de Concha Espin a 6
MADRID 16

11051 VAQUERO R., A., Dr - Geologist
Geoexperts
Av. del General Peron 22
MADRID

7499 VAZQUEZ R., E., Dr - Chemist
Bertran 123-125
BARCELONA

1944 VIDAL F., J., Dr - Geologist
Balmes 316
BARCELONA

11052 VILLANUEVA P., P., Mr - Road Engr
Comandante Zorita, 2
MADRID 20

1486 VILLA, P., J.M., Mr - Road Engr
Hidroelectrica Espanola, S.A.
Hermosilla 1
MADRID

1408 VIRGILI R., C., Dr - Geologist
Facultad de Ciencias
Universidad Complutense
MADRID

1443 ZAPATA T.,F., Mr - Road Engr
PILSON, S.A.
Fernan Gonzalez 44
MADRID

1444 ZARAGOZA G., G., Mr - Road Engr
\+ Servicio Geologico de O.P.
Av. de Portugal 81
MADRID 11

1487 ZULUETA A., A., Mr - Road Engr
Castilla 7
(Colonia Sto. Domingo) - Algete
MADRID

13572 Representative of:
ADARO
Empresa de Investigaciones Mineras
Serrano 116
MADRID

1609 Representative of:
AUXINI
Dep. Construccion
Pl. de Salamanca, 8
MADRID

1493 Representative of:
CIMENTACIONES ESPECIALLES S.A.
RODIO
Av. del Generalisimo 20
MADRID

1615 Representative of:
CUBIERTAS Y TEJADOS, S.A.
Velazquez, 47
MADRID

1496 Representative of:
DRAGADOS Y CONSTRUCCIONES, S.A.
Av. America 32
MADRID

1498 Representative of:
ENTRECANALES Y TAVORA, S.A.
Juan de Mena 8
MADRID

7475 Representative of:
GEOTECNIA Y CIMIENTOS, S.A.
Orense 11
MADRID

13568 Representative of:
HUARTE Y CIA
Avda. Generalisimo, 12
MADRID

1634 Representative of:
IBERDUERO, S.A.
Gardoqui 8
BILBAO

1912 Representative of:
IBERING
Muntaner, 270
BARCELONA

1463 Representative of:
INTECSA
Condesa de Venadito, 1
MADRID 27

13569 Representative of:
OCISA
Princesa, 3
MADRID

SUPPORTING MEMBERS

582 DRAGADOS Y CONSTRUCCIONES S.A.
P. Alameda de Osuna, S/N
Apartado de Correos nº 13054
MADRID 22

13044 GRUPO ESPECIALISADO DE GEOMECANICA DE LA ASOCIACION NACIONAL DE INGENIEROS DE MINAS
Rios Rosas, 21
MADRID 3

SWEDEN

SWEDISH NATIONAL COMMITTEE FOR ROCK MECHANICS

Royal Swedish Academy of Engineering Sciences, Box 5073, S-102 42 STOCKHOLM

Chairman: Dr S. Bjurström
Secretary General: Mr T. Franzen

Total Number of Ordinary Members - 113
Total Number of Supporting Members - 13

ORDINARY MEMBERS

1281 ABRAHAMSSON, E., Dr - Engr
Research Engr, Head of Dept
The Royal Fortifications Administration
Fack
S-104 50 STOCKHOLM

10837 AHLSTEDT, J., Mr - Engr
Engr - Development
Ytong AB
S-692 02 KUMLA

12030 + AKESSON, J.-A., Mr - Geologist
Consultant
BERGAB - Berggeologiska Undersökningar AB
S:t Pauligatan 33 A
S-416 60 GÖTEBORG

1282 ALBERTS, C., Mr - Civil Engr
Tech. Manager
Stabilator AB
S-182 25 DANDERYD

1283 ALESTAM, M., Mr - Engr
Managing Engr
Sydkraft AB
Fack
S-200 70 MALMO

10838 * ALSKÄR, A., Mr - Engr
Engr - Geotech. Section
Stockholm Public Works Office
Fack
S-104 20 STOCKHOLM

10839 ANDERSSON, S., Dr - Mining Engr
AB Atomenergi
Studsvik
Fack
S-611 01 NYKÖPING

10840 ANDERSSON, S., Mr - Civil Engr
Consultant
Vattenbyggnadsbyran, VBB
Box 5038
S-102 41 STOCKHOLM

1293 BÄCKSTRÖM, S., Mr - Civil Engr
Contractor/Consultant
Stabilator AB
S-182 25 DANDERYD

10841 BENEDIK, R., Mr - Mining Engr
Tyrens
Fack
S-103 40 STOCKHOLM

10842 BERGDAHL, S.G., Mr - Mining Engr
Managing Director
The Swedish Mining Association
P.O. Box 5501
S-114 85 STOCKHOLM

7572 + BERGMAN, M., Mr - Civil Engr
Consultant
Hagconsult AB
Banergatan 37
S-115 22 STOCKHOLM

1288 * BERGMAN, S.G.A., Dr
Consultant
Föreningsvägen 19
S-182 74 STOCKSUND

6025 + BJURSTRÖM, S., Dr - Engr
Research Director
The Swedish Rock Mechanics Research Foundation
P.O. Box 5501
S-114 85 STOCKHOLM

1290 BOMAN, J., Mr - Mining Engr
Managing Dir, ret.
The Swedish Mining Association
P.O. Box 5501
S-114 85 STOCKHOLM

10843 BRANEFORS, Y., Mr - Engr
Harbour Engr
Gävle Hamnförvaltning
Hamntorg 4
S-803 50 GÄVLE

7573 BRÄNNFORS, S., Mr - Civil Engr
Contractor
Skanska Cementgjuteriet
S-182 25 DANDERYD

1291 BROMS, B., Prof. - Civil Engr
* U. Prof.
The Royal Institute of Technology
S-100 44 STOCKHOLM

10845 CARLSSON, A., Dr - Geologist
* Research Engr
The Swedish State Power Board
S-162 87 VÄLLINGBY

10846 CARLSSON, B., Mr - Civil Engr
Dept Manager
John Mattson Byggnads AB
P.O. Box 8744
S-402 75 GÖTEBORG

10844 CEDERBERG, H., Mr - Civil Engr
Research Engr
The Swedish State Power Board
S-162 87 VALLINGBY

10847 CHRISTENSSON, K., Mr - Civil Engr
Contractor
Armerad Betong Vägförbättringar, ABV
S-112 88 STOCKHOLM

10848 CORP, L.G., Mr - Mining Engr
Department Head
The National Board of Occupational
Safety and Health
Fack
S-100 26 STOCKHOLM

10849 DAHLIN, O., Mr - Civil Engr
* Chief Consultant
Vattenbyggnadsbyran, VBB
P.O. Box 5038
S-102 42 STOCKHOLM

3386 ELG, S., Mr - Engr
Research Engr
The Academy of Engineering Sciences
P.O. Box 5073
S-102 41 STOCKHOLM

1298 EMMELIN, L.O., Mr - Civil Engr
* Chief Consultant
+ Vattenbyggnadsbyran, VBB
P.O. Box 5038
S-102 41 STOCKHOLM

10850 ERIKKSON, A., Dr - Geologist
* Consultant
Allmänna Ingenjörsbyran AB
P.O. Box 5511
S-114 85 STOCKHOLM

6027 ERIKSSON, K., Mr - Civil Engr
Chief Consultant
Vattenbyggnadsbyran, VBB
P.O. Box 5038
S-102 41 STOCKHOLM

7576 EURENIUS, J., Mr - Civil Engr
* Consultant
+ Vattenbyggnadsbyran, VBB
P.O. Box 5038
S-102 41 STOCKHOLM

7577 FAGERSTROM, H., Mr - Civil Engr
* Chief Consultant
+ Vattenbyggnadsbyran, VBB
P.O. Box 5038
S-102 41 STOCKHOLM

10851 FARESTVEIT, S., Mr - Civil Engr
Stockholm Public Works Office
Underground Section
Fack
S-104 20 STOCKHOLM

10852 FELLDIN, S., Mr - Engr
Contractor
SIAB-Byggen AB
P.O. Box 53018
S-400 14 GÖTEBORG

10853 FINKEL, M., Mr - Engr
Research Engr
The Royal Institute of Technology
Dep. of Mining & Excavation
S-100 44 STOCKHOLM

8005 FISCHER, H. Ch., Dr
* U. Prof.
University of Uppsala
Institute of Technology
P.O. Box 534
S-751 21 UPPSALA

10854 FRANZEN, T., Mr - Civil Engr
Research Engr
The Swedish Rock Mechanics Research
Foundation
P.O. Box 5501
S-114 85 STOCKHOLM

10855 FREDRIKSSON, A., Mr - Civil Engr
* Research Engr
The Royal Institute of Technology
Dep. of Soil and Rock Mechanics
S-100 44 STOCKHOLM

10856 GRAHN, G., Mr - Engr
Consultant
CEGE Consult
Gränsvägen 39
S-137 00 VASTERHANINGE

10857 GRANLUND, N., Dr
Consultant Geologist
Hagconsult AB
Banergatan 37
S-115 22 STOCKHOLM

1303 GRANSTRÖM, S., Dr
Chief Consultant
Tyrens
Fack
S-103 40 STOCKHOLM

7583 HAHN, T., Dr
\+ Research Engr - Geologist
The Royal Fortification Administr-
ation
S-104 50 STOCKHOLM

10058 HALLENIUS, T., Mr - Civil Engr
President
WP- System AB
P.O. Box 339
S-851-03 SUNDSVALL

1312 HANSAGI, I., Mr - Mining Engr
Tech. Manager
Luossavaara-Kiirunavaara AB - LKAB
Fack
S-100 41 STOCKHOLM

10859 HEINER, A., Dr - Civil Engr
* Consultant
\+ Vattenbyggnadsbyran, VBB
P.O. Box 5038
S-102 41 STOCKHOLM

1316 HELFRICH, H., Prof.
* Consultant - Engng Geology
\+ Brunnsvägen 5
S-181 10 ENEBYBERG

10860 HELLGREN, A., Mr - Civil Engr
* Tech. Manager
Stockholm Public Works Office
Construction Section - Geotechnique
Fack
S-104 20 STOCKHOLM

1319 HILTSCHER, R., Dr
Senior Research Scientist (ret.)
Hasseluddsvägen 170
S-132 00 SAUTSJO-BOO

10861 HOLMBERG, R., Mr - Engr
Research Engr
Swedish Detonic Research Foundation
P.O. Box 32058
S-126 11 STOCKHOLM

10862 HOLMGREN, J., Dr
Research Engr
The Royal Fortification Administr-
ation
Fack
S-104 50 STOCKHOLM

10863 HULTIN, S.A., Mr - Civil Engr
Vice President - Head of Civil
Engineering Department
The Swedish State Power Board
S-162 87 VÄLLINGBY

10864 INGLER, D., Mr - Civl Engr
Chief Consultant - Mining
Vattenbyggnadsbyran, VBB
P.O. Box 5038
S-102 41 STOCKHOLM

10865 ISACSSON, C., Mr - Civil Engr
Managing Director
Stabilator AB
S-182 25 DANDERYD

1324 JANELID, I., Prof. - Mining Engr
U. Prof.
The Royal Institute of Technology
Dept of Mining & Excavation
S-100 44 STOCKHOLM

10866 JANSSON, B., Mr - Architect
Chief Consultant
Vattenbyggnadsbyran, VBB
P.O. Box 5038
S-102 41 STOCKHOLM

10867 JERNELIUS, I., Mr - Civil Engr
Chief Consultant
Vattenbyggnadsbyran, VBB
P.O. Box 5038
S-102 41 STOCKHOLM

13029 JOHANSSON, S., Mr - Engr
Engr
Göteborgs Byggmastartjanst AB
Föreningsgatan 2
S-411 27 GÖTEBORG

1329 KANGAS, K., Mr - Mining Engr
Chief Engr - Tech. Manager
Luossavaara-Kiirunavaara AB - LKAB
Fack
S-971 00 MALMBERGET

10868 KARLSSON, L., Mr - Civil Engr
Consultant - Civil Engr
Bergsäker Konsult AB
Folkvisegatan 12
S-422 41 HISINGS BACKA

10869 KIHLSTRÖM, B., Mr - Engr
Chief Consultant
Nitro Consult AB
P.O. Box 32058
S-126 11 STOCKHOLM

1331 KNUBB, E., Mr - Mining Engr
Chief Mine Inspector
Södra Bergmästardistriktet
Asgatan 35
S-791 00 FALUN

10870 KOLSRUD, B., Mr - Mining Engr
Head of Engng Dept
Boliden Metall AB
S-930 50 BOLIDEN

6311 KRAULAND, N., Mr - Mining Engr
Research Engr
Boliden Metall AB
S-930 50 BOLIDEN

10871 LAGERGREN, S., Mr - Civil Engr
Consultant
Vattenbyggnadsbyran, VBB
P.O. Box 5038
S-102 41 STOCKHOLM

13030 LARSSON, B., Mr - Mining Engr
Dept Manager
Nitro Nobel AB
S-710 30 GYTTORP

3846 LARSSON, I., Prof. - Geologist
+ U. Prof.
The Royal Institute of Technology
Dept of Land Improvement & Drainage
S-100 44 STOCKHOLM

13031 LINDAHL, T., Mr - Civil Engr
Executive Secretary
The Swedish Association of General
Contractors and House-Builders
SBEF
P.O. Box 27029
S-102 51 STOCKHOLM

13032 LINDBLOM, U., Dr
* Research Geologist - Consultant
Hagconsult AB
Postgatan 28
S-411 06 GOTEBORG

1338 LINDER, S., Mr - Civil Engr
Consultant
Vattenbyggnadsbyran, VBB
P.O. Box 5038
S-102 41 STOCKHOLM

13033 LINDSKOG, R., Mr - Civil Engr
Chief Consultant
Vattenbyggnadsbyran, VBB
P.O. Box 5038
S-102 41 STOCKHOLM

13034 LINDSTRÖM, A., Mr - Civil Engr
Consultant
Vattenbyggnadsbyran, VBB
P.O. Box 5038
S-102 41 STOCKHOLM

13037 LÖNNQVIST, G., Mr - Mining Engr
Mine Planning
Gränges International Mining
P.O. Box 16316
S-103 26 STOCKHOLM

13035 LUDVIG, B., Dr - Geologist
+ Assistant Prof. - Research Engr
University of Lulea
Div. of Rock Mechanics
S-951 87 LULEA

1341 LUNDBORG, N., Dr
Head of Research Laboratory
The Swedish Detonic Research
Foundation
P.O. Box 32058
S-126 11 STOCKHOLM

360 LUNDSTRÖM, R., Mr - Civil Engr
Head of Soil and Mech. Dept
Ingenjörsfirma Orrje & Co AB
Fack
S-102 60 STOCKHOLM

13036 LYSEN, L., Mr - Civil Engr
President
Göteborgs Vatten-och Avloppsverk
Fack
S-401 10 GÖTEBORG

13038 MASSARSCH, K. R., Dr
* Consultant - Civil Engr
Vattenbyggnadsbyran, VBB
P.O. Box 5038
S-102 41 STOCKHOLM

13039 MATTSON, F., Mr - Constr. Engr
Construction Engr
AB Karlshamns Oljefabriker
S-292 00 KARLSHAMN

1344 MORFELDT, C.O., Dr - Geologist
President
Hagconsult AB
Brnergatan 37
S-115 22 STOCKHOLM

13040 NILSSON, G., Mr - Engr
Project Engr
Gränges International Mining
P.O. Box 16316
S-103 26 STOCKHOLM

12005 NORD, B., Mr - Constr. Engr
* Contractor
+ BINAB
Fack
S-102 20 STOCKHOLM

12006 NORD, G., Mr - Civil Engr
Contractor
Skanska Cementgjuteriet
S-182 25 DANDERYD

12007 NORDGREN, I., Mr - Constr. Engr
Managing Director - Chief Consultant
Viking Engineering
Mölndalsvägen 85
S-412 85 GÖTEBORG

12008 OLSSON, L., Mr - Civil Engr
* Research Engr
The Royal Institute of Technology
Dept of Soil & Rock Mechanics
S-100 44 STOCKHOLM

12009 OLSSON, T., Dr - Geologist
+ Research Engr
The Geological Survey of Sweden
Bredgränd 4
S-753 20 UPPSALA

1352 OTTOSON, L., Mr - Mech. Engr
Tech. Director - Long Range Planning
Atlas Copco MCT AB
Fack
S-102 60 STOCKHOLM

12010 OUCHTERLONY, F., Dr - Physicist
Research Engr
The Swedish Detonic Research
Foundation
P.O. Box 32058
S-126 11 STOCKHOLM

12011 PAGANUS, T., Mr - Engr
Luossavaara-Kiirunavaara AB - Ikab
Regionkontoret
Fack
S-981 01 KIRUNA

12012 PALMQVIST, K., Dr
+ Consultant
BERGAB - Berggeologiska Undersökningar AB
S:t Pauligatan 33 A
S-416 60 GÖTEBORG

12013 PERSSON, P.A., Dr
President
Swedish Detonic Research
P.O. Box 32058
S-126 11 STOCKHOLM

12014 PETERSSON, L., Mr - Engr
Consultant
Nitro Consult AB
P.O. Box 32058
S-126 11 STOCKHOLM

12015 PRECHT, K., Mr - Engr
Sales Manager
AB Svenska Sika
P.O. Box 6009
S-175 06 JÄRFÄLLA

12016 REHBINDER, G., Dr - Physicist
Research Engr
Atlas Copco MCT AB
Fack
S-102 60 STOCKHOLM

12017 ROCKSTRÖM, H., Mr - Civil Engr
Managing Director
Svensk Petroleumlagring Tre AB
Sveavägen 45
S-111 34 STOCKHOLM

1359 ROOSAAR, H., Dr - Geologist
+ Senior Consultant
Tyrens
P.O. Box 7852
S-103 99 STOCKHOLM

6035 ROSEN, B., Dr - Geologist
Consultant
Ingenjörsbyra G. Nabholz AB
Tranebergsvägen 41
S-161 32 BROMMA

1360 ROSENSTRÖM, S., Mr - Civil Engr
+ Consultant
Vattenbyggnadsbyran, VBB
P.O. Box 5038
S-102 41 STOCKHOLM

12018 RUNDGREN, L., Mr - Civil Engr
Chief Consultant
Vattenbyggnadsbyran, VBB
P.O. Box 5038
S-102 41 STOCKHOLM

12019 SANDELL, B., Mr - Constr.Engr
President
AB BESAB
Tegenevägen 7
S-422 47 HISINGS BACKA

1366 SANDSTRÖM, H., Mr - Engr
Managing Director
Innovation AB
Adler Salviusväg 59
S-146 00 TULLINGE

6036 SCHERMAN, K.A., Mr - Civil Engr
* Tech. Manager
The Swedish State Power Board
S-162 87 VÄLLINGBY

12019 SCHMIDT, C.A., Mr - Civil Engr
Consultant
Vattenbyggnadsbyran, VBB
P.O. Box 5038
S-102 41 STOCKHOLM

12021 SJÖBERG, C., Mr - Engr
Chief Consultant
Nitro Consult AB
P.O. Box 32058
S-126 11 STOCKHOLM

1373 STANFORS, R., Prof. - Geologist
+ U. Prof.
University of Lund
Dept of Geology
Sölvegatan 13
S-223 62 LUND

6037 STEPHANSSON, O., Prof. - Geologist
+ U. Prof.
University of Lulea
Div. of Rock Mechanics
S-951 87 LULEA

12020 STILLE, H., Prof.-Civil Engr
* U. Prof., Consultant
Hagconsult AB
Banergatan 37
S-115 22 STOCKHOLM

1375 SUND, B., Dr - Geologist
+ Senior Consultant
Tyrens
P.O. Box 7852
S-103 99 STOCKHOLM

12022 THOREN, H., Mr - Geologist
* Research Engr
+ Statens Väg-och Trafikinstitut
Fack
S-581 01 LINKÖPING

12023 THURNER, H., Prof. - Civil Engr
* Prof.
+ Geodynamik AB
Regeringsgatan 111
S-111 39 STOCKHOLM

12024 WALTER, O., Mr - Civil Engr
Consultant
Vattenbyggnadsbyran, VBB
P.O. Box 5038
S-102 41 STOCKHOLM

12025 WESSLEN, A., Dr - Geologist
Senior Consultant
Viak AB
Fack
S-162 10 VÄLLINGBY

12026 WETTLEGREN, G., Mr - Civil Engr
Tech. Manager
Evenska Väg AB
S-171 57 SOLNA

12027 WIJK, G., Dr - Physicist
Research Engng Div. Head
Atlas Copco MCT AB
Fack
S-102 60 STOCKHOLM

12028 WIKING, T., Mr - Civil Engr
Chief Consultant
Vattenbyggnadsbyran, VBB
P.O. Box 5038
S-102 41 STOCKHOLM

12029 ZIEMELIS, O., Mr - Geologist
* Consultant
\+ Vattenbyggnadsbyran, VBB
P.O. Box 5038
S-102 41 STOCKHOLM

SUPPORTING MEMBERS

522 ATLAS COPCO AB
S-105 23 STOCKHOLM

13030 BESAB Betongsprutnings AB
GOTEBORG

6991 HAGCONSULT AB
Banergatan 37
S-115 22 STOCKHOLM

13031 L K A B
Fack
S-100 41 STOCKHOLM

7226 NATIONAL SWEDISH ROAD AND
TRAFFIC RESEARCH INSTITUTE - VTI
Fack
S-581 01 LINKOPING

7551 NITRO CONSULT AB
P.O. Box 32058
S-126 11 STOCKHOLM

7450 SKANSKA CEMENTGJUTERIET DANDERYD
S-182 25 DANDERYD

7451 SWEDISH DETONIC RESEARCH FOUNDATION
SVE DE FO
P.O. Box 32058
S-126 11 STOCKHOLM

7227 SWEDISH STATE POWER BOARD
S-162 87 VALLINGBY

13032 THE SWEDISH ROCK MECHANICS RESEARCH
FOUNDATION - BEFO
P.O. Box 5501
S-114 85 STOCKHOLM

13034 TYRENS
P.O. Box 7582
S-103 99 STOCKHOLM

13033 UNIVERSITY OF LULEA
Fack
S-100 41 LULEA

13035 VBB AB
P.O. Box 5038
S-10241 STOCKHOLM

SWITZERLAND

SCHWEIZERISCHE GESELLSCHAFT FÜR BODEN-UND FELSMECHANIK (SGBF)

Institut für Grundbau und Bodenmechanik, ETH-Hönggerberg 8093 ZÜRICH

President: Prof. E. Recordon
Secretary: Mr R. Wulliman

Total Number of Ordinary Members - 143
Total Number of Supporting Members - 4

ORDINARY MEMBERS

650 ABDEL-RAHAM, N., Dr. sc. techn. ETH
* Dipl. Bauing
Hauptstr. 76, 11. rechts
4132 MUTTENZ

12072 AEBERLI, U., Dr. sc. nat. Geologe
* Assistant ETH
Alte Landstr. 185
8800 THALWIL

656 ALLET, B., Ing. civ. dipl. EPF
* General-Werdmüllerstr. 13
8804 AU ZH

658 AMBERG, G., Dipl. Masch.-Ing. ETH
* Adjunkt
Tuschgenweg 111
8041 ZÜRICH

660 AMSLER, P., Ing. civ. dipl. EPF
* Directeur adjoint
17C, ch des Buclines
1224 CHENE- BOUGERIES

12073 ANDRASKAY E., Dipl. Bauing. ETH
Säntisstr. 14
8133 ESSLINGEN

12074 BALISSAT M., Ing. civil dipl. EPF
Chef Groupe Mecanique des sols
c/o Motor-Columbus AG
Parkstr. 27
5401 BADEN

671 BANZIGER, D.J., Dipl. Bauing. ETH/ASIC/SIA
* Inhaber Ing.-Büro
Engimattstr. 11
8027 ZÜRICH

12075 BÄR, H., Ing.-Techn. HTL
* Poststr. 4
8700 KÜSNACHT

12076 BIRRER, H., Dipl. Bauing. ETH
* Rosengartenhalde 4
6000 LUZERN

8021 BRENNER, R.P., Dr. Dipl. Bauing.ETH
* Rosenstr. 8
8570 WEINFELDEN

3297 BRUGMAN, B.J., Dipl. Bauing. ETH/SIA
+ Beethovenstr. 11
3073 GÜMLIGEN

713 BÜCHI, H.F., Dipl. Bauing. ETH/SIA
* Inhaber Ing.-Büro
Rotfluhstr. 47
8702 ZOLLIKON

729 CLEUSIX, C., Ing. civ. dipl. EPF/SIA
* Directeur
c/o Compagnie d'etudes et de
Realisations Techniques SA
Av. Ritz 35
1950 SION

732 COMTE, C.-L., Ing. dipl. EPF
* Direktor Solexperts AG
Rebstr. 1
8703 ERLENBACH

12077 COSTOPOULOS, S., Ing. civ. Dr. an
LEGEP, EPF
Directeur
Saranta Pjhou 51, Lysabette
ATHENS (GREECE)

735 CUENOD, Y., Geologue dipl. GE
+ Chef du groupe geologie
Motor-Columbus Ing.-cons.
Klosterfeld
5649 STETTEN AG

737 DAVAUD, G., Ing. civ. dipl. EPF
* Les "Scarabees B"
+ 1837 CHATEAU D'OEX

3303 DESCOEUDRES, F., Prof. EPF
Laboratoire de Geotechnique
* 1049 SULLENS

742 DESCOUEDRES, J., Ing.-Conseil
ETH/SIA
* Hohenbühlstr. 8
8032 ZÜRICH

746 DEVAY, L., Project civ. engineer
US Army Eng. Waterways
Experiment Station
P.O. Box 1072
VICKSBURG, MISS. 39180 (U.S.A.)

751 DUBAS, C., Dr. es. sc. tech.,
* Ing. civ. dipl. ETH
Adm.-del. et directeur gen.
des ACMV SA
Au Sichoz 39
1814 LA TOUR-DE-PEILZ

12078 DURET, J. J., Geologue
* 12, rue des Cordiers
1207 GENEVE

756 DYSLI, M., Ing. civ. dipl. EPF
* Chef de section
Labor. Geotechn. EPF
22, av. du Parc de la Rouvraie
1018 LAUSANNE

12079 ECKARDT, P.M., Dr.phil. II Geologe
Wabergstr. 13
8624 GRÜT.

761 EINSTEIN, H.H., Dr. sc. tech. ETH
* Assoc. Prof. of Civil Engng
M.I.T. Room 1-330
CAMBRIDGE, MASS. 02139 (U.S.A.)

12080 FECHTIG, R., Dipl. Ing. ETH/SIA
* Oberingenieur
Sempacherstr. 52
8032 ZÜRICH

12081 FEDERER, O.A., Dipl. Bauing, ETH
Haini-Rennhas-Str. 13
9403 GOLDACH SG

775 FREIMANN, M., Dipl. Bauing. ETH/SIA
* Prokurist
Parkweg 5
8800 THALWIL

778 FRIEDLI, P., Dipl. Bauing. ETH/SIA
* Büro f. Geotechnik
Kreuzbühlstr. 8
8008 ZÜRICH

12082 GAST, H., Bauing. SIA
Chef Bureau techn.
27, route du Rawyl
1950 SION

785 GAUTSCHI, M.A., Dipl. Bauing.
ETH/SIA/ASIC
* Mitinh. Geotech. Büro
Dr. von Moos AG
Zürichstr. 39
8122 PFAFFHAUSEN

8116 GICOT, O., Ing.civ.dipl.
EPF/SIA/ASCE
* Copropr. bur. Soltechnique
Rte de Bertigny 11
1700 FRIBOURG

791 GILG, B., Dr. sc. techn
* Mitgl. der Dir. Elektrowatt AG
Zeisenberg
8911 RIFFERSWIL

3311 GROB, H., Dipl.Ing. ETH/SIA
* Prof. ETH u. Rektor, Institut f.
Strassen, Eisenbahn-u.Felsbau
ETH-Hönggerberg
8093 ZÜRICH

12083 GRUBER, N., Dipl.Ing. für Bauwesen
Wiss. Angest. an der TU
München
Irisstr. 9
8031 GRÖBENZELL (FRG)

801 von GUNTEN, H., Dipl.Bauing. ETH/
SIA
* Geschäftsleiter Geotechn.
Inst. AG
Kappelenring 38 b
3032 HINTERKAPPELEN

12084 GUERPILLON, Y., Ing. Ecole Centrale
de Paris
9, ch. des Tuilieres
1028 PREVERENGES

806 HALTER, H., Dr.,Dipl. Bauing. ETH/
SIA
* Berat. Ing.
Gubelstr. 32
8050 ZÜRICH

3312 HORISBERGER, W., Dipl.Bauing. ETH/
SIA
Teilh. HGS Bauing, AG, Bern
Melchtalstr. 17
3014 BERN

831 HUDER, J., Prof. Dr. sc. tech.
* o. Prof. ETH
Greifenseestr. 31
8603 SCHWERZENBACH

12085 HÜNERWADEL, M.F., Dipl. Bauing ETH/
SIA
* Direktor
Spechtweg 1
4103 BOTTMINGEN

12086 ITEN, H., Bauing. Tech. W'thur
* Teilh. Ing.-Büro
Gfennstr. 7
8603 SCHWERZENBACH

835 JAECKLIN, F.P., Dr.Dipl.Bauing ETH/
SIA
* Inhaber Ing.-Büro
Geissbergstr. 59
5400 ENNETBADEN

1258 JUHOLA, M.O., Dipl. Ing.
* Direktor Pohjavahvistus Oy.
Lounaisväyla 2 a. 4
HELSINKI (FINLAND)

12087 KAJEL, L., Dipl. Geologe
Alpenblickstr. 23
3052 ZOLLIKOFEN

3316 KAPP, H., Dr.phil II, Geologe
* Teilh.Grundbauberatung AG
\+ Teufenerstr. 25
0000 St. GALLEN

3394 KARAKAS, I., Ing.civ.dipl. MSCE/SIA
* Bd de Grancy 19a
1006 LAUSANNE

12088 KOLLBRUNNER, A., Dipl.Ing. ETH, MS
* Schlossbergstr. 21
8702 ZOLLIKON

8213 KOLLBRUNNER, C.F., Dr., Dr.h.c.,
Dipl. Bauing. ETH
* Ing. conseil
Witellikerstr. 50
8702 ZOLLIKON

3317 KOVARI, K., Dr.sc.techn.,Dipl.
\+ Bauing. ETH
Adjunkt, ISETH
ETH-Hönggerberg
8093 ZÜRICH

8220 KREUZER, H.L., Dr.tech., Dipl.Bauing.
Ahornweg 14
5416 KIRCHDORF

570 KRÖTZL, J., Dipl.Verm.Ing./SIA
Inhaber Ing.-Büro
Wingertweg 11
7000 CHUR

874 KRUMDIECK, A., Dipl.Bauing. FIUC
* Katzenrütistr. 63
8153 RÜMLANG

882 LANG, H.-J., Prof. Dipl. Ing. ETH
* Vorsteher Institut f. Grundbau
und Bodenmechanik
ETH-Hönggerberg
8093 ZÜRICH

8235 LASRAM, A., Ing.civ.dipl.EPF/SIA
* 10, av. du Gros Chene
1213 ONEX

12089 LEONI, R., Ing.civ.dipl.ETH/SIA
* Consultente geotecnico
Studio Ing. A. Linder
Via alla Campagna 2
6924 SORENGO

892 LOCHER, H.G., Dipl.Bauing.ETH/SIA
* Gumpisbühlweg 17
\+ 3067 BOLL

12090 LOCHER, T., Dr. phil. II
* Beratender Geologe
\+ Waldstr. 4
8122 PFAFFHAUSEN

12091 MARCACCI, H., Ing.civ.dipl. EPF
Fonde de pouvoir
6, rue des Vollandes
1207 GENEVE

907 MAYOR, H., Dr. es sc.Geologue et
* Ing. civ. dipl. EPF
\+ 1141 MONTRICHER

8267 MEIA, J., Dr., Geologue
* 2, ch. des Villarets
\+ 2036 CORMONDRECHE

12092 MEIER, P., Dipl. Bauing. ETH
Riedthofstr. 78
8105 REGENSDORF

910 MENGIS, R., Dipl.Bauing. ETH/SIA
* Teilh.R.Mengis + R.Schmidt
Mühlehofstr. 14
6030 EBIKON

3326 von MOOS, A., Dr.phil
* Beratender Geologe
\+ Bachofnerstr. 5
8037 ZÜRICH

919 MOTTIER, J.L., Dipl.Bauing. EPF/SIA
* Prokurist
Huttenackerstr. 13
5452 OBERROHRDORF

920 MUGGLIN, G., Dipl.Bauing. ETH/SIA
* Inhaber Ing.-Büro
Bolleystr. 29
Postfach
8006 ZÜRICH

12093 MÜLLER, E.R., Dipl.phil. Geologe SIA
* Teilh. Ing.-Büro
Weingartenstr.
8557 FRUTHWILEN

932 MULLER, H.W., Dipl.Bauing.ETH/SIA
* Teilh. Grundbauber. AG
Fähnernstr. 23 a
9000 St. GALLEN

933 MÜLLER, H., Dipl. Bauing. ETH/SIA
Vize-Direktor
Boglerenstr. 46
8700 KÜSNACHT

12094 MÜLLER, R.W., Dipl.Bauing.ETH
Geotechniker
General-Guisanstr. 42
5000 AARAU

12095 NOHER, H., Dipl.Bauing.ETH
c/o Fam. E. Nohev
Scheuermattweg 4
3007 BERN

947 NORBERT, J., Geologue-conseil/SIA
* Propr. Bur. Techn. Geol.
+ Rue Enning 6
1003 LAUSANNE

3330 PEDROZZI, P., Ing.geol.dipl.ETH/SIA
* Tit. Studio di Geologia
+ Via Ceresio 8
6963 PREGASSONA

962 PIRCHER, W., Dr., Techn.Dipl.Ing.
Prokurist
Tiroler Wasserkraftwerke AG
Landhausplatz 2
6020 INSBRUCK (AUSTRIA)

8346 PUGIN, L., Geologue
* Professeur
+ Institut de Geologie de l'Universite
1700 FRIBOURG

970 RACINE, C., Dipl.Bauing. ETH/SIA
* Bergstr. 125
8032 ZÜRICH

971 RECORDON, E., Ing.civ.dipl.EPF/SIA
* Prof. a l'EPF
+ Lab. de geotechnique EPF
Ecublens
1015 LAUSANNE

12096 RITZ, P., Dr.sc.techn., Dipl.Bauing.ETH
Teilh. Ing.-Büro
Nordstr. 11 a
3900 BRIG

12097 de RIVAZ, B., Geologue SIA
* Dir. bur. d'etudes
Les Combes
1961 GRIMISUAT

976 ROBERT, F., Ing. dipl. EPF/SIA
* Dir. de Tecnoconsult SA
1249 CHOULEX-GENEVE

8361 ROOSE, K., Dipl.Bauing.ETH/SIA
* 25, Joaquinstreet
San Lorenzo Village
Makati
METRO MANILA (PHILIPPINES)

12098 ROSSET, F., Ing. civ. EPF
Ing. en chef
4, chemin de Chandieu
1006 LAUSANNE

12099 RÜEGGER, R., Dipl.Bauing. ETH/SIA
Filialleiter
Säntisstr. 8 a
9034 EGGERSRIET

993 SCHAERER, C., Dipl.Bauing. ETH/SIA
* Sektionschef IGB, ETHZ
Buchenweg 2
8116 WÜRENLOS AG

1002 SCHELLENBERG M., Dipl.Bauing.
ETH/SIA
* Teilh. Ing.-Büro
Gubelstr. 28
Postfach 26
8050 ZÜRICH

1004 SCHIESS, J.M.S., Dipl.Bauing.ETH/SIA
* Direktor BBL Baulab. AG
Langmattstr. 14
4132 MUTTENZ

1007 SCHINDLER, C.M., Dr.sc.nat.ETH,
Geologe
Mitinh. Dr. von Moos AG
Schwerzestr. 7
8618 OETWIL AM SEE

13000 SCHLEGEL, H., Dipl.Bauing.ETH/SIA
Handlungsbevollmachtigter
Gruner AG, Basel
Bruderholzallee 169
4059 BASEL

1009 SCHMASSMANN, H., Dr. Geologe/SIA
* Inhaber Ing.-Büro
+ Langhagstr. 7
4410 LIESTAL

1013 SCHMIDHALIER, P., Dipl.Bauing.
ETH/SIA
* Teilh. Ing.-Buro
Nordstr. 11 a
3900 BRIG

13001 SCHMIDT, R., Dipl.Bauing.ETH/SIA
* Dipl. Ing.-Geologe ETH
Professor
Bühlmattöhe 11
6045 MEGGEN

13002 SCHNEIDER, T., Dr.sc.nat.ETH
* Beratender Geologe, Büroinh.
+ Rütihofstr. 53
8713 UERIKON ZH

13003 SCHNELLER, A., Dipl.Bauing. ETH
Lindenstr. 55
8307 EFFRETIKON

3340 SCHNITTER, G., Prof.Dr.h.c.,
Dipl.Bauing ETH/SIA Im Ruhestand
* Im Kalender 3
8700 KÜSNACHT

8409 SCHWARZ, H.R., Dr.Geologe
* Inhaber Ing.-Büro
+ Baumgartenrain 1
3600 THUN

8413 SIEGENTHALER, R., Dipl.Bauing.ETH/
SIA
* Teilh. Dr Staudacher
Siegenthaler AG
Germaniastr. 19
8006 ZÜRICH

1032 SINNIGER, R., Prof.Ing. ETH/SIA
* Vize-Direktor Elektrowatt
Sonnenbergstr. 36
8708 MÄNNEDORF

1039 STAMM, C., Dipl. Bauing.ETH
* Benkenstr. 36
4054 BASEL

13004 STEINER, W., Dipl. Bauing.ETH, SMCE
* c/o Consorzio Galleria della Piumogna
Lotto 259, c.p. 42
6760 FAIDO

7948 STÖCKLIN, F., Bauing. HTL
* Prokurist, Chefauslanddel
Rutistr. 28
4103 BOTTMINGEN

1053 STREICH, W., Dipl.Bauing.ETH/SIA/ASIC
* Teilh. Ing.-Büro
Firststr.3
8125 ZOLLIKERBERG

1055 STUCKI, E., Dipl.Bauing.ETH/SIA
* Teilh. Ing.-Büro
Engweg 7
8006 ZÜRICH

1063 SUTER, R., Ing.civ.dipl.EPF/SIA
* Propr. bur. d'ing.
Av. Eugene-Rambert 15
1005 LAUSANNE

1070 THUT, A., Dr., Dipl.Bauing. ETH
c/o Solexperts SA
Postfach
8034 ZÜRICH

8476 VIRET, R., Ing. civ. EPF
* Chef de projet
Bucholzstr. 25
8053 ZÜRICH

1087 VOLLENWEIDER, U., Dr.,Dipl.Bauing. ETH
\+ Hegarstr. 22
8032 ZÜRICH

1091 WACKERNAGEL, A., Dipl.Bauing. ETH/SIA
Prokurist
Sissacherstr. 61
4052 BASEL

13005 WANNER, H., Dr.sc.nat., Geologe SIA
* Grossweid
8607 AATHAL-SEEGRÄBEN

13006 WEBER, G., Dipl.Ing.ETH/SIA
* Dir.Schweiz. Wasserwirte
Wehrenbachhalde 59
8053 ZÜRICH

1105 WENK, P., Dipl.Ing.ETH/SIA/ASIC
* Teilh. Ing.-Büro
Zähner + Wenk
Hauptbahnhof, Büro 109
9000 St. GALLEN

3345 WEYERMANN, W., Dipl.Bauing.ETH/SIA
* Ruhestand
Rütistr. 6
8032 ZÜRICH

8511 WIENER, G., Dr.phil.nat., Geologe SIA
* Inhaber Geol + Hydrol. Büro
\+ Gruenhagweg 2
4410 LIESTAL

1115 WILHELM, J., Ing.civ.dipl.SIA
* Fonde de pouvoir
\+ Av.C.-F. Ramuz 5
1009 PULLY

8523 ZÄHNER, A., Bauing. SIA
* Inhaber Ing.-Büro
"Aesch"
9114 HOFFELD

12112 Representative of:
ALLEMAND JEANNERET SCHMID SA
Ingenieurs civ. dipl. EPF/SIA
4, rue du Musee
2001 Neuchatel

13107 Representative of:
AUTOBAHNAMT DES KANTONS BERN
Schermenweg 7
Postfach 1042
3001 BERN

13108 Representative of:
BAUDIREKTION DES KANTONS ZUG
Poststr. 18
Postfach 203
6301 ZUG

13113 Representative of:
BONNARD & GARDEL, INGENIEURS-CONSEILS SA
Av. de Cour 61
Case Postale
1001 Lausanne

13106 Representative of:
BUNDESAMT FÜR STRASSENBAU
Monbijoustr. 40
3003 Bern

13133 Representative of:
ED. ZUBLIN & CIE AG
Bauingenieure und-Bauunternehmer
Okenstr. 4-6
8037 ZÜRICH

13116 Representative of:
FRUTIGER SÖHNE AG
Bauunternehmung
Frutigenstr 37
3601 THUN

13124 Representative of:
Dr ING. G. LOMBARDI
Bureay d'etudes
Via A. Ciseri 3
6601 LOCARNO

13118 Representative of:
GEOTECHNISCHES INSTITUT AG
Gartenstr. 13
3007 BERN

13119 Representative of:
GEOTEST AG
Bodenuntersuchung und Fundationsberatung
Birkenstr. 15
3052 ZOLLIKOGEN

13127 Representative of:
J.C. OTT
Ingenieur Conseil
8, ch. Deux-Communes
1226 THONEX GE

13109 Representative of:
KANTONALES TIEFBAUAMT GRAUBÜNDEN
Grabenstr. 30
7000 CHUR

13120 Representative of:
KISSLING & ZBINDEN
Ingenieurbüro für Hoch-und Tiefbau
Seftingenstr. 22
3000 BERN

13121 Representative of:
LABORATOIRE DE GEOTECHNIQUE
H.B. de Cerenville SA
Av. de Rumine 11
1005 LAUSANNE

13122 Representative of:
LABORATOIRE ET ETUDES GEOTECHN.
SA (LEG)
Bureau d'ingenieurs-geotechniciens
1931 BOVERNIER

13126 Representative of:
LOUCHER & CIE AG
Bauingenieure und Bauunternehmer
Pelikanplatz 5
Postfach
8022 ZURICH

13126 Representative of:
MOTOR-COLUMBUS INGENIEURUNTERNEHMUNG AG
Parkstr. 27
Postfach
5401 BADEN

13127 Representative of:
POLYDRO, INGENIEURS CONSEILS SA
Rue du Tunnel 1
1005 LAUSANNE

13129 Representative of:
REIFLER & GUGGISBERG ING.AG
Bauunternehmung
Länggasse 9
2500 BIEL

13115 Representative of:
S. FACCHINETTI & CIE., TRAVAUX PUBLICS
Genie civil, Carrieres
Gouttes d'Or 78
2000 NEUCHATEL 8

13125 Representative of:
SA UFFICIO D'INGEGNERIA MAGGIA
Via Stefano Franscini 5
6600 LOCARNO

13132 Representative of:
SOCIETE ANONYME CONRAD ZSCHOKKE
Entreprise de travaux publics et du batiment
Rue du 31-Decembre 2
1211 GENEVE 6

13130 Representative of:
SOCIETE GENERALE POUR L'INDUSTRIE
Ingenieurs-conseils
71, av. Louis-Casai
1216 COINTRIN GE

13131 Representative of:
SOCIETE ROMANDE D'ELECTRICITE
1815 CLARENS-MONTREAUX

13110 Representative of:
TIEFBAUMAMT DES KANTONS ZÜRICH
Walchetor
Postfach
8090 ZÜRICH

13111 Representative of:
TIEFBAUMT DER STADT ZÜRICH
Werdmühleplatz
Postfach
8023 ZÜRICH

13114 Representative of:
DR ULRICH P. BÜCHI SIA
Geologische Expertisen & Forschungen AG
Bodenacherstr. 79
8121 BENGLEN

13117 Representative of:
W. GALLI, INGENIEURBURO GOLTA
A. Sandacker 6
8052 ZÜRICH

SUPPORTING MEMBERS

6826 BERNOLD AKTIENGESELLSCHAFT
im Hasenberg
8880 WALENSTADT

9008 ELECTROWATT
Bellerivestrasse 36
8022 ZÜRICH

9012 FGU (Specialized Group of SIA for Underground Construction)
Selnaustrasse 16
8039 ZÜRICH

9007 INSTITUT FÜR STRASSEN-EISENBAHN-UND FELSBAU (ISETH)
ETH - Honggerberg
8093 ZÜRICH

UNITED KINGDOM

BRITISH GEOTECHNICAL SOCIETY

Institution of Civil Engineers, Great George Street, Westminster, LONDON SW1P 3AA

Secretary: Mr J. G. Watson

Total Number of Ordinary Members - 280
Total Number of Supporting Members - 8

ORDINARY MEMBERS

1971 AITKEN, P. L., Mr - Civil Engr
* Hydro-Electric Engineer
17 Craigleith View
EDINBURGH EH4 3J2

8538 AJAZ, A., Dr
* 30-H School Road
F 6/3 ISLAMABAD (PAKISTAN)

1973 ALDERSLADE, R.
* 32 Tenterbalk Lane
ADWICK-LE-STREET DN6 7EF

13315 AL-HASHIMI, K., Mrs
* 3 Tennyson Road
STRATFORD-UPON-AVON, Warwick

8539 ALLEN, A. C., Mr - Civil Engr
* MIStructE AMIWE FGS
\+ Roseleigh
Dalginross
COMRIE PH6 2ED

13316 AL-MOSAWY, N.M., Dr
* c/o Saudi Letco
P.O. Box 3975
JEDDAH (SAUDI ARABIA)

13328 ALONSO, E. E., Prof.
* Professor of Civil Engineering
ETS Ingenieros de Caminos
Jorge Girona Salgado 31
BARCELONA 17 (SPAIN)

13329 ARAYA, R. I., Mr
* Qta Frentavila
Av. San Felipe
La Castellana
CARACAS (VENEZUELA)

1989 ARTHUR, J.R.F., Dr
U. Teacher
34 Chipstead Park
SEVENOAKS, Kent

1988 ARROWSMITH, E.J., Mr - Civil Engr
Government Engineer
St. Mary's House
MOSS SIDE
nr. Lytham FY8 4LZ

1990 ASHBEE, R.A., Mr - Engr
* Civil Engineer
'Silverwood'
Skinners Lane
ASHTEAD KT21 2LY

1992 ATTEWELL, P.B., Dr - Civil Engr
* U. Teacher
50 Archery Rise
Nevilles Cross
DURHAM DH1 4LA

1997 BAKER, C.O., Mr
Flat 8. Parkview
Alexandra Road
EPSOM, Surrey

13330 BALASUBRAMANIAM, A.S., Dr - Civil Engr
* GE Divn, Asian Institute of Technology
P.O. Box 2754
BANGKOK (THAILAND)

13331 BANERJEE, P.K., Dr - Civil Engr
* U. Teacher
2 Cefn Coed Avenue
CYNCOED, Cardiff

13332 BARLEY, A.D., Mr
* 34 Monks Drive
Withnell
CHORLEY, Lancs

13333 BARNES, G.R., Mr - Civil Engr
* 23 Windale
Worsley
MANCHESTER M28 5SR

2003 BARTHOLOMEW, R.F., Mr - Civil Engr
* Manager
Frankipile Ltd
Davis House, High Street
CROYDON CR9 1PN

2004 BARTON, M.E., Dr.
* U. Lecturer
Dept of Civil Engineering
The University
SOUTHAMPTON SO9 5NH

13334 BASSETT, R.H., Dr - Engr
* The Thatch Cottage
Hill Road
HEMPSTEAD
nr Saffron Waldron, Essex

2010 BEAVAN, G.C.G., Mr - Civil Engr
* Consulting Engineer
Binnie & Partners
Artillery House
Artillery Row
LONDON SW1P 1RX

13335 BELTRAM-MORENO, L., Mr
* Apartado Aereo 30420
BOGOTA (COLUMBIA)

2015 BENT, M.K., Mr
* Spring Hill Farm
Molesworth
HUNTINGDON PE18 OQF

13336 BERTHELSEN, O.J.
* 13 Lawn Road
Hampstead
LONDON NW3

13337 BEVAN, D.J., Mr - Civil Engr
* 30 Oak Hill Crescent
WOODFORD GREEN, Essex

2020 BINNIE, G.M., Mr
* Binnie & Partners
Artillery House
Artillery Row
LONDON SW1P 1RX

13338 BLACKFORD, S., Mr - Civil Engr
* Civil Engr
P.O. Box 360
DUBAI (UNITED ARAB EMIRATES)

13339 BODEN, J.B., Mr
* Transport & Road Research Laboratory
CROWTHORNE, Berks

2024 BONNEY, J.E., Mr
14 Neath Abbey Road
NEATH SA10 7BD

3928 BOOTH, A.R., Mr - Civil Engr
* 67 Argyle Way
DUNBLANE FK15 9DY

13340 BRADY, B.H.G., Mr - Engr
* c/o Imperial College of
Science & Technology
South Kensington
LONDON SW7 2AZ

2032 BRAND, E.W., Dr - Civil Engr
* U. Teacher
c/o Geotechnical Control Office
Public Works Department
Murray Building
HONG KONG

13341 BRIAN-BOYS, K.C., Mr - Civil Engr
* Flat 3
6 Mansfield Road
HONG KONG

2038 BRITTAIN, E.W., Dr - Engr
* Engineer in Charge
The County Highways Laboratory
Wharf House
GNOSALL
nr Stafford

13342 BROWN, E.T. Dr
* Dept of Mining & Mineral Technology
Imperial College
Imperial Institute Road
LONDON SW7 2AZ

2051 BURLAND, J.B., Dr - Civil Engr
* Research Engineer
13 Barton Road
Wheathampstead
ST. ALBANS, Herts

13343 CARRINGTON, R.W.M., Mr - Civil Engr
* Sir Alex Gibb & Partners
P.O. Box 30020
NAIROBI (KENYA)

13344 CARTER, T.G., Mr
76 Chatsworth Road
CROYDON, CRO 1HB

13345 CASANOVAS-RODRIGUEZ, J.
* Valls y Taberner 2
BARCELONA 6 (SPAIN)

13346 CAWSEY, D.C., Dr
School of Natural Sciences
The Hatfield Polytechnic
HATFIELD, Herts

2074 CHAPLOW, R., Mr
c/o Sir Alex Gibb & Partners
Earley House
427 London Road
Earley
READING RG6 1BL

13347 CHAPMAN, G.T., Mr
* 41 Rafford Way
BROMLEY BR1 3EN

13348 CHARTRES, F.R.D., Mr
* Hurst Wood
Great Woodcote Drive
PURLEY CR2 PL3

2076 CHAUDHARI, M.A., Mr
* Soil Engineer
PWD P.O. Box 3
ABU DHABI (ARABIAN GULF)

1349 CHEETHAM, J.A., Dr - Civil Engr
* Managing Director
4 Moorfield Drive
WILMSLOW SK9 6DL

13350 CHEW, Y.K., Mr
* 315 Geylang Road
SINGAPORE 14 (SINGAPORE)

2873 CHILD, K.T., Mr
* Managing Director
2 Sandhill Grove
LEEDS 17

3180 CHOWDHURY, R.N., Dr - Civil Engr
* Senior Lecturer
The University of Wollongong
WOLLONGONG NSW 2500 (AUSTRALIA)

2889 CLARKE, A.M.J.K., Mr - Geologist
Chief Geologist
12 Glenhurst Avenue
LONDON NW5 1PS

13351 CLARKE, T.J.C.P., Mr - Civil Engr
* Stirling Astaldi (Nig) Ltd
P.O. Box 2067
SANA'A (YEMEN ARAB REPUBLIC)

2083 COCKSEDGE, J.E., Mr - Civil Engr
* Consulting Engr
38 Sycamore Crescent
ALDERSHOT GU13 ONN

13352 COGAN, K., Mr - Civil Engr
* Cogan & Associates
68A High Street
WARGRAVE, Berks

8543 COLBACK, P.S.B., Mr - Engr
* Geotechnical Engineer
Binnie & Partners
Artillery House
Artillery Row
LONDON SW1P 1RX

13353 CONSTANT, D., Mr - Engr
3211 74th Street
EDMONTON Alta (CANADA)

13354 COOK, A.W.
* c/o Mr C. W. Bartlett
Foundation Engineering Ltd
109 Blackfriars Road
LONDON SE1 8HW

2094 CORMIE, W.M., Mr - Civil Engr
* c/o Crouch & Hogg
18 Woodside Crescent
Charing Cross
GLASGOW G3 7UU

13355 COSTA FILHO, L. de M., Mr - Engr
* Consulting Engineer
Dept of Civil Engineering
Imperial College
Imperial Institute Road
LONDON SW7 2BU

2830 COTTINGTON, J., Mr - Civil Engr
* Consulting Engineer
3 Downs Street
BRIDPORT DT6 3JR

2009 COUMOULOS, D.G., Dr - Engr
* Consulting Engineer
10 Krisila Street
Pagrati
ATHENS 502 (GREECE)

13356 CRAIG, C., Mr - Civil Engr
* 3 Withyholt Park
Charlton Kings
CHELTENHAM GL53 9BP

2106 CRUTCHLOW, S.B., Miss
Engineer Public Service
DOE Room 4/55
St Christopher House
Southwark Street
LONDON SE1

13357 DAHLGREN, J.A., Mr
* c/o Golder Associates Inc
10628 NE 38th Place
KIRKLAND (Seattle) WA 98033 (USA)

13358 DARAMOLA, O.
* Dept of Civil Engineering
Imperial College
Imperial Institute Road
LONDON SW7 2BU

6765 DAVIES, M.J., Mr
26 Bellsburn Avenue
LINLITHGOW EH49 7LB

2833 DAVIES, P., Mr - Engr
* 104 Buckingham Road
KLOOF, Natal (S.AFRICA)

8547 DAW, G.P., Dr
* Research Scientist
Cementation Research Ltd
Denham Way
Maple Cross
RICKMANSWORTH WD3 2SW

13359 DAWES, B.J.
* 9 Milton Lawns
Chesham Bois
AMERSHAM, Bucks

13360 DAWS, G., Dr
64 Shelton Avenue
off Nottingham Road
HUCKNALL, Notts

13361 DE ABREU, F.L.R., Mr
* c/o Hidroservice Ltda (UA-MS)
R. Afonso Celso 235 4
04119 SAO PAULO, SP (BRAZIL)

6050 DEARMAN, W.R., Prof.
* U. Teacher
Dept of Geology, Drummond Building
The University
NEWCASTLE-UPON-TYNE NE1 7RU

13362 DEPPNER, W.
* Director
The White House
Loggerheads
Market Drayton
SALOP, Lancs

13363 DOUGLAS, T.H., Mr - Civil Engr
J. Williamson & Partners
219 St Vincent Street
GLASGOW

13364 DOWSE, B.E.W., Mr - Civil Engr
* Consulting Engineer
17 Hunt Club Drive
LONDON, ONT. (CANADA N6H 3Y2)

2890 EARLY, K.R., Mr
* 4 Marlpit Lane
COULSDON, Surrey

2148 EASTWOOD, W., Prof.-Engr
* Consulting Engineer
Bowmaker House
23 Kingfield Road
SHEFFIELD S11 9AS

3182 EGERTON, R.H.L., Mr
* 57 High Street
+ GODSTONE RH9 8LT

2151 ELLIOTT, A.J.
* 10 Montpellier Villas
CHELTENHAM GL50 2XE

2152 ELLIOTT, S.G., Mr - Civil Engr
* Scott Wilson Kirkpatrick & Partners
1720 Star House
Salisbury Road
HONG KONG

13365 EL-RUWAYIH, A., Mr
* College of Engineering
P.O. Box 800
RIYDAH (SAUDI ARABIA)

2154 ELSON, W.K., Dr
* Lecturer
Orchard End
7 Finches Court
LINDFIELD
nr Haywards Heath, Sussex

2155 ENDICOTT, L.J., Dr
* Maunsell Consultants Asia
14th Floor
1 Kowloon Park Drive
Tsim Sha Tsui, Kowloon
HONG KONG

2835 EVANS, B.M., Mr - Civil Engr
* Flat 1, Burnham Court
32 Ashley Road
ALTRINCHAM WA14 2DW

13366 EVANS, N.D., Mr
* c/o Dames & Moore
123 Mortlake High Street
LONDON SW14 8SN

13367 FARMER, I.W., Dr
* Dept of Mining Engineering
The University
NEWCASTLE-UPON-TYNE NE1 7RU

2166 FARRAR, N.S., Mr - Civil Engr
22 The Close
REIGATE, Surrey

13368 FERGUSON, G.A., Mr
Dale House
Belmont Vale
MAIDENHEAD SL6 6BH

2172 FINEY, J.T., Mr - Civil Engr
* Kimberley
20 Albion Crescent
CHALFONT ST GILES HP8 4EU

8554 FLINTOFF, W.T., Mr - Civil Engr
* Consulting Engineer
Geotechnical Control Office
Public Works Department
Murray Building
Garden Road
HONG KONG

2179 FOOKES, P.G., Dr
* Winters Wood
47 Crescent Road
CATERHAM CR3 6LH

2182 FOX, D.A., Mr - Civil Engr
* 11 Hillside
Wimbledon
LONDON SW19 4NH

2187 FUSHBERGER, M.
* General Manager
c/o ICOS (Gt Britain)
25/28 Buckingham Gate
LONDON SW1E 6LD

2188 FUTCHER, R.G., Mr - Civil Engr
* U. Teacher
Dept of Civil Engineering
University of Technology
LOUGHBOROUGH LE11 3TU

8555 GALLACHER, K.A., Mr - Civil Engr
* Research Engineer
23 Princes Avenue
WATFORD WD1 7RR

13369 GAMMON, J.R.A., Mr - Engr
* Consulting Engineer
27 Gorse End
HORSHAM RH12 4XW

2191 GANENDRA, P., Mr - Civil Engr
* Malaysia International Consults
P.O. Box 2263
14 Road
20/16A Paramount Gardens
Petaling Jaya
SALANGOR (MALAYSIA)

2195 GEDDES, J.D., Dr - Civil Engr
Dept of Civil Engineering &
Building Technology
U.W.I.S.T.
King Edward VII Avenue
Cathay's Park
CARDIFF CF1 3NU

13370 GENS, A., Mr - Civil Engr
* Dept of Civil Engineering
Imperial College
Imperial Institute Road
LONDON SW7 2AZ

3186 GILL, S.A., Dr - Civil Engr
* Chief Engineer
Soil Testing Services Inc
111 Pfingsten Road
P.O. Box 266
NORTHBROOK IL. 60062 (USA)

2201 GOLDBERGER, M., Mr - Engng Geologist
* 11 Hazanhanim Street
GIVATAYIM (ISRAEL)

2202 GOLDER, H.Q., Dr - Civil Engr
* 4 Links Court
GROUVILLE, Jersey C.I.

13371 GONANO, L.P., Dr
c/o Golder Associates
5 Forlease Road
MAIDENHEAD SL6 1RP

13372 GOSSCHALK, E.M., Mr - Civil Engr
* Consulting Engineer
Church Farm House
Park Lane
FORD CHIPPENHAM SN14 8RN

2207 GRATEROL, M.J.
Aptdo 80912
CARACAS 108 (VENEZUELA)

2209 GRAY, D.A., Mr
46 Bonnersfield Lane
HARROW HA1 2LE

2212 GREEN, F.M., Mr - Engr
* Consulting Engineer
William J. Green & Associates
Brook House
449 Brook Lane
Kings Heath
BIRMINGHAM B13 OBT

2213 GREEN, G.E., Dr - Engr
* Consulting Engineer
Shannon & Wilson Inc
1105 N. 38th Street
SEATTLE WA. 98103 (USA)

2217 GREENWOOD, D.A., Mr
* Technical Director
49 St Huberts Close
Fulmer Road
GERRADS CROSS BL9 7EN

13373 GRIFFIN, A.R., Mr
* 'Markswood', Townsend Grove
WANTAGE OX12 OAR

2222 GWYNN, J.D., Mr - Civil Engr
* Consulting Engineer
49A Lytton Grove
LONDON SW15 2HD

2223 HADDAD, P., Mr - Civil Engr
* 61 Truman Road
WILLOWDALE Ont. (CANADA M2L 2L7)

13374 HADEN, G.P., Mr - Civil Engr
* 16 Lon Cedwyn
CWMGWYN, Swansea SA2 OTH

13375 HARDING, C., Mr
* c/o Foundation Engineering (Nigeria) Ltd
P.O. Box 2100
LAGOS (NIGERIA)

8559 HARPER, T.R.
Dames & Moore
123 Mortlake High Street
LONDON SW14 8SN

2234 HARRIS, J.S.
* 6 Willerby Road
WOODTHORPE NG5 4PB

8560 HAWKINS, A.B., Dr
* U. Lecturer
Dept of Geology
University of Bristol
Queen's Building
University Walk
BRISTOL BS8 1TR

13376 HAWORTH, G.W.
7 St Vincents
Weathercock Lane
WOBURN SANDS MK17 8NR

13377 HELMY, R.H., Dr - Engr
* The City University
Dept of Civil Engineering
St John Street
LONDON EC1V 4PB

13378 HENRY, K.N., Mr
5C Peck Hay Road
SINGAPORE 9 (SINGAPORE)

13379 HICKIN, C.R., Mr
* Anglican Engineering & Survey International
Survey House
Earls Street
THETFORD, Norfolk

2251 HO, C.S., Mr
* Flat B. 17th Floor
Yee Shan Mansion
Taikoo Shing
HONG KONG

2252 HOBBS, N.B.
Soil Mechanics Ltd
Foundation House
Eastern Road
BRACKNELL, Berks

2254 HODGES, W.G.H., Mr - Engr
* Lecturer
18 Abshot Close
Titchfield Common
FAREHAM, Hants

3940 HOLT, D.N., Mr - Civil Engr
* 57 The Gardens
WATFORD, Herts

13380 HOLTZ, R.D., Dr
School of Civil Engineering
Purdue University
WEST LAFAYETTE IN 47907 (USA)

13381 HORSWILL, P., Mr
36 Kelthorpe Close
Geeston Ketton
STAMFORD, Lincs

2262 HOUSDEN, R.T.
* 23 Wilton Grove
Wimbledon
LONDON SW19 3QU

2264 HOWELLS, D.A., Mr - Civil Engr
* Consulting Engineer
63 Victoria Drive
LONDON SW19 6HN

13382 HOWLAND, A.F., Mr
506 Fellow Court
Hackney Road
LONDON E2

8561 HUDSON, J.A., Dr
7 The Quadrangle
WELWYN GARDEN CITY AL8 6SG

13383 HUDSON, R.R., Mr - Engr
22 Amity Grove
LONDON SW20 OLJ

2266 HUGHES, F.H., Mr - Civil Engr
* Director
88 Blacketts Wood Drive
CHORLEYWOOD, Bucks

2265 HULME, T.W., Mr - Civil Engr
4 Bernard Gardens
Wimbledon
LONDON SW19 7BE

2268 HUMPHREYS, A.P., Mr
* Kerves Wale
Kerves Lane
HORSHAM RH13 6ES

13385 IFEDIGBO, R.E., Mr
Flat 1A
12 York Terrace East
LONDON NW1 4PT

13384 ILLANES, J., Mr - Civil Engr
* The Limes
123 Mortlake High Street
LONDON SW 14

13386 JAMES, A.N., Dr - Engr
* Consulting Engineer
1 Woodcroft Road
CHESHAM, Bucks

3191 JARRETT, P.M., Mr
Teacher
Dept of Civil Engineering
Royal Military College
KINGSTON Ont. (CANADA K7L 2W3)

2283 JENNINGS, R.A.J., Mr
68 Coleherne Court
LONDON SW5 OEF

13387 JOHNSTON, I.W., Dr
* Dept of Civil Engineering
Monash University
CLAYTON Vic. (AUSTRALIA 3168)

13388 JONES, D.L., Mr
* Robert Leslie & Partners
P.O. Box 4585
CAPE TOWN 8001 (S. AFRICA)

13389 JONES, R.H., Dr - Civil Engr
U. Lecturer
70 Arundel Drive
Bramcote Hills
BEESTON NG9 3FQ

8563 JORDAN, P.G., Mr - Civil Engr
* Consulting Geologist/Engineer
6 The Garth
HAMPTON TW12 1SR

2290 KALAUGHER, P.G.
* U. Lecturer
Dept of Engineering Science
Applied Science Building
North Park Road
EXETER EX4 4QF

13390 KARUNARATNE, G.P., Dr - Engr
* U. Lecturer
Faculty of Engineering
University of Singapore
Kent Ridge Campus
SINGAPORE 5 (SINGAPORE)

2291 KATHIRGAMAN, K., Mr - Civil Engr
* P.O. Box 2990
KAMPALA (UGANDA)

3944 KENNETT, P.
Geophysicist
84 Copse Avenue
WEST WICKHAM, Kent

13391 KIEFER, M.L., Mr
* Dames & Moore
1550 Northwest Highway
PARK RIDGE IL 66068 (USA)

2309 KNILL, J.L., Prof.-Engng Geologist
* Dept of Geology
\+ Royal School of Mines
Prince Consort Road
LONDON SW7 2BP

13392 KOWALCZYK, A.K.
* c/o Wimpey Laboratories Ltd
P.O. Box 1811
DUBAI (UNITED ARAB EMIRATES)

2315 KUTTER, H.K., Prof.
* U. Professor
Institut für Geologie
Ruhr Universität
D-4630 BOCHUM (GERMANY, FR)

13393 LARSEN, H., Mr - Engr
* Research Engineer
Geodan
Gribskovvej 2
2100 COPENHAGEN (DENMARK)

2886 LEACH, B.A., Mr - Civil Engr
* Chartered Engineer
2 Tytherington Park Road
MACCLESFIELD, Cheshire

13394 LEE, K., Mr
* Dept of Engineering Science
University of Oxford
Parks Road
OXFORD OX1 3PJ

8566 LEGGE, T.F.H., Mr - Civil Engr
Watermeyer, Legge, Piesold & Uhlmann
35 Station Road
ASHFORD TN23 1PP

13395 LEONARDS, G.A., Dr
Grissom Hall
Purdue University
LAFAYETTE IN. 47906 (USA)

2330 LITTLE, A.L., Mr - Civil Engr
* Consulting Engineer
4 Front Street
Mendlesham
STOWMARKET IP14 5CX

2336 LONGWORTH, C.R., Mr - Civil Engr
* Consulting Engineer
P.O. Box 785
NORTH SYDNEY NSW 2060 (AUSTRALIA)

2888 LONGWORTH, T.I., Mr
10 Dalby Close
Brookview Estate
Leagrave High Street
LUTON LU4 OXF

13396 LUPINI, J.F.B.
* 26 Hartswood Road
LONDON W12

13397 McDERMOTT, R.J.W., Mr
Principal Planning Officer
Plas Heulog
Mountain Lane
PENMAENMAWR, Gwynedd

2355 McKINLAY, A.C.M., Mr
c/o Geological Survey of Papua New Guinea
P.O. Box 778
Port Moresby
PAPUA (NEW GUINEA)

2356 McKINLAY, D.G., Dr
* U. Teacher
51 Norwood Park
BEARSDEN G61 2RZ

13398 MALDEN, P.J.
Central Laboratories
English Clays Lovering Pochin & Co Ltd
Pentewan Road
ST AUSTELL, Cornwall

2364 MALKIN, A.B., Mr - Engr
* Consulting Engineer
Spring View Chase Lane
STOKE-ON-TRENT ST12 9HH

13399 MARRIOTT, D.L., Mr - Engng Geologist
* Engineering & Power Development Consultants Ltd
Marlowe House
SIDCUP, Kent DA15 7AU

2371 MARSLAND, A., Mr
* 7 Hyburn Close
Bricket Wood
ST ALBANS, Herts

13400 MASTER, J.
* Laboratory Technician
42 Victoria Avenue
HILLINGDON, Middx

2896 MAURENBRECHER, Mr - Civil Engr
* Groot Hertoginnelann 15
2517EA S-GRAVENHAGE (NETHERLANDS)

2377 MEIGH, A.C., Mr - Civil Engr
* Managing Director
Fishponds Farm House
West Woodhay
NEWBURY, Berks

13401 MILLIGAN, G.W.E., Dr - Civil Engr
* Consulting Engineer
Hillrise
Haughurst Hill
Baughurst
BASINGSTOKE, Hants

2846 MITCHELL, P.B., Mr - Civil Engr
* Rofe Kennard & Lapworth
2/4 Sutton Court Road
SUTTON SM1 4SS

13402 MOH, Z-C., Mr
6-1 5th Floor, Lane 137
Yen Chi Street
TAIPEI (CHINA)

13403 MORI, R.T.
* c/o Hidroservice Ltda
Rua Afonso Celso, 235 - 4º andar
04119 SAO PAULO SP (BRAZIL)

2810 MUIR WOOD, A.M., Dr - Engr
* Sir Wm. Halcrow & Partners
Newcombe House
45 Notting Hill Gate
LONDON W11

2400 NASH, J.K.T.L., Prof - Civil Engr
* Kings College
+ Strand
LONDON WC2R 2LS

13404 NEGRO, A., Mr - Engr
* Consulting Engineer
Rue Barao de Tatui
594 Apt 52
01226 SANTA CECILIA, SP (BRAZIL)

2405 NEWBURY, J., Dr - Geologist
* Consulting Geologist
Howard Humphreys & Sons
Thorncroft Manor
Dorking Road
LEATHERHEAD KT22 8JB

13405 NGO, H.H.
* Block 108 No. 49
Jalan Hang Jebat
SINGAPORE 5 (SINGAPORE)

13406 NICHOLSON, D.P., Mr - Engr
* Consulting Engineer
36 Gladstone Road
COLCHESTER COT 2RA

13407 NUNN, J.O.H., Mr - Engr
* P. Bag 7712. Causeway
SALISBURY (ZIMBABWE)

6810 OVER, V.G., Mr
* 49 Seaford Road
Harwood
BOLTON BL2 4BX

13408 OWEN, R.C., Dr
10 Badgers Croft
Stone Road
ECCLESHALL ST21 6DS

6363 PACHAKIS, M.D., Mr - Engr
* Consulting Engineer
76 Metonos Street
Cholargos
ATHENS (GREECE)

2898 PAGELLA, J.F., Mr
* 4 Carlyle Road
EAST CROYDON, Surrey

2605 PAINE, N., Mr - Civil Engr
* 2 Yew Tree Gardens
Fordwich
CANTERBURY CT2 ODQ

2606 PALMER, D.J., Mr - Civil Engr
* Director
6 Ember Lane
ESHER KT10 8ER

13409 PAN, Y.F., Mr
* Hidroservice Ltda (UA-MS)
Rua Afonso Celso 235
04119 SAO PAULO SP (BRAZIL)

13410 PASSARIS, E.K.S., Dr
29 Rosewood Gardens
Kenton
NEWCASTLE-UPON-TYNE NE3 3DH

13411 PATELY, D.R.
* Editor
8 Junction Road
BRENTWOOD, Essex

2613 PATERSON, A.C., Mr - Civil Engr
* Consulting Engineer
F R Bullen & Partners
125 High Street
CROYDON CR9 IPT

13412 PAU, M.K., Mr - Civil Engr
* Civil Engineering Office PWD
Murray Building, Garden Road
HONG KONG

2616 PEAKE, D.J., Mr - Civil Engr
* Head of Dept of Civil & Structural
Engineering
Hong Kong Polytechnic
Hung Hom, Kowloon
HONG KONG

13413 PENNA, A.S.D.
* Rua Jacurici 166
Apto 81.B.3.
01453 SAO PAULO SP (BRAZIL)

13414 PERELBERG, S., Mr
* Room 603
Lillian Penson Hall
Talbot Square
LONDON W2 ITT

2620 PERONIUS, A.N.
* 9 de la Warr Road
East Grinstead
SUSSEX RH19 3BS

2624 PIGEON, Y., Mr - Engr
Consulting Engineer
406 Lockhart
TOWN OF MOUNT ROYAL PQ (CANADA H3P 1Y5)

8577 PINE, R.J., Mr - Engr
* Geotechnical Engineer
c/o Golder Hoek & Associates
5/7 Forlease Road
MAIDENHEAD SL6 1RP

2626 PITT, L.C., Mr
* 'Valhalla'
Sandhills Lane
VIRGINIA WATER, Surrey

13415 POPE, R.G., Mr - Civil Engr
* Technical Engineer
18 The Ferns
TETBURY GL8 8JE

123 POTTS, E.L.J., Prof
\+ Dept of Mining Engineering
Drummond Building
The University
NEWCASTLE-UPON-TYNE NE1 7RU

2632 PRICE, D.G., Mr
* Delft University of Technology
Dept of Mining
Mijnbouwstraat 20
DELFT (NETHERLANDS)

13416 PURI, R., Mr - Engr
* Soil Engineer
U.G.M.S. - Stategeol
606 Black Hawk Way
SALT LAKE CITY UT 84108 (USA)

13417 QURESHI, M.A., Mr
* 3 - J Wahdat Colony
Ferozepur Road
LAHORE (PAKISTAN)

13418 RANKILOR, P.R., Mr - Civil Engr
* Consulting Engineer
Manstock Engineering Co Ltd
1 North Parade
PARSONAGE M3 2FB

13419 RASHID-SALEHY, M., Dr
* Dept of Geology
Ferdowsi University
MASHHAD (IRAN)

2641 REDHOUSE, I.A.
* Director
39 Sherborne Road
NEWARK PE14 4RS

2647 RICHARDS, A.F., Prof
U. Professor
Marine Geotechnical Laboratory
Lehigh University
BETHLEHEM PA. 18015 (USA)

2655 ROBERTSHAW, J., Mr - Civil Engr
* Director
Buck Cottage
45 Buck Lane
LONDON NW9 OAP

8581 ROBERTSON, N.F., Mr - Civil Engr
Materials Branch
Main Roads Dept GPO 1412
BRISBANE (AUSTRALIA 4001)

13419 RODRIGUEZ-GUILLEN, H.
* Apartado 66.302
Las Americas 106
CARACAS (VENEZUELA)

2659 ROGERS, S.H., Mr - Civil Engr
* 'Casanene'
The Green
Badby
DAVENTRY NN11 6AF

13420 ROMAYA, J.M.
* Hmood Building
Nidhal Street
BAGHDAD (IRAQ)

3200 ROSS-BROWN, D.M.
* 14453 W Third Avenue
GOLDEN CO. 80401 (USA)

13421 ROTHWELL, G.H., Mr - Civil Engr
* P.O. Box 91 Padang
WEST SUMATRA (INDONESIA)

2664 ROWE, J.R., Mr - Civil Engr
42 Hill Street
ST ALBANS, Herts

2666 ROWLANDS, G.O., Mr
* Principal Lecturer
16 Birchwood Road
Penylan, CARDIFF

4015 SAPPEH, D.K., Mr
* 33 Lisburne Road
Hampstead
LONDON NW3 2NS

2676 SAYNOR, F.
* Frank Saynor & Associates
5 Broughton Market
EDINBURGH EH3 6NU

13422 SCHOVILLE, J.A., Mr
* 919 Pawnee Road
WILMETTE IL. 60091 (USA)

2679 SCOTT, R.A., Dr
* Research Consultant
Cementation Specialist Holdings Ltd
Denham Way, Maple Way
RICKMANSWORTH WD3 2SW

3203 SCRIVEN, J., Mr - Civil Engr
Edgar Morton & Partners
Alpha House
126/130 Beech Lane
MACCLESFIELD SK10 2EY

2684 SEARLE, W.J., Mr - Civil Engr
* 23 Radley House
Park Road
Regents Park
LONDON NW1

13423 SENNER, D.W.F., Mr
* c/o McClelland Engineers
319 Pinner Road
HARROW HA1 4XX

13424 SEOW, Y-C., Mr - Engr
* Senior Soil Engineer
68 Pesiaran Zaaba
Tun Dr Ismail
Damansara
KUALA LUMPUR (MALAYSIA)

2691 SHARMAN, F.A., Mr - Civil Engr
* 10 Erskine Hill
Hampstead Garden Suburb
LONDON NW11

13425 SHERMAN, W.C., Mr
Soils Division
Waterways Experiment Station
VICKSBURG MS 39180 (USA)

2698 SHERRELL, F.W., Dr - Civil Engr
* Consulting Engineer Geologist
66 West Street
TAVISTOCK PL19 8AJ

2700 SIMM, K.F., Mr - Civil Engr
* 26 Claremont Road
Ealing
LONDON W13 ODQ

2707 SKINNER, A.E., Dr - Civil Engr
* 56 Alfriston Road
Clapham Common
LONDON SW11 6NW

2713 SMITH, P.H.J.
* 14 York Road
MAIDENHEAD, Berks

3962 SMITH, W.F., Mr - Civil Engr
* 231 North Gower Street
LONDON NW1

2715 SOCARRAS, M.
* Apartado Aereo No. 21332
BOGOTA DE (COLOMBIA)

2720 SPENCER, P.G., Mr - Civil Engr
* Director & General Manager
Essex Water Company
342 South Street
ROMFORD RM1 2AL

13426 STEENFELT, J.S., Dr
Churchill College 42B
CAMBRIDGE CB3 ODS

13427 STOKFISZ, F.J.
* R. Baronesa de Itu 858
01231 SAO PAULO SP (BRAZIL)

8587 SUMMERS, P.V., Mr
* Diamant 30
HEERHUGOWAARD (NETHERLANDS)

2732 SUTHERLAND, H.B., Prof.-Civil Engr
* U. Professor
Dept of Civil Engineering
The University
GLASGOW G12 8LT

2733 SUTTON, J.L.E., Mr - Civil Engr
* 12 Littlebrook Close
HADFIELD
VIA HYDE SK14 8AW

3964 SWEENEY, M., Mr - Civil Engr
* 135 Waller Road
LONDON SE14 5LX

2736 SZEKI, A.
34 Brackenfield Road
Gosforth
NEWCASTLE-UPON-TYNE NE3 4DY

2741 TAYLOR, R.K., Mr - Civil Engr
* U. Teacher
\+ University of Durham
Dept of Geology
South Road
DURHAM DH1 3LE

13428 THIA, H.J., Mr
* Room No.5. Al-Attas Hottle
P.O. Box 1299
JEDDAH (SAUDI ARABIA)

2756 TRIGG, C.F., Mr - Civil Engr
* Dept of Civil Engineering
University College
Newport Road
CARDIFF CF2 1TA

13429 TRUMAN-DAVIES, C.
* General Manager
Ground Anchors Ltd
63A High Street
REIGATE RH2 9AE

13430 TURNER, M.J., Mr
28 St Peters Road
KINETON, Warwick

2905 TYE, P.F., Mr - Engr
* Consulting Engineer
Rofe Kennard & Lapworth
Rafferty House
2-4 Sutton Court Road
SUTTON SM1 4SS

2762 TYSON, R.E., Mr - Civil Engr
* New Lodge
\+ Links Drive
CHELMSFORD CM2 9AW

13431 VADGAMA, N.J., Mr - Civil Engr
* Civil Engineer
78 Windmill Lane
GREENFORD, Middx

2765 VAIL, A.J., Mr - Civil Engr
* Consulting Engineer
1A Man Kei Toi
Pak Sha Wan
Sai Kung
New Territories
HONG KONG

13432 VALENZUELA, L., Mr - Engr
* Project Engineer
c/o THEMAG
Largo do Arouche 24, 10º Andar
01219 SAO PAULO SP (BRAZIL)

2766 VANNER, M.J., Mr
* 9 Elgin Crescent
CATERHAM CR3 6ND

2767 VAUGHAN, P.R., Dr - Civil Engr
* U. Teacher
41A Eton Place
Eton College Road
LONDON NW3

2775 WARDELL, K., Mr
K. Wardell & Partners
Lancaster Building
High Street
Newcastle, STAFFS

8589 WAREHAM, B.F., Mr - Civil Engr
15 Cypress Way
Derby Green
Blackwater
CAMBERLEY, Surrey

2863 WATKINS, D.J., Dr - Civil Engr
* Project Engineer
1400 Solano No. 12
ALBANY CA. 94706 (USA)

13433 WATSON, S.H., Mr
24 Valley View
Jesmond Dene
NEWCASTLE-UPON-TYNE NE2 2JS

2783 WATT, G.A., Mr - Civil Engr
62 Grand Drive
Raynes Park
LONDON SW20 9DY

13434 WATTERS, R.J., Mr - Geologist
Mackay School of Mines
University of Nevada
RENO NV. 89557 (USA)

13435 WAYGOOD, C.M.P., Mr - Civil Engr
* 9 Duffield Road
Walton-on-the-Hill
TADWORTH, Surrey

2864 WEX, B.P., Mr - Civil Engr
* Consulting Engineer
Freeman Fox & Partners
Devon House
12-15 Dartmouth Street
LONDON SW1H 9BL

8590 WHITE, D.A., Mr
* Lecturer
2 Wood Leigh
Aldershot Road
FLEET GU13 9NL

8591 WHITTAKER, B.N., Dr
* U. Teacher
Mining Department
University of Nottingham
NOTTINGHAM NG7 2RD

13436 WICKSON, J.L., Mr - Civil Engr
* Geotechnical Engineer
Ostlers Cottage, The Green
Dinton
AYLESBURY HP17 8UR

8592 WIDDIS, T.F., Mr
* 19 The Coppins
Brighton Road
Foxrock
DUBLIN (IRELAND)

2799 WILLIAMS, G.M.J., Mr - Civil Engr
* 81 Woodsford Square
LONDON W14

8593 WILLIAMS, O.T., Mr - Civil Engr
* 18 Little Gaddesden
BERKHAMSTED HP4 1PA

2802 WILSON, A.H., Mr
Scientist
91 Ashby Road
BURTON-ON-TRENT DE15 OPP

2803 WILSON, C.D.V., Dr
U. Teacher
Dept of Geology
The University
LIVERPOOL L69 3BX

8954 WILSON, L.C., Mr - Engr
Consulting Engineer
Ninham Shand & Partners
P.O. Box 6101
ROGGEBAAI 8012 (S.AFRICA)

2807 * WILSON, R.L., Mr - Engr
Consulting Engineer
R. Travers Morgan & Partners
136 Long Acre
LONDON WC2E 9AD

8595 WOOD, I.R., Mr - Engr
Soil Engineer
15 Brookshill Avenue
HARROW WEALD, Middx

13437 * WOOD, L.A., Dr - Engr
Consulting Engineer
43 Taylors Crescent
CRANLEIGH GU6 7EN

13438 * WRIGHT, J.A., Mr - Engr
Turner Wright & Partners
Aston House
High Street
WEST WYCOMBE, Bucks

2817 WRIGHTMAN, J., Mr - Civil Engr
1 Old Bury Hill House
Milton Heath, Westcott
DORKING, Surrey

2819 * WYKEHAM, G.B.
Consultant
Merrileas
Gay Lane
HOLYPORT
nr Maidenhead, Berks

13440 * WYNN, P.G.L., Mr
27 Wordsworth Avenue
MALDON, Essex

13441 * YU, P.L.K., Mr
461-463 Chatham Road
Ground Floor, Kowloon
HONG KONG

2822 * YUNG, E-H., Mr - Civil Engr
Company Director
Goodyear Management
(Malaysia) SDN BHD
Bangunan UMBC, Cround Floor
Jalan Suleiman
KUALA LUMPUR (MALAYSIA)

13442 * ZACLIS, E.
Rua Piracicaba - 895 Apt 71
Campo Belo
04618 SAO PAULO SP (BRAZIL)

13443 ZEITLEN, J.G., Mr
Faculty of Civil Engineering
Israel Institute of Technology
Technion City
HAIFA (ISRAEL)

2825 ZIENKIEWICZ, O.C., Dr - Civil Engr
School of Engineering
University College of Swansea
Singleton Park
SWANSEA SA2 8PP

SUPPORTING MEMBERS

3838 BRITISH LIBRARY (LENDING DIVISION)
Boston Spa
WETHERBY LS23 7BQ

7234 ENGINEERING LABORATORY EQUIPMENT LTD
Eastman Way
HEMEL HEMPSTEAD HP2 7HB

13192 GEO RESEARCH LTD
Brighton Road
Heaton Norris
STOCKPORT SK4 2Be 214

8596 PERGAMON PRESS LTD
Headington Hill Hall
OXFORD OX3 OBW

12031 REES INSTRUMENTS LTD
Westminster House
OLD WOKING GU22 9LF

10672 ROBERTSON RESEARCH ENGINEERING
SERVICES LTD
Back York Road
Deganwy
GWYNEDD, N. Wales

8012 ROCK INSTRUMENTS LTD
Bell Lane
UCKFIELD TN22 1QL

586 SOIL MECHANICS LTD
Eastern Road
BRACKNELL RG12 2UZ

UNITED STATES OF AMERICA

US NATIONAL COMMITTEE FOR ROCK MECHANICS (USNC/RM)

National Research Council, NAS-NAE, 2101 Constitution Avenue, N.W., WASHINGTON, DC 20418

Chairman: Dr D.C. Banks
Secretary: Mr R.L. Bangert

Total Number of Ordinary Members - 222
Total Number of Corresponding Members - 3
Total Number of Supporting Members - 12

ORDINARY MEMBERS

551 ABEL, Jr., J.F., Prof.
Professor
Department of Mining
Colorado School of Mines
GOLDEN, CO 80401

5534 ABOU-SAYED, A.S.
Manager
Terra Tek, Inc.
420 Wakara Way
SALT LAKE CITY, UT 84108

5403 ADAMS, Jr, W.R.
Chief, Tests and Geotechnical Engineering Section
Allegheny County Department of Engineering and Construction
942 Mirror Street
PITTSBURGH, PA 15217

5511 AGAPITO, J.F.T.
President
Agapito & Associates
715 Horizon Drive
Suite 340
GRAND JUNCTION, CO 81501

5568 AGBABIAN, M.S.
President
Agbabian Associates
250 North Nash Street
EL SEGUNDO, CA 90245

5105 APPL, F.C., Prof.
Professor
Mechanical Engineering Dept
Kansas State University
MANHATTAN, KS 65506

5569 ARNOLD, A.B., Mr - Geologist
Chief Geologist
Bechtel Corporation
50 Beale Street
SAN FRANCISCO, CA 94119

5366 ARZI, A.A.
80 Krinizi Street, Suite 23
RAMAT-GAN (ISRAEL)

375 ATCHISON, T.C., Prof.
Visiting Professor
Dept of Civil and Mineral Engineering
University of Minnesota
221 Church Street, S.E.
MINNEAPOLIS, MN 55455

5363 ATKINSON, R.H.
Vice President
Atkinson-Noland & Associates, Inc.
2619 Spruce Street
BOULDER, CO 80302

5172 AUGHENBAUGH, N.B., Prof.
Department Chairman & Professor
Dept Mining, Petroleum & Geological Engineering
University of Missouri-Rolla
ROLLA, MO 65401

5418 AUSTIN, W.G.
Head, Rock Mechanics Section
Water and Power Resources Service
Box 25007 DFC
DENVER, CO 80225

5507 AVASTHI, J.M., Mr
Research Engineer
Gulf Research & Development Co.
P.O. Drawer 2038
PITTSBURGH, PA 15230

5342 AVERY, R.T., Mr - Engr
Staff Senior Engineer
University of California
Lawrence Berkeley Laboratory,
Building 46-189
BERKELEY, CA 94720

5549 BANGERT, R.L.
Executive Secretary
USNC/Rock Mechanics
National Research Council
2101 Constitution Ave., N.W.
WASHINGTON, D.C. 20418

5338 BANKS, D., Dr
Chief, Engineering Geology and
Rock Mechanics Division
U.S.A.E. Waterways Experiment Station
P.O. Box 631
VICKSBURG, MS 39180

5591 BAUER, R.A., Mr - Geologist
Assistant Geologist
Illinois State Geological Survey
Natural Resources Bldg
URBANA, IL 61801

5510 + BECKWITH, G.H.
Senior Vice President
Sergent, Hauskins & Beckwith
3940 W. Clarendon Avenue
PHOENIX, AZ 85019

5608 + BECKWITH, Jr., W.B.
Dravo Engineers and Constructors
Civil, Mining and Marine Division
One Oliver Plaza, 33rd Floor
PITTSBURG, PA..15222

5560 BELESKY, R.M.
Graduate Student
The Pennsylvania State University
104 Mineral Science Bldg
UNIVERSITY PARK, PA 16802

5489 BENNETT, W., Mr - Engr
Geotechnical Engineer
New Mexico State Highway Department
Materials Laboratory and Research
Bureau
P.O. Box 1149
SANTA FE, NM 87501

472 BIENIAWSKI, Z.T., Prof.
U. Professor
104 Mineral Sciences Bldg
The Pennsylvania State University
UNIVERSITY PARK, PA 16802

5367 BLACKSTONE, R., Mr - Engr
Chief Engineer
Union Metal Manufacturing Co.
American Concrete Division
908 Sir John Drive
LAKE VILLA, IL 60046

5193 BLAKE, W., Mr - Mining Engr
Consulting Mining Engineer
P.O. Box 928
HAYDEN LAKE, ID 83835

5422 BRANDON, J.R.
6028 Owens Street
ARVADA, CO 80004

5368 BRANTHOOVER, G.L.
Vice President
Branthoover & Richards, Inc.
P.O. Box 6074
HARRISBURG, PA 17112

5423 + BREKKE, T.L., Prof.
U. Professor
Department of Civil Engineering
434B Davis Hall
University of California - Berkeley
BERKELEY, CA 94720

5227 BRIERLEY, G.S.
Manager
H & A of New York
50 Chestnut Plaza
ROCHESTER, NY 14604

5117 BRUHN, R.W.
Project Engineer
GAI Consultants, Inc.
570 Beatty Road
PITTSBURGH, PA 15239

520 + BUKOVANSKY, M.
President
GEO-HYDRO Consulting, Inc.
9485 West Colfax 304
DENVER, CO 94720

5522 BUR, T.R., Mr - Mining Engr
100 W. Crystal Lake Road
BURNSVILLE, MN 55337

5389 BUREAU, G.
Geotechnical Consultant
Civil Systems, Inc.
2450 Washington Avenue
SAN LEANDRO, CA 94577

5570 BUTTERS, S.A.
Vice President
Terra Tek, Inc.
420 Wakara Way
SALT LAKE CITY, UT 84108

5558 CELESTINO, T.B.
Research Assistant
University of California - Berkeley
Dept of Civil Engineering
BERKELEY, CA 94720

5536 CHAE, Y.S., Prof.
U. Professor
Dept of Civil Engineering
Rutgers University
P.O. Box 909
PISCATAWAY, NJ 08854

5599 CHAN, L.-Y.
Research Assistant/Teaching Assistant
440 Davis Hall
University of California-Berkeley
BERKELEY, CA 94720

5254 CHASE, G.H., Mr - Hydrologist
U.S.Geological Survey
National Center MS 760
RESTON, VA 22092

5425 CHOI, D.S.
Senior Research Scientist
CONOCO
Drawer 1267
PONCA CITY, OK 94601

5508 CHONG, K.P., Prof.
U. Professor
Dept of Civil Engineering
University of Wyoming
LARAMIE, WY 82071

5480 CIKANEK, E.M., Mr - Engr
Lead Engineer
Harza Engineering Company
150 South Wacker
CHICAGO, IL 60606

5274 CLARK, B.R.
Director of Rock Mechanics
Leighton and Associates, Inc.
17975 Sky Park Circle
IRVINE, CA 92714

5546 CLEARY, M.P.
Associate Professor
Massachusetts Institute of Technology
77 Massachusetts Avenue
CAMBRIDGE, MA 02139

5101 COGAN, J., Mr - Geotech. Engr
International Engineering Co.
220 Montgomery Street
SAN FRANCISCO, CA 94104

5426 COLLINS, J.P.
Partner
James P. Collins & Associates
642 Kreag Road
PITTSFORD, NY 14534

5610 COOK, N.G.W.
Department of Materials Science and Mineral Engineering
University of California, Berkeley
BERKELEY, CA 94720

5307 COOLEY, W.C.
President
Terraspace Inc.
304 North Stonestreet Avenue
ROCKVILLE, MD 20850

560 COULSON, J.H., Mr - Civil Engr
Principal Civil Engineer
Tennessee Valley Authority
400 Commerce Avenue W3D228
KNOXVILLE, TN 37902

5199 COURSEN, D.L.
Research Fellow
E.I. Du Pont de Nemours & Company, Inc.
Potomac River Laboratory
P.O. Box 761
MARTINSBURG, WV 25401

401 CRADDOCK, C., Prof.
U. Professor
University Wisconsin
MADISON, WI 53706

5166 DAEMEN, J.J.K.
Assistant Professor
University of Arizona
Dept of Mining and Geological Engineering
TUCSON, AZ 85721

424 D'APPOLONIA, E.
\+ D'Appolonia Consulting Engineers,Inc.
1177 McCully Drive
PITTSBURGH, PA 15235

5268 DAVIS, A.H.
Los Alamos Scientific Laboratory
MS 665, Group G-6
LOS ALAMOS, NM 87545

203 DEERE, D.U., Prof.
Consultant
6834 S.W. 35th Way
GAINESVILLE, FL 32601

5520 DENDROU, B., Mr - Engr
Research Engineer
Agbabian Associates
250 North Nash
EL SEGUNDO, CA 90245

5159 DESAI, C.S., Prof.
* U. Professor
Dept of Civil Engineering
Virginia Polytechnic Institute and State University
BLACKSBURG, VA 24061

5503 DOE, T.
Staff Scientist
31 Hopkins Court
BERKELEY, CA 94706

5571 DONATH, F.A., Prof.
U. Professor
University of Illinois
URBANA, IL 61801

5224 DOWDING, C.H.
Associate Professor
Department of Civil Engineering
Northwestern University
EVANSTON, IL 60201

5611 DZIUBEK, J.A.
Manager, Geotechnical Department
Michael Baker, Jr., Inc.
4301 Dutch Ridge Road, Box 280
BEAVER, PA 15009

5013 EINSTEIN, H.H.
Associate Professor
Massachusetts Institute of Technology
Room 1-330
CAMBRIDGE, MA 02130

5407 EVANS, D.A.
President
D.A. Evans, Inc.
23011 Moulton Parkway
Suite j-10
LAGUNA HILLS, CA 92653

5547 FELLNER, R.W., Mr - Engr
Agbabian Associates
250 N. Nash Street
EL SEGUNDO, CA 90245

5533 FOLWELL, W.T.
Manager, Rock Mechanics
Dresser Industries, Inc. -
Mining Services & Equipment Division
P.O. Box 24647
DALLAS, TX 75224

5512 FONT, R.G.
Associate Professor
Department of Geology
Baylor University
WACO, TX 76703

5561 FRIEDMAN, M., Prof.
U. Professor
Center for Tectonophysics
Texas A&M University
COLLEGE STATION, TX 77843

5513 FUH, G.F.
Research Scientist
Continental Oil Company
M&T Lab/Mining Research Division
P.O. Box 1267
PONCA CITY, OK 74601

5537 GALLET, A.S., Mr - Geotech.Engr
Senior Geotechnical Engineer
Sargent & Lundy Engineers
55 East Monroe Street
CHICAGO, IL 60603

5217 GATES, R.H.
LTC, US Army
8709 Kerry Lane
SPRINGFIELD, VA 22152

5600 GERCEK, H.
Graduate Student
The Pennsylvania State University
104 Mineral Science Bldg
UNIVERSITY PARK, PA 16802

5209 GIGER, M.W., Mr
Doctor of Engineering
TAMS
Consulting Engineers
695 3rd Avenue
NEW YORK, NY 10017

5572 GILL, J.D.
Head, Civil/Geotechnical Departments
Acres American
900 Liberty Bank Building
BUFFALO, NY 14202

5433 GOETTLE, III, R.J.
President
Richard Goettle, Inc.
12071 Hamilton Avenue
CINCINNATI, OH 45231

5014 GORTON, Jr., Q.A., Mr - Geologist
Consulting Civil Engineer and
Geologist
2571 Moraga Drive
PINOLE, CA 94564

338 GRAY, R.E.
* Vice President
\+ GAI Consultants, Inc.
570 Beatty Road
MONROEVILLE, PA 15146

5482 GREEN, S.J.
President
Terra Tek, Inc.
420 Wakara Way
SALT LAKE CITY, UT 84108

5573 GRONSETH, J.M., Mr - Engr
Project Engineer
Terra Tek, Inc.
420 Wakara Way
SALT LAKE CITY, UT 84108

5106 HAAS, C.J., Prof.
U. Professor
Rock Mechanics & Explosives
Research Center
University of Missouri-Rolla
ROLLA, MO 65401

5562 HADLEY, K., Mr - Geologist
Senior Supervisory Geologist
Exxon, U.S.A. - Offshore District
P.O. Box 23704
HARAHAN, LA 70183

5015 HAIMSON, B.C., Prof.
U. Professor
University of Wisconsin
1509 University Avenue
MADISON, WI 53706

5059 HAMEL, J.V., Mr - Engr
Consulting Engineer
Hamel Geotechnical Consultants
1992 Butler Drive
MONROEVILLE, PA 15146

5000 HANDIN, J., Prof.
U. Professor
Center for Techtonophysics
Texas A&M University
COLLEGE STATION, TX 77843

5145 HANSEN, D.E.
President
Lachel Hansen & Associates, Inc.
P.O. Box 741
BELLEVUE, WA 98009

5435 HANSMIRE, W.H., Mr - Geotech. Engr
Senior Geotechnical Engineer
Parsons, Brinckerhoff, Quade & Douglas
One Penn Plaza
NEW YORK, NY 10001

5016 HARDY, Jr., H.R., Prof.
U. Professor
Department Mineral Engineering
Room 117, MS Building
Penn State University
UNIVERSITY PARK, PA 16802

5163 HATHEWAY, A.W., Mr - Geologist
Vice President and Chief Geologist
Haley & Aldrich, Inc.
238 Main Street
CAMBRIDGE, MA 02142

5436 HAVERLAND, M.L., Mr - Engr
Supervisory Structural Engineer
Water & Power Resources Service
Code D-225, Denver Federal Center
DENVER, CO 80225

5550 HAYS, T.
Department of the Army
New England Division, Corps of Engineers
424 Trapelo Road, Building 100A
WALTHAM, MA 02142

351 HESS, J.D.
President
Structural Behavior Engineering Labs, Inc.
P.O. Box 23167
PHOENIX, AZ 85063

5574 HOCKING, G.
Advanced Technology Group
Dames & Moore
44 Mall Road
BURLINGTON, MA 01803

5551 HOFMANN, R.
Scientist/Manager
Science Applications, Inc.
2450 Washington Avenue 120
SAN LEANDRO, CA 94577

5019 HOSKINS, E.R., Prof.
U. Professor
Department of Geophysics
Texas A&M University
COLLEGE STATION, TX 77843

5594 HUNT, S.R., Mr - Geologist
Research Geologist
Exxon Production Research Company
Box 2189
HOUSTON, TX 77001

5575 INGRAFFEA, A.R.
Assistant Professor and Manager of Structural Research
Dept of Structural Engineering
School of Civil and Environmental Engineering
371 Hollister Hall
Cornell University
ITHACA, NY 14853

5003 ISRAELSEN, O.A.
Executive Secretary
Council on Natural Disasters
National Academy of Sciences
2101 Constitution Avenue, N.W.
WASHINGTON, D.C. 20418

5557 JEFFERIS, R.G.
Graduate Assistant
Department of Geology
Penn State University
308 Deike Building
UNIVERSITY PARK, PA 16802

5576 JONES, A.H.
Executive Vice President
Terra Tek, Inc.
420 Wakara Way
University Research Park
SALT LAKE CITY, UT 84108

5563 JONES, R.S.
Geology Department
Magma Copper Co.
P.O. Box M
SAN MANUEL, AZ 85631

5382 JORGENSEN, D., Dr - Geologist
Senior Staff Geologist
U.S. Gypsum Company
BR 1
SALTVILLE, VA 24370

286 JUDD, W.R., Prof.
* U. Professor
+ School of Civil Engineering
Purdue University
W. LAFAYETTE, IN 47907

5442 KALKANI, E.C.
Sevastoupoleos 50
ATHENS 608 (GREECE)

5601 KARMIS, M.
Assistant Professor
Mining & Minerals Engineering Dept
Virginia Technology
Holden Hall 213
BLACKSBURG, VA 24061

5577 KENYON, W.E.
Scientific Consultant
Schlumberger-Doll Research Center
P.O. Box 307
RIDGEFIELD, CT 06877

5277 KIESEWETTER, C.H.
Instructor
Dept of Mechanical/Civil Technology
Middlesex County College
EDISON, NJ 08817

5501 KIM, C.M.
Manager, Engineering Research
and Development
BJ-Hughes, Inc
1133 Avenue "G" East
ARLINGTON, TX 76011

5491 KIM, K.
Associate Professor
Mining Engineering Department
Michigan Technological University
HOUGHTON, MI 49931

5210 KIM, K.S.
Assistant to the Dean
MacKay School of Mines
University of Nevada
RENO, NV 89557

5552 KLAUBER, W.
Graduate Student, Mining
University of California, Berkeley
1105 F 6th Street No. 34
ALBANY, CA 94710

5136 KO, K.C.
President
Kenneth C. Ko & Associates, Inc.
6911 South Yosemite No. 108
ENGLEWOOD, CO 80112

5612 KOELSCH, T.A.
EXXON Production Research Co.
P.O. Box 2189
HOUSTON, TX 77001

5220 KUHN, A.K.
D'Appolonia Consulting Engineers, Inc.
2340 Alamo SE, Suite 306
ALBUQUERQUE, NM 87106

5198 KULHAWY, F.H.
Associate Professor
School of Civil and Environmental Engineering
Cornell University
ITHACA, NY 14853

5516 LACHEL, D.J.
Vice President
Lachel Hansen & Associates, Inc.
16027 West 5th Avenue
GOLDEN, CO 80401

5445 LAWRENCE, R.J.
Member of the Technical Staff
Sandia Laboratories - Organisation 400
P.O. Box 5800
ALBUQUERQUE, NM 87185

5161 LEONARD, R.J., Prof.
* U. Professor
University of Kansas
LAWRENCE, KS 66045

5479 LIENHART, D.A.
Director
Ohio River Division Laboratory
U.S. Army Corps of Engineers
5851 Mariemont Avenue
CINCINNATI, OH 45227

5528 LIFRIERI, J.J.
Environmental Engineering
744 Mountain Boulevard
WATCHUNG, NJ 07060

5192 LOWE, III, J.
* Partner
Tippetts-Abbett-McCarthy-Stratton
655 Third Avenue
NEW YORK, NY 10017

5291 LUNDQUIST, R.G.
Mining Research Consultant
35 Mohegan Road
ACTON, MA 01720

5108 LYTLE, J.D.
* Chief, Instrumentation and
Evaluation Section
U.S. Army Corps of Engineers
210 Tucker Boulevard North
Attn: ED-FI
ST. LOUIS MO 63101

5548 MA, M.S.
Research Assistant
Dept of Metalurgical & Mineral Engineering
University of Wisconsin
MADISON, WI 53706

5067 MacPHERSON, H.H.
* Graduate Student
Dept of Civil Engineering
Rm 2206 CEB
University of Illinois
URBANA, IL 61801

5068 MARLIAVE, B.H., Dr - Geologist
Consulting Geologist
67 Grandview Place
WALNUT CREEK, CA 94595

5352 MARSHALL, G.S.
Franki Foundation Co.
203 N. Live Oak Drive
MONCKS CORNER, SC 29461

5496 McCARTER, M.K., Prof.
U. Professor
111 Browning Building
University of Utah
SALT LAKE CITY, UT 84112

5109 McCLAIN, W.C.
Executive Vice President
Re/Spec. Inc.
P.O. Box 725
RAPID CITY, SD 57701

5408 McCUSKER, T.G.
7 Echo Court
WAYNE, NJ 07470

5578 McHUGH, S., Dr - Geophysicist
SRI International
333 Ravenswood Avenue - A286
MENLO PARK, CA 94025

5541 MERRITT, J.L.
President
Merritt Cases Inc.
710 Brookside Avenue
P.O. Box 1206
REDLANDS, CA 92373

5602 MILLER, S.M., Mr - Geol.Engr
4025 North 4th Avenue
TUCSON, AZ 85705

5041 MISTEREK, D.L.
Head, Instrumentation Section
Water & Power Service - Code 225
P.O. Box 25007
DENVER, CO 80225

5579 MONTGOMERY, C.T.
Research Specialist
Dowell Division of the Dow Chemical Company
1151 N Rockford, Building 8
TULSA, OK 74106

5529 MORLAN, L.E., Mr - Engr
Senior Engineer
Foundation Sciences, Inc.
1630 S.W. Morrison
PORTLAND, OR 97205

5603 MRUGALA, M.
Graduate Student
Mining Department
The Pennsylvania State University
104 Mineral Sciences Bldg
UNIVERSITY PARK, PA 16802

5358 MYSTKOWSKI, W.
Scientific Programmer B
Colorado Department of Highways
626 S. Oakland Street
AURORA, CO 80012

5556 NELSON, J.W., Mr - Geotech.Engr
D'Appolonia Consulting Engineers, Inc.
7400 South Alton Court
ENGLEWOOD, CO 80112

5556 NELSON, R.A.
AMOCO Production Co.
Research Center
Box 591
TULSA, OK 74102

5098 NESBITT, R.H., Dr - Engng Geologist
7225 Morningside Lane
ALEXANDRIA, VA 22308

5454 NICHOLAS, D.E.
5225 E. Pima, Suite B.
TUCSON, AZ 85712

5110 NORMAN, C.E.
Department of Geology
University of Houston
HOUSTON, TX 77004

5072 NOVICK, D.A.
Senior Vice President
Lester B. Knight & Associates, Inc.
549 West Randolph Street
CHICAGO, IL 60606

5200 OSTERWALD, F.W., Dr - Geologist
U.S. Geological Survey
Box 25046 MS 903
DENVER, CO 80225

5185 OUTMANS, H.D.
Research Associate
Union Oil Company of California
Union Science & Technology Division
376 South Valencia Avenue
P.O. Box 76
BREA, CA 92621

5457 PALMER, M.W.
Civil/Geotechnical Substation Supervisor
Acres Consulting Services
5259 Dorchester Road
Niagara Falls
ONTARIO (CANADA LZE 6W1)

5073 PANEK, L.A.
Senior Technical Advisor, Rock Mechanics
U.S. Bureau of Mines
Bldg 20 Federal Center
DENVER, CO 80225

5564 PARATE, N.S.
Managing Director/President
Parates International Inc.
504 Terrace Drive
NEW CUMBERLAND, PA 17070

5177 PARISEAU, W.G., Prof.
U. Professor
315 WBB
University of Utah
SALT LAKE CITY, UT 84112

5580 PARK, D.-W.
Assistant Professor
College of Mineral and Energy Resources
West Virginia University
117 White Hall
MORGANTOWN, WV 26506

5325 PECK, R.B., Mr - Civil Engr
* 1101 Warm Sands Drive, South East
ALBUQUERQUE, NM 87123

1650 PENG, S.S.
Mining Engineering Dept
West Virginia University
MORGANTOWN, WV 26506

349 PINCUS, H.J.
Department of Geological Sciences &
Civil Engineering
University of Wisconsin-Milwaukee
P.O. Box 413
MILWAUKEE, WI 53201

5181 PODNIEKS, E.R.
Research Supervisor, Surface Mining
Twin Cities Mining Research Center
P.O. Box 1660
TWIN CITIES, MN 55111

5030 PRATT, H.R.
Director, Applied Geosciences
Terra Tek Inc.
420 Wakara Way
SALT LAKE CITY, UT 84108

5581 PRICE, R.H.
Research Assistant/Graduate Student
Center for Tectonophysics
College of Geosciences
Texas A&M University
COLLEGE STATION, TX 77843

5285 PUGLIESE, J.M.
Environmental Specialist (Mining)
U.S. Bureau of Mines
P.O. Box 1660
TWIN CITIES, MN 55111

5565 PYKE, R.
Geotechnical Consultant
Civil Systems Inc.
2450 Washington Avenue
SAN LEANDRO, CA 94577

5517 RADER, D.
Manager, Sensor Development
NL Petroleum Services /DST
P.O. Box 1473
HOUSTON, TX 77001

5031 RAGAN, D.M., Prof.
U. Professor
Arizona State University
TEMPE, AZ 85281

5553 RAJARAM, V., Mr - Mining Engr
E.K. Lehmann & Associates, Inc.
1409 Willow Street
MINNEAPOLIS, MN 55403

240 REED, J.J., Mr - Mining Engr
Consultant
P.O. Box 126
ANACORTES, WA 98221

5543 REN, N.-K., Mr - Engr
Research Engineer
P.O. Box 21
TULSA, OK 74102

5155 RICHARDS, D.P., Mr - Engr
Senior Engineer/Associate
Jenny Engineering Corporation
16027 West 5th Avenue
GOLDEN, CO 80401

5076 RICHARDS, J.W.
President
Branthoover & Richards, Inc.
P.O. Box 6074
HARRISBURG, PA 17112

5092 ROBBINS, R.J.
President
The Robbins Company
650 South Orcas Street
SEATTLE, WA 98108

5582 ROBERDS, W.J.
Instructor
University of Texas
Room 62 ECJ
AUSTIN, TX 78712

5583 ROBERTSON, E.C., Dr - Geophysicist
U.S. Geological Survey
922 National Center
RESTON, VA 22092

5160 ROLLINS, R.R.
Associate Professor
Rock Mechanic Research Centre
University of Missouri-Rolla
ROLLA, MO 65401

5032 ROSENBLAD, J.L.
* Manager, Geotechnical Division
Stone & Webster Engineering Corp.
P.O. Box 5406
DENVER, CO 80217

RUDNICKI, J.W.
Department of Theoretical and
Applied Mechanics
216 Talbot Lab.
University of Illinois
URBANA, IL 61801

5490 RUGGIERO, J.G.
Project Coordinator
N.Y.C. Department of Environmental
Protection
3927 Victory Blvd.
STATEN ISLAND, NY 10314

5222 RUSSEL, J.E., Prof.
U. Professor
Texas A&M University
Dept of Petroleum Engineering
COLLEGE STATION, TX 77843

5146 SAIDMAN, M., Mr - Engr
Senior Consulting Engineer
1845 West Point Drive
CHERRY HILL, NJ 08003

5559 SANCIO, R.T.
Graduate Student
University of California-Berkeley
M39 The Alameda
BERKELEY, CA 94707

5384 SANGREY, D.A., Prof.
* U. Professor
School of Civil and Environmental Engineering
Cornell University
ITHACA, NY 14853

5584 SANYAL, S.K.
Manager
Stanford University Petroleum Research Institute
Lloyd Noble Building
STANFORD, CA 94305

5562 SAVELY, J.P., Mr - Geologist
7138 East Paseo San Andres
TUCSON, AZ 85710

562 SCHMIDT, B.
Project Manager
Parsons, Brinckerhoff, Quade & Douglas, Inc.
165 Post Street
SAN FRANCISCO, CA 94108

5476 SCHMIDT, R.A.
Member of Technical Staff
Organization 4732
Sandia Laboratories
ALBUQUERQUE, NM 87185

5605 SCHWARTZ, C.W.
Assistant Professor
Dept of Civil Engineering
University of Maryland
COLLEGE PARK, MD 20742

5567 SCOTT, G.A., Mr - Civil Engr
Water and Power Resources Service
P.O. Box 25007
DENVER, CO 80225

5500 SCOTT, J.C., Mr - Geotech. Engr
* Geotechnical Engineer
Woodward-Clyde Consultants
2909 West 7th Avenue
DENVER, CO 80204

5226 SEEGMILLER, B.L.
Principal
Seegmiller Associates
447 East 200 South
SALT LAKE CITY, UT 84111

5521 SHARMA, R.P., Mr - Mining Engr
Duval Corporation
4715 East For Lowell Road
TUCSON, AZ 85712

397 SINGH, M.M.
President
Engineers International, Inc.
2514 WISCONSIN Avenue
DOWNERS GROVE, IL 60515

5377 SLEBIR, E.J., Mr - Engr
Structural Engineer
U.S. Bureau of Reclamation
Water & Power Resources Service
Bldg 56
Division of Research, Code 1543
Denver Federal Center
DENVER, CO 80225

5530 SOTTILLE, A.L.
P.O. Box 28292
WASHINGTON, D.C. 20005

5301 SOWERS, G.F.
Senior Geotechnical Consultant
Law Engineering Testing Co.
2749 Delk Road, S.E.
MARIETTA, GA 300067

5616 SPRUNT, E.S.
Mobil Research & Development Corp.
P.O. Box 900
DALLAS, TX 75221

5566 ST. JOHN, C.M.
Assistant Professor
Department of Civil & Mineral Engineering
University of Minnesota
MINNEAPOLIS, MN 55455

5523 STRANSKY, T., Dr - Geologist
U.S. Army Corps of Engineers
Ohio River Division Laboratories
5851 Mariemont Avenue
MARIEMONT, OH

5585 STRICKLAND, F.G., Mr - Engr
Research Engineer
Dowell Division of Dow Chemical
Box 21
TULSA, OK 74102

5236 STRID, J.A., Mr - Mining Engr
Chief Mining Engineer
Carbon County Coal Co.
P.O. Box 733
HANNA, WY 82327

5897 STUDEBAKER, I.G., Mr - Mining Engr
Occidental Oil Shale
2090 Sequoia Road
GRAND JUNCTION, CO 81503

5532 SUNDARAM, P.N.
Assistant Professor
University of Wisconsin-Milwaukee
P.O. Box 784
MILWAUKEE, WI 53201

5898 SWEENEY, N.F., Mr - Geotech. Engr
Pacific Gas & Electric Company
77 Beale Street, Rm 1382
SAN FRANCISCO, CA 94106

5211 TARKOY, P.J.
Geotechnical Consultant
102 N. Main Street
SHERBORN, MA 01770

5036 THIRUMALAI, K., Mr - Mining Engr
Mine Safety and Health Administration
Office of Technical Support
4015 Wilson Boulevard
ARLINGTON, VA 22030

5586 THOMPSON, T.W.
Assistant Professor
Department of Petroleum Engineering
University of Texas - Austin
AUSTIN, TX 78731

5504 THOMS, R.L., Prof.
U. Professor
Institute for Environmental Studies
Louisiana State University
BATON ROUGE, LA 70803

5472 TOCKSTEIN, C.D., Mr - Civil Engr
Tennessee Valley Authority
400 Commerce Avenue, W3D222
KNOXVILLE, TN 37902

5606 TOUSEULL, J.A., Mr - Engr
Structural Engineer
1037 South Miller Way
LAKEWOOD, CO 80226

5344 UEBLACKER, H., Mr - Engr
Consulting Engineer
Ueblacker & Associates
7230 West Archer Place
LAKEWOOD, CO 80226

5533 UNAL, E., Mr - Mining Engr
Graduate Assistant
The Penn State University
104 MS Bldg
UNIVERSITY PARK, PA 16802

5401 VAN EECKHOUT, E.M.
Assistant Professor
Montana Technology
BUTTE, MT 59701

367 VAN ZELST, T.W.
* Chairman
Soiltest, Inc.
1213 Wagner Road
GLENVIEW, IL 60025

5082 VARNES, D.J., Dr - Geologist
\+ U.S. Geological Survey
MS 903 Box 25046
DENVER, CO 80225

5505 VELTROP, J.A., Mr - Engr
Vice President and Chief Engineer
Harza Engineering Company
150 South Wacker Drive
CHICAGO, IL 60076

5083 VERSAW, R.E., Mr - Engr
Specialist Engineer
Golder Associates
Suite 180
1658 Cole Boulevard
GOLDEN, CO 80401

5487 VISCA, P.J., Mr - Geol.Engr
Dept of Mining & Geological
Engineering
University of Arizona
TUCSON, AZ 85721

5412 WAWERSIK, W.R.
Member of Technical Staff
Org. 5532, Geomechanics Research
Division
Sandia Laboratories
ALBUQUERQUE, NM 87115

5346 WEAKLY, L.A.
Director of Mining Research
St. Joe Minerals Corporation
P.O. Box 500
VIBURNUM, MO 65566

5607 WEISS-MALIK, R.F., Dr - Engng
Geologist
D'Appolonia Consulting Engineers,
Inc.
M.R. 522-B
CHESTERON, IN 46304

5339 WILLARD, R.J., Dr - Geologist
U.S. Bureau of Mines
P.O. Box 1660
TWIN CITIES, MN 55111

5587 WILLIAMS, J.R.
Geotechnical Consultant
Dames & Moore, Advanced Technology
Group
44 Mall Road
BURLINGTON, MA 01803

340 WILLIS, C.L.
Vice President and Chief Geologist
Harza Engineering Company
150 South Wacker Drive
CHICAGO, IL 60606

5588 WONG, I.H., Mr - Engr
Consulting Engineer
Ebasco Services, Inc.
91st Floor
2 World Trade Center
NEW YORK, NY 10048

5340 YEN, B.C., Prof.
* U. Professor
Dept of Civil Engineering
California State University
LONG BEACH, CA 90840

5609 YEUNG, D.
Graduate Student
Mining Engineering Department
The Pennsylvania State University
UNIVERSITY PARK, PA 16802

5186 ZACCHEO, P.F., Mr - Geotech. Engr
* Tippetts-Abbett-McCarthy-Stratton
655 Third Avenue
NEW YORK, NY 10017

5589 ZAVODNI, Z.M., Mr - Mining Engr
Kennecott Minerals Company
Engineering Center
1515 Mineral Square
SALT LAKE CITY, UT 84147

5360 ZEIGLER, T.W., Mr - Civil Engr
U.S. Department of Energy
Carbondale Mining Technology Center
P.O. Box 2587
CARBONDALE, IL 62901

5617 ZEUCH, D.H.
Department of Materials Science
and Engineering
Cornell University
ITHACA, NY 14853

5488 ZOBACK, M.D., Dr - Geophysicist
U.S. Geological Survey
345 Middlefield Road
MENLO PARK, CA 94025

CORRESPONDING MEMBERS

557 MERRITT, A.H., Dr
7726 SW 36 Avenue
GAINESVILLE, FL 32601

456 MURPHY, V.J., Dr - Consulting
Geophysicist
* c/o Weston Geophysical Engineers, Inc.
P.O. Box 550
WESTBORO, MA 01581

10633 STRACHAN, G.C., Dr - Engng Geologist
2615 N.W. Mulkey
CORVALLIS, OR 97330

SUPPORTING MEMBERS

13036 ACKER DRILL COMPANY, INCORPORATED
P.O. Box 830
SCRANTON, PA 18501

13037 AGBABIAN ASSOCIATES
250 N. Nash Street
EL SENGUNDO, CA 90245

13038 FENIX & SCISSON, INC.
1401 South Boulder
TULSA, OK 74119

8621 GEOLOGY LIBRARY YALE UNIVERSITY
Box 1301 A Yale Station
NEW HAVEN CT 06520

8194 HAMEL GEOTECHNICAL CONSULTANTS
1992 Butler Drive
MONROEVILLE, PA 15146

7163 HARZA ENGINEERING COMPANY
150 South Wacker Drive
CHICAGO, IL 60606

13039 IRAD GAGE
14 Parkhurst St
LEBANON, NH 03766

13040 SLOPE INDICATOR COMPANY
3668 Albion Place North
SEATTLE, WA 98103

6848 SOILTEST INTERNATIONAL, INC.
2205 Lee Street
EVANSTON, IL 60202

13041 STRUCTURAL BEHAVIOR ENGINEERING
LABORATORIES, INC.
P.O. Box 23167
PHOENIX, AZ 85063

13042 TERRAMETRICS, INC.
Division of Teledyne Geotech
16027 W. 5th Avenue
GOLDEN, CO 80401

13043 TERRATEK
University Research Park
420 Wakara Way
SALT LAKE CITY, UT 84108

UNION OF SOVIET SOCIALIST REPUBLICS

SOVIET COMMITTEE FOR THE PARTICIPATION IN INTERNATIONAL SOCIETY FOR ROCK MECHANICS (SC ISRM)
"Hydroproject" Institute, Volokolamskoe shosse 2, MOSCOW A-80, 125812, U.S.S.R.

President: Mr N. Ivantsov
Secretary: Dr V. Rechitski

Total Number of Ordinary Members - 45

ORDINARY MEMBERS

10654 AITMATOV, I.G., Dr
Director
Institute of Physics and Mechanics of Rocks
Kirghizian Academy of Sciences
FRUNZE

10655 BABAIANTS, G.M., Dr
Chief Specialist
Ministry of Ferrous Metallurgy
Technical Department
MOSCOW

10656 BACKLASHOV, I.V., Dr
Prof.
Moscow Mining Institute
MOSCOW

10657 * BOULYCHEV, N.S., Dr
Prof.
Leningrad Mining Institute
LENINGRAD

10658 DIAKOVSKY, V.B., Dr
Deputy Director
"Unipromed" Institute
SVERDLOVSK

10659 DMITRIEV, M.V.
Director
Hydrospetzstroy,
Ministry of Power and Electrification of the USSR
MOSCOW

10660 DOTSENKO, T.P., Mr
Chief Engineer
"Hydroproject" Institute
MOSCOW

10661 ERZHANOV, Z.S., Mr
Academician Kazakh Academy of Sciences
Director
Seismology Institute
ALMA-ATA

10662 FADDEEV, A.B., Dr
Chief of Laboratory
VNIMI Institute
LENINGRAD

10663 FISHMAN, J.A., Dr
Chief of Dept
Rock Foundations Dept
"Hydroproject" Institute
MOSCOW

10664 GAMALEYA, B.M., Dr
Scientific Secretary
VSEGINGEO Institute
MOSCOW

10665 * + GAZIEV, E.G., Dr
Chief Specialist in Rock Mechanics
"Hydroproject" Institute
MOSCOW

10666 IVANTSOV, N.M., Mr - Engr
Vice-Minister
Ministry of Power and Electrification of the USSR
MOSCOW

10667 * KANDAUROV, I.I., Dr
Prof.
Railway Engineers Institute (LIIZT)
LENINGRAD

10668 KATIN, K.P., Dr
Chief of Laboratory
Laboratory of Physical and Mechanical Tests of Rocks
VNII-tsvetmet Institute
UST-KAMMENOGORSK

10669 * KHRISTOFOROV, V.S., Dr
Prof. in Soil Mechanics
Gosstroy USSR
MOSCOW

10670 KUDOYAROV, L.I., Dr
Deputy Chief
Glavniiproject
Ministry of Power and Electrification of the USSR
MOSCOW

10671 KUZNETSOV, S.V., Dr
Prof., Chief of Laboratory
Rock Pressure Laboratory
Academy of Sciences of the USSR
MOSCOW

10672 LYKOSHIN, A.G., Dr
* Deputy Chief Engineer
\+ "Hydroproject" Institute
MOSCOW

10673 MARCHUK, A.N., Dr
Deputy Chief Engineer
Glavgidroenergostroy
Ministry of Power and Electrification of the U S S R
MOSCOW

10674 MASLOV, N.N., Dr
* Prof.
\+ Moscow Anto-Road Institute
MOSCOW

10675 MGALOBELOV, Y.B., Dr
Chief Specialist
"Hydroproject" Institute
MOSCOW

10676 MICHAILOV, L.P., Dr
Director
"Hydroproject" Institute
MOSCOW

10677 MOSTKOV, V.M., Dr
Prof., Chief of Dept
Dept of Underground structures
Research Centre of "Hydroproject" I Institute
MOSCOW

10678 NOSOV, N.V., Mr - Engr
Chief of Technical Dept
GPTUS, Ministry of Power and Electrification of the U S S R
MOSCOW

10679 RECHITSKI, V.I., Dr
Chief of Group
Dept of Rock Foundations
"Hydroproject" Institute
MOSCOW

10680 RUPPENEIT, K.V., Dr
Prof. NIIOSP Institute
MOSCOW

10681 SAPEGUIN, D.D., Dr
* Chief of Foundation Laboratory
VNIIG Institute
LENINGRAD

10682 SAVICH, A.I., Dr
* Chief Geophysicist
\+ Dept of Geophysics
"Hydroproject" Institute
MOSCOW

10683 SHANGUIN, V.S., Mr
Chief Engineer
Naryngidroenergostroy
Ministry of Power and Electrification of the U S S R
MOSCOW

10684 SHAUMIAN, L.V. (Mrs.), Dr
Senior Specialist
Geological Faculty
Moscow State University
MOSCOW

10685 SHIRIAEV, R.A., Dr
* Chief of Section
Foundation Laboratory
VNIIG Institute
LENINGRAD

10686 SKVORTSOV, G.G., Dr
Chief of Laboratory
VSEGINGEO Institute
Settlement Zeleniy Noginsk District
Moscow Region
MOSCOW

10687 SOKOLOV, I.B., Dr
Deputy Director
VNIIG Institute
LENINGRAD

10688 SOROKIN, A.A., Dr
Chief Specialist
"Hydroproject" Institute
LENINGRAD

10689 STEPANOV, A.P., Mr
Chief Engineer
Glavgidroenergostroy
Ministry of Power and Electrification of the U S S R
MOSCOW

10690 TER-MARTIROSIAN, Z.G., Dr
* Prof. in Soil Mechanics
MISI Institute
MOSCOW

10691 TIZDEL, R.R., Mr
Chief Geologist
Rock Foundations Dept
"Hydroproject" Institute
MOSCOW

10692 TOROPOV, L.N., Mr
Deputy Chief Engr
Glavvostokgidroenergostroy
Ministry of Power and Electrification of the U S S R
MOSCOW

10693 TSYTOVICH, N.A., Prof.
* Academician, Prof. in Soil Mechanics
\+ MISI Institute
MOSCOW

10694 TURCHANINOV, I.A., Dr
Prof. Director
Mining Institute, Kola Branch of Academy of Sciences of the U S S R
APATITY

10695 UKHOV, S.B., Dr
* Prof in Soil Mechanics
MISI Institute
MOSCOW

10696 VLOKH, N.P., Dr
Prof.
Mining Institute
SVERDLOVSK

10697 VORONKOV, O.K., Dr
Foundation Laboratory
VNIIG Institute
LENINGRAD

10698 YUPHIN, S.A., Dr
MISI Institute
MOSCOW

VENEZUELA

SOCIEDAD VENEZOLANA DE MECANICA DE SUELO E INGENIERIA DE FUNDACIONES

Apartado 4074, CARACAS 101

President: Mr F.H. Tinoco

Total Number of Ordinary Members - 193
Total Number of Corresponding Members - 1

ORDINARY MEMBERS

10221 ABREU CH., E., Mr - Civil Engr
Calle El Carmen, Qtª Elminelmer
Urb. Prados del Este
CARACAS 108

10222 ACEVEDO, J., Mr - Civil Engr
* Apartado 75.709
CARACAS 107

10223 AGUILERA L.,L., Mr - Mining Engr
* Apartado 68.509
CARACAS 106

10224 ALCAINO A.,E., Dr - Geologist
* Av. Los Jardines, Qtª La Natera
Urb. Prados del Este
CARACAS 108

10225 ALVAREZ S., J., Mr - Civil Engr
* Qtª Alvalar, Calle El Bosque
Urb Caracas Country Club
CARACAS 106

10226 ANGARITA, J., Mr - Civil Engr
* C.V.P. Apartado 425, El Tigre
ESTADO ANZOATEGUI

10227 ARDILA U., R., Mr - Civil Engr
* Calle El Bosque, 0-99, Urb. Stª Maria
Merida
ESTADO MERIDA

10228 ARELIANO M., A., Mr - Civil Engr
* Resd. Las Carolinas, Apto. 2-C, Av.Ppal.
Santa Fe Sur,
CARACAS 108

10229 ARMADA DE RAVEN, N., Mr - Civil Engr
* Calle Veracruz, Resd. La Hacienda
Apto. B-13
Urb. Las Mercedes
CARACAS 106

10230 ARNAL M., M., Mr - Civil Engr
Av. Ppal. Las Mercedes
Edf. Irune - Of. 4
CARACAS 106

10231 ASCANIO T.,G., Dr - Geologist
* Colinas de Santa Monica, Ruta 7
Resd. Las Rosas Apto. 7-1
CARACAS 104

10232 AYALA C., M., Mr - Civil Engr
Av. Ppal. de Los Chorros, No. 13-01
CARACAS 107

10233 BAJETTI, B., S., Dr - Geologist
* Av. Manaure, Seccion "K" Qtª Dora
Urb. Macaracuay
CARACAS 107

10234 BARON J., E., Mr - Civil Engr
* Calle Las Flores con Esq.Av.Los
Jabillitos
Edf. Los Jabillitos Pent-House
Sabana Grande
CARACAS 105

10235 BENARROCH, R., Mr - Civil Engr
* Conjunto Sans-Souci, Resd. El Parrill
Apto. 64 - Chacaito
CARACAS 106

10236 BENAVIDES F., R., Mr - Civil Engr
* Edif. Roraima, Piso 1º Of.B-1,
Av. Francisco de Miranda
Chacaito
CARACAS 106

10237 BEZIC G., J., Mr - Civil Engr
* Plaza Las Delicias, Edf.Caribe
Apto. 2-A
Sabana Grande
CARACAS 105

10238 BISI, A., Mr - Civil Engr
* Edf. Cediaz, Of. E-74
Av. Casanova
Sabana Grande
CARACAS 105

10239 BORGES O., C., Dr - Engng Geologist
* Av. Solano, Resd. Sans-Souci
Edf. Mari-Mari
Apto. 102
El Bosque
CARACAS 106

10240 BRACHO S., W., Mr - Civil Engr
* Av. 15-1 No. 56-35
Urb. La Trinidad-Maracaibo
ESTADO ZULIA

10241 BRICENO M., L., Mr - Civil Engr
* Calle La Iglesia, Qtª Maria Luisa
Urb. Prados del Este
CARACAS 108

10413 BRICENO R., F., Mr - Civil Engr
* Centro Villasmil, Piso 8º
Ofc. 817, Esq. Puente Victoria
CARACAS 101

10242 BURGOS A.,J., Dr - Geologist
* Av. La Trinidad, Edf.Uracoa
Apto. 5-B
Urb. Santa Paula
CARACAS 106

10243 CABRERA DE B., T., Mr - Civil Engr
* Calle Rio Orinoco, Edf. Canaima
Apto. 1-A
Cumbres de Curumo
CARACAS 108

10244 CALLIMAN P., S., Mr - Civil Engr
* 1ª Av. Los Palos Grandes, Edf.Monte Alona
4º Piso, Apto. 41-A
CARACAS 106

10245 CAMPILONGO, G., Mr - Civil Engr
* Apartado 51.649
CARACAS 105

10246 CARABALLO DE P., E., Dr - Engng Geologist
* Calle Transv.40, Resd. Don Juan
Piso 11, Apto. 11-B (frente al
Automercado TIA)
Montalban - La Vega
CARACAS 102

10247 CARCIENTE, J., Mr - Civil Engr
* Av. Lecuna, Parque Central
Edf. Mohedano
Pasillo 1, Of. 1-G
CARACAS 101

10248 CARRERO S., J., Mr - Civil Engr
* Av. Ppal. La Castellana
Resd. La Castellana
Apto. 302
CARACAS 106

10249 CARRILLO P., P., Mr - Civil Engr
* Apartado 61.621
CARACAS 106

10250 CARRILLO, V., Mr - Civil Engr
* Edf. La Vera, Resd. Sans-Souci
Apto. 124
Av. Francisco Solano
El Bosque
CARACAS 106

10251 CENTENO W., R., Mr - Civil Engr
* Av. Sojo, Qtª Chichi
El Rosal
CARACAS 106

10252 CONTRERAS, F., Mr - Civil Engr
* Apartado 440 Merida
ESTADO MERIDA

10253 CORTES M., F., Mr - Civil Engr
* Calle Central, 411-406
Qta. Peregrina
Urb. Santa Cecilia
CARACAS 107

10254 CORDOLIANI F., N., Mr - Civil Engr
INAPLICA, Apartado 375 Ciudad Bolivar
ESTADO BOLIVAR

10255 CORRALES, G., Mr - Civil Engr
* Colinas de Bello Monte, Calle
Casiquiare,
Qta. Tere
CARACAS 105

10256 CRESPO T., C., Mr - Civil Engr
* Qta. Tin. Calle Urape
Colinas de La Trinidad
CARACAS 108

10257 DABOIN R., C., Mr - Civil Engr
* Calle del Valle, Qta. Senda
Country Club
CARACAS 106

10258 DE ELGUEZABAL O., J., Mr - Ind. Engr
* Av. 11 entre 7a y 8a Transv.
Qta. Aldape
Altamira
CARACAS 106

10259 D'ESCRIVAN G., E., Mr - Civil Engr
* Av. Orinoco, entre Av. Caroni y
Calle Tacagua
Qta. Gam, Bello Monte de Sabana
Grande
CARACAS 105

10260 DE FRIES, K., Mr - Civil Engr
* Av. Libertador, Edf. Fertec
\+ Piso 1º - Of. A.
Chacao
CARACAS 106

10261 DE LA CRUZ M., L., Dr - Geologist
* Calle Roraima, Qta. Floriela
Urb. Chuao
CARACAS 106

10262 DE SOLA, O., Dr - Geologist
* MISECA - Apartado 4369
CARACAS 101

10263 DIAZ D., A., Mr - Civil Engr
* Qta. Los Pinos, cruce Av. Andres
Bello con Av. La Salle, Los Caobos
CARACAS 105

10264 DIAZ M., A., Mr - Civil Engr
* Av. Buenos Aires, Edf. Naiguata
Apto. 11
Los Caobos
CARACAS 105

10265 DIAZ Q., A., Dr - Geologist
* Urb. Piedra Azul,Calle Casiquiare
\+ Quinta Los Silvestres
CARACAS 105

10266 DISILVESTRO, C., Mr - Civil Engr
* Urb. Alta Florida, Av.Florida Este
Quinta Los Silvestres
CARACAS 105

10267 DOMINGUEZ Z., C., Mr - Civil Engr
* Urb. Altamira, Av. San Juan Bosco
Edf. 5, Apto.2
Sur La Plaza Altamira
CARACAS 106

10268 ENGLERT K., D., Mr - Civil Engr
* Calle Junin con Calle Ayacucho
Qta. Alborada
El Rosal
CARACAS 106

10269 ESCOVAR M., R., Dr - Geologist
* Qta. Farallones, Calle Paso Real,
Urb. Prados del Este
CARACAS 108

10270 ESPANA F., N., Mr - Civil Engr
* Av. Lazo Marti, Edf. Delta-Torre B
Apto. 4-A
Urb. Santa Monica
CARACAS 104

10271 ESPINAL V., R., Mr - Civil Engr
* Apartado 60.419
CARACAS 106

10272 FALCON A., M., Mr - Civil Engr
* U.C.V. Facultad de Ingenieria
Laboratorio de Hidraulica - 3º Piso
CARACAS 104

10273 FARIAS M., F., Mr - Civil Engr
* Av. Pichincha Chacaito (frente a la
\+ Cortina)
Edf. Askain, 7º Piso Of 26
CARACAS 106

10274 FEBRES-CORDERO, E., Mr - Civil Engr
* Av. Universidad, Edf. Murachi A-1
Merida
ESTADO MERIDA

10275 FERRER F., D., Mr - Hydr.Engr
* Av. Tamanaco, Qta. More
Urb. Macaracuay
CARACAS 107

10276 FERRER R., O., Mr - Civil Engr
* Av. Cordillera de Los Andes, Edf.
Orion
Apto. 2-D
Urb. Cumbres de Curumo
CARACAS 108

10277 FERNANDEZ F., L., Mr - Civil Engr
Av. Francisco Lazo Marti, Edf.
Arcos
3º Piso, 31
Santa Monica
CARACAS 104

10278 FINOL F., J., Mr - Civil Engr
* Apartado 16.285
CARACAS 101

10279 FLORES C.,C., Dr - Geologist
* Apartado 4639
CARACAS 101

10280 FORTUNA M.,A., Mr - Civil Engr
* Resd. Los Techos Rojos, Qta.Alpa
Colinas de la California Sur
CARACAS 107

10281 GALAVIS M.,L., Mr - Civil Engr
* Apartado 68.651
Altamira
CARACAS 106

10282 GALLOVICH P., E., Dr - Geologist
* Apartado 68.509
\+ CARACAS 106

10283 GARCIA G., J., Mr - Civil Engr
* Qta. Romelia, Comercio Residencial
Boleita
Calle A, 57 Los Ruices Norte
CARACAS 107

10284 GARCIA I., L., Mr - Civil Engr
* Av. Don Bosco, Qta. Urupagua, 19
La Florida
CARACAS 105

10285 GARCIA R.. R., Dr - Engng Geolog
ist
Urb. El Parque 6 Los Teques
EDO. MIRANDA

10286 GIFFONNI L., C., Mr - Civil Engr
* Urb. Nueva Segovia, Calle 2
Qta. Gilu-Barquisimeto
ESTADO LARA

10287 GIL RUIZ - T., F., Dr - Engng
Geologist
* Av. Ingenieria, Edf. Adan, Piso
3º, 11
CARACAS 104

10288 GOMEZ C.,J., Mr - Civil Engr
* Boulevar El Cafetal
Resd. Macairal Apto 8-C
CARACAS 106

10289 GOMEZ R.,J., Dr - Engng Geologist
* Conjunto Residencial El Saman
Torre B-2, 32
Urb. Valle Abajo
CARACAS 104

10290 GOMEZ, V., L., Mr - Civil Engr
Resd. La Magnolia, Apto.1-A,
3ª Avenida
Los Palos Grandes
CARACAS 106

10291 GONZALEZ D.,T., Mr - Civil Engr
* Av. Rio de Janeiro, Edf. Guayamuri
Penthouse B
Urb. Chuao
CARACAS 106

10292 GRATEROL M., J., Mr - Civil Engr
* Calle San Miguel, Qta. Luisa
Urb. Sorocaima
La Trinidad
ESTADO MIRANDA

10293 GUEVARA B., R., Mr - Civil Engr
* Av. Francisco de Miranda
Torre La Primera
Piso 14, Of. 14-C
Chacaito
CARACAS 106

10294 HEGEDUS G., L., Mr - Civil Engr
* Ruta 6, Qta. Piluca
Colinas de Sta. Monica
CARACAS 104

10295 HERNANDEZ T., Mr - Civil Engr
* Av. Los Jabillos, No. 33-40
Apto. 31
La Florida
CARACAS 105

10296 HERNANDEZ S., J., Mr - Civil Engr
* Edf. Everest A, Piso 3º
Apto. 33 cruce Calle Calcano con
Calle Simon Plana, Sta Monica
CARACAS 104

10297 HERRERA A., T., Mr - Civil Engr
* Av. Los Castanos, Edf. Dunas
Palace, Planta Baja
La Florida
CARACAS 105

10298 HERRERO N., J., Dr - Engng Geologist
* Urb. La Fundacion A-20, Qta.Terepe
Ciudad Bolivar
ESTADO BOLIVAR

10299 HIEDRA-LOPEZ, J., Mr - Civil Engr
* Calle La Iglesia, Qta. La Hiedra
Prados del Este
CARACAS 108

10300 IRAZABAL R., A., Mr - Civil Engr
* Vista Alegre, Calle 5,
Qta. Giomah
CARACAS 102

10301 ISAVA C., M., Mr - Civil Engr
* Qta. Camia, Calle C-7
Urb. Caurimare
CARACAS 106

10302 LAFOREST L., R., Dr - Geologist
* Apartado 1745
CARACAS 101

10303 LANZ L., L., Mr - Civil Engr
Av. Ppal. Alto Prado
Qta. La Maranona
CARACAS 106

10304 LAPADULA, D., Mr - Civil Engr
* Calle Araure, Qta. Mi Gorda
El Marques
CARACAS 107

10305 LEAL, R., Mr - Civil Engr
* Av. A. Qta. Estrella, 405-1130
\+ La Carlota
CARACAS 107

10306 LEON M., V., Dr - Engng Geologist
* Calle Upata, Qta. Chela
\+ El Cafetal
CARACAS 106

10307 LINARES E., G., Mr - Civil Engr
* Av. Ppal. Los Chorros, 61
Qta. Lourdes
CARACAS 107

10308 LOPEZ A., J., Mr - Civil Engr
* Av. Paez, Edf. Villa Madrid
Apto. 54
El Paraiso
CARACAS 102

10309 LOPEZ C., A., Mr - Civil Engr
* Apartado 6390
CARACAS 101

10310 LOPEZ, E., Mr - Civil Engr
* Av. San Felipe, 45, Qta. Inaqui
La Castellana
CARACAS 106

10311 LOPEZ H., M., Mr - Civil Engr
* Apartado 51.649
CARACAS 105

10312 MANRIQUE, B., Mr - Civil Engr
* Consorcio Estudios Hidroelectrico
Edf. Torreon, Piso 10, Calle Veracruz
Las Mercedes
CARACAS 106

10313 MARQUEZ P., E., Mr - Civil Engr
* Calle Guaicaipuro, Edf. El Mirador
Apto. 71
El Marques
CARACAS 107

10314 MARTINEZ, L., Mr - Civil Engr
* Calle Corama, Av. Motatan
Colinas de Bello Monte
CARACAS 105

10315 MARTINEZ C., R., Mr - Civil Engr
* Apartado 6731 Carmelitas
CARACAS 101

10316 MARTINEZ DE LA P., C., Mr - Civil Engr
* Av. Las Palmas, Qta. Tibidabo
La Florida
CARACAS 105

10317 MARTINEZ F., J., Mr - Civil Engr
* Apartado 70.367
CARACAS 107

10318 MARTINEZ G., J., Mr - Civil Engr
* Qta. Shantiniketan, Calle La Mesa
Prados del Este
CARACAS 108

10319 MEDINA, F., Mr - Civil Engr
* Av. Lopez Mendez, Edf. Paraiso
2º Piso
Apto. 5
San Bernardino
CARACAS 101

10320 MEDINA U., C., Mr - Civil Engr
* Calle Carabobo, Qta. Coromoto
El Rosal
CARACAS 106

10321 MEJIA P.,T., Dr - Geologist
* Boulevard El Cafetal, Resd. Valy-Ana
Cuerpo B, Apto. 41-B, Piso 4º
CARACAS 106

10322 MELA S., L., Mr - Civil Engr
* Av. Leonardo Da Vinci
Edif. Pigalle - Of. 41
Colinas de Bello Monte
CARACAS 105

10411 MENDEZ G., F., Mr - Civil Engr
Av. 19 Abril, Edif.Consuelo, Apto. 2-1
San Cristobal
ESTADO TACHIRA

10323 MERENFELD, J., Mr - Civil Engr
* 3a. Av., 2.11.3609
Los Palos Grande
CARACAS 106

10324 MIBELLE D., O., Mr - Civil Engr
* Centro Villasmil, Piso 8º Of. 817 ESq.
Puente Victoria
CARACAS 101

10325 MILA DE LA ROCA, A., Mr - Civil Engr
* Apartado 68.271
CARACAS 106

10326 MILLAN T., D., Mr - Civil Engr
* Calle Suapure, Qta. Ana Maria
El Marques
CARACAS 107

10327 MOGOLLON, J., Mr - Civil Engr
* Calle 4, 11-6, La Concordia
San Cristobal
ESTADO TACHIRA

10328 MOLINA V., H., Mr - Civil Engr
* Edf. Claret, Apto.2,Av.3
Merida
ESTADO MERIDA

10329 MORALES C., F., Mr - Civil Engr
* Av. Cecilio Acosta, 14, Qta.Rosita
San Bernardino
CARACAS 101

10330 MONTAUTI P., H., Mr - Civil Engr
Av. Rio Manapire, Resd. Guacharo
Planta - Baja - B, Parque Humboldt
Prados del Este
CARACAS 108

10331 MOREAU, J., Mr - Civil Engr
Av. Rio de Janeiro, Resd. Robledar,
Apto. 32
Urb. Caurimare
CARACAS 106

10332 MORREO L., F., Mr - Civil Engr
Urb. El Placer, Calle Norte 6,
131.1010
CARACAS 108

10333 MUNOZ J., N., Dr - Geologist
U.C.V. Escola de Geologia
Apartado 50.926
CARACAS 105

10334 NEGRO C., P., Mr - Civil Engr
* Calle Apure, Qta. Chucha,
Las Palmas
CARACAS 105

10335 NIETO G., E., Mr - Civil Engr
* Calle Los Alpes, Qta. Maria-Cristina
Prados del Este
CARACAS 108

10336 OLIVARES R., P., Mr - Civil Engr
* Apartado 6495 Carmelitas
CARACAS 101

10337 OLIVARES S., C., Mr - Civil Engr
* Av. Urdaneta, Edf. El Candil, Piso 6º
Of. C y D
La Candelaria
CARACAS 101

10338 OLIVEROS R., L., Mr - Civil Engr
* 2a. Av. con 2a. Transv.
Los Palos Grandes
Resd. Don Manuel, Apto. 13-A
CARACAS 106

10339 OLIVEROS, O., Mr - Engng Geologist
Calle 18, 11-17
San Cristobal
ESTADO TACHIRA

10340 PADILLA, D., Dr - Geologist
* Calle El Limon, 43, Qta. Maira
\+ El Cafetal
CARACAS 106

10341 PAIVA H., A., Dr - Geologist
* Vereda 74, 20
Urb. Delgado Chalbaud
Coche
CARACAS 104

10342 PANALOZA R., N., Mr - Civil Engr
Urb. Santa Fe, Calle Santa Ines
Qta. Nepalba
CARACAS 108

10343 * PEREZ-GUERRA, G., Mr - Civil Engr
Apartado 60.316
CARACAS 101

10344 * PEREZ-GUERRA, J., Mr - Civil Engr
Apartado 60.316
CARACAS 101

10345 * PEREZ-LA SALVIA, H., Mr - Civil Engr
Av. El Parque, Qta. Carhu
Prados del Este
CARACAS 108

10412 * PEREZ M., H., Mr - Geophys. Engr
Calle Transv. 40, Resd. Don Juan
Piso 11, Apto. 11-B (frente al
Automercado TIA)
Montalban - La Vega
CARACAS 102

10346 * PEREZ M., J., Mr - Civil Engr
9a Transv. entre 2a y 3a Av.
Qta. Chapaleta
Altamira
CARACAS 106

10347 * PEREZ R., I., Mr - Civil Engr
Calle Carabobo, Qta. Coromoto
El Rosal
CARACAS 106

10348 * PERICHI D., L., Mr - Civil Engr
Calle A, Qta. Elsa
Urb. Santa Marta
CARACAS 106

10349 * PESTI J., A., Mr - Civil Engr
Av. Santa Teresa, Qta. Trinidad 255
Planta Alta - La Floresta
CARACAS 106

10350 POMBROL S., M., Mr - Civil Engr
1a Calle de Bello Monte
Edf. Doru, 301
CARACAS 105

10351 * POPOFF M., CH., Mr - Civil Engr
Av. 2-B, 1104,
La Carlota
CARACAS 107

10352 * PORCARELLI G., F., Mr - Civil Engr
Qta. Gladys, Ruta 5-A,
Parcela 97-P
Colinas de Santa Monica
CARACAS 104

10354 PRUSZA V., Z., Mr - Civil Engr
Apartado 62.413
CARACAS 106

10353 * PUCHE N., A., Mr - Civil Engr
Av. La Alfareria, Qta. Coingri 9-29
Colinas de Los Ruices
CARACAS 107

10355 * QUINTANA U., A., Mr - Civil Engr
Av. Isla de Margarita, Qta.Arigain
Cumbres de Curumo
CARACAS 108

10356 RADINA, A., Mr - Civil Engr
Urb. Santa Fe, Av. Jose Maria Vargas
Resd. Julio - Apto. 12
CARACAS 106

10357 * RAMIREZ C., C., Dr - Geologist
M.E.M. Div. de Geotecnia
Edf. Gran Via,
Mezz. Esq. Cruz Verde a Zamuro
CARACAS 101

10358 * RAMIREZ C., L., Mr - Civil Engr
Edf. Centro Andres Bello Of. 82-E
Avenida Andres Bello
CARACAS 105

10359 * RAMIREZ O., O., Mr - Civil Engr
Diaz Moreno, Edf. Don Pelayo
Piso 12 - Of. 6
Valencia
EDO. CARABOBO

10360 RANGEL J., T., Mr - Civil Engr
Av. 2-B, 1104
La Carlota
CARACAS 107

10361 * RENGEL F., J., Dr - Geologist
Av. Arturo Michelena, Qta. Rengel
Santa Monica
CARACAS 104

10362 * RIOS L., J., Mr - Civil Engr
Av. Rufino Blanco Fombona, Edf. Beta
Torre B, Apto. 12-C
Santa Monica
CARACAS 104

10363 * RIVERO M., H., Mr - Civil Engr
Edf. Caruachi, Apto. 14-B
Urb. Santa Sofia
CARACAS 106

10364 * ROBLES T., J., Dr - Geologist
Av. Berriveitia, Qta. Tabu
El Paraiso
CARACAS 102

10365 * RODRIGUEZ A., P., Mr - Civil Engr
Calle Corozo, Resd. Yurubi 2
1º Piso, Apto. 1-4
Colinas de Los Ruices 1
CARACAS 107

10366 * RODRIGUEZ C., O., Mr - Civil Engr
Av. Bolivar, 192
Ejido
EDO.MERIDA

10367 RODRIGUEZ D., N., Mr - Civil Engr
Av. Ppal. Las Mercedes
Edf. Irune - Of. 4
CARACAS 106

10368 RODRIGUEZ G., H., Mr - Civil Engr
Calle Auyantepuy, Qta. Limonero
\+ Chuao
CARACAS 106

10369 RODRIGUEZ T., M., Mr - Civil Engr
* Edf. Doral, PH,
4a Av. Los Palos Grandes
CARACAS 106

10370 ROMERO R., I., Mr - Civil Engr
* Av. Ppal. San Luis, Edf. Atlanta
Apto. 7-A
CARACAS 106

10371 ROMERO Z., H., Dr - Engng Geologist
Av. Libertador,Edf. Residencias Florida
7º Piso, Apto. 32
CARACAS 105

10372 ROSARIO, R., Mr - Civil Engr
* Av. Norte 23-1, Resd. Irene
1º Piso - No. 2, Mariperez
CARACAS 105

10373 SALAME R., L., Mr - Civil Engr
* Calle Sta. Sofia Sur, Qta. Yo no Se
Urb. Santa Sofia
CARACAS 106

10374 SALCEDO R., D., Dr - Engng Geologist
* Av. El Retiro y Boyaca, Qta. Orion
El Rosal
CARACAS 106

10375 SANANEZ, J., Mr - Civil Engr
* Av. Leonardo Da Vinci, Centro Comercial
Piso 4º, Letra F
Bello Monte
CARACAS 105

10376 SANCHO F., I., Mr - Civil Engr
* Calle Caroni, Resd. Jardin de las
Mercedes, 4-C
Colinas de Bello Monte
CARACAS 105

10377 SARDI S., V., Mr - Civil Engr
* Av. Las Palmas, Qta. Tibidabo
La Florida
CARACAS 105

10378 SARRIA H., L., Mr - Civil Engr
* Calle A-3, Qta. Pelusal
Urb. La Lagunita
El Hatillo
CARACAS 108

10379 SEGUELA F., J., Mr - Geod Engr
* Qta. Ana Maria, Calle El Arenal
Urb. La Trinidad
Baruta
CARACAS 108

10380 SELVI B., M., Mr - Civil Engr
* Calle Honduras, 16
Ejido
EDO. MERIDA

10381 SEMINARIO M., J., Mr - Civil Engr
Urb. San Luis, Calle Comercio
Edf. Tajamar, 10-4
CARACAS 106

10382 SEVA C., A., Mr - Civil Engr
* Calle C, Resd. Las Margaritas
Apto. 4-A
Santa Rosa de Lima
CARACAS 106

10383 SILVA C., M., Mr - Civil Engr
* Qta. Querecual, Av. Guatemala
Las Acacias
CARACAS 104

10384 SOTO V., A., Mr - Civil Engr
* Av. 11-B, 52-49
Urb. Cantaclaro
Maracaibo
ESTADO ZULIA

10385 STALHUTH, W., Mr - Civil Engr
* Calle Venezuela, 40
Qta. San Onofre
Terrazas del Club Hipico
CARACAS 108

10386 SUAREZ F., R., Mr - Civil Engr
* Urb. Montalban, Cruce de la 4 con
la 2
Resd. Olga, Piso 11, Apto. 111
CARACAS 102

10387 TAPIA G., M., Mr - Civil Engr
* la. Transv., 21-12
Urb. Monte Cristo
CARACAS 107

10388 TEPPA A., J., Mr - Civil Engr
* Qta. Mama-Diosa, Av. Lago Maracaibo
Cumbres de Curumo
CARACAS 108

10389 TINOCO A., F., Mr - Civil Engr
* Calle Canaima, la. Colina de Chuao
Quinta La Nonina
CARACAS 106

10390 TOMAT K., S., Dr - Engng Geologist
* Qta. Edelweiss, Calle San Luis
Urb. San Luis
El Cafetal
CARACAS 106

10391 TROCONIS, F., Mr - Civil Engr
* Misericordia a Miguelacho, 30
CARACAS 101

10392 UCAR N., R., Mr - Mining Engr
* Apartado 289 Merida
ESTADO MERIDA

10393 UNDA R., J., Mr - Civil Engr
* LAGOVEN - Apartado 889
CARACAS 101

10394 VALERO B., I., Mr - Civil Engr
* Av. Libertador frente Centro
Commercial Los Cedros
Edf. Samson, Apto. 8-B
CARACAS 105

10395 VALLE-RODAS, R., Mr - Civil Engr
* Apartado 60.105
CARACAS

10396 VARGAS P., J., Mr - Civil Engr
* Rivas 96, Los Teques
ESTADO MIRANDA

10397 VASQUEZ R., G., Dr - Geologist
* Av. Gral, Paez, Resd. Gral. Paez
Apto A-33
El Paraiso
CARACAS 102

10398 VEGAS S., J., Mr - Civil Engr
* Calle Taborda, Qta. Rosario
San Roman
CARACAS

10399 VELASCO G., J., Mr - Civil Engr
* Av. Orinoco, Qta. Praga
Las Mercedes
CARACAS 106

10400 VELASCO G., N., Mr - Civil Engr
* Av. Orinoco, Qta. Praga,
Las Mercedes
CARACAS 106

10401 VELASQUEZ M., L., Dr - Engng Geologist
Edf. Los Roques, Apto. 43
Parque Humboldt
Prados del Este
CARACAS 108

10402 VIDAL G., I., Mr - Civil Engr
* Edf. Vilma, Apto. 122
Calle Maria-Auxiadora
Parcelamiento Don Bosco
Los Dos Caminos
CARACAS 107

10403 VIGNIERI, L., Mr - Civil Engr
* Av. Paris con Calle Caroni
Edf. La Isla, Apto 1
Las Mercedes
CARACAS 106

10404 VILLALTA, D., Dr - Geologist
* Carrera 22, 12-125
\+ San Cristobal
ESTADO TACHIRA

10405 VILLAMIZAR P., C., Mr - Civil Engr
* Calle Santa Sofia Sur, 72
Urb. Santa Sofia
CARACAS 106

10406 WEISER B., C., Mr - Civil Engr
* Terrazas de Santa Ines
Resd. Carenero, Apto. 4-B
CARACAS 106

10407 YACKOVLEV, V., Mr - Civil Engr
* Apartado 51.566
CARACAS 105

10408 YANEZ F., M., Mr - Civil Engr
* Av. Rio Orinoco, Resd. Kioto
Apto. 1-B
Urb. Cumbres de Curumo
CARACAS 108

10409 YEPEZ C., F., Mr - Civil Engr
* Av. La Salle, Resd. Tony,
Apto. 24
Colinas de Los Caobos
CARACAS 105

10410 ZAGALA T., M., Mr - Civil Engr
* Av. Upata, Edif. Tiuna
Apto. "D"
Ciudad Bolivar
ESTADO BOLIVAR

CORRESPONDING MEMBERS

10202 PEREZ E., M., V., Mr - Civil Engr
2da. Avenida - Urb. Monte Cristo
Esq. 2da. Transversal Qta.
TENSAVEN (Fte a Ferreteria
Moreste)
CARACAS

YUGOSLAVIA

JUGOSLOVENSKO DRUSTVO ZA MEHANIKU STENA I PODZEMNE RADOVE

Institut za vodoprivredu "J. Cerni", Bulevar vojvode Misica 43, BEOGRAD

President: Prof. B. Kujundzic
Secretary: Mr O. Markovic

Total Number of Ordinary Members - 197
Total Number of Supporting Members - 2

ORDINARY MEMBERS

2975 ABRAMOVIC, V., Mr - Mining Engr
Pjerottijeva 6
41000 ZAGREB

2976 ACIMOVIC, M., Prof.-Civil Engr
Generala Zdanova 80/IV
11000 BEOGRAD

13192 ADAMIC, T., Mrs - Civil Engr
Ljubljanska 34
61000 LJUBLJANA

2977 AKSENTIJEVIC, D., Mr - Engng Geologist
Drugi bulevar 189/6
11070 NOVI BEOGRAD

13193 ANDJELKOVIC, V., Mr - Geophys. Engr
Rudarsko-geoloski fakultet
Djusina 7
11000 BEOGRAD

2979 ANTONIC, C., Mr - Civil Engr
Bul. Lenjina 3/84
11070 NOVI BEOGRAD

7452 BECIROVIC, H., Mr - Mining Engr
Rudarsko-geoloski fakultet
Rudarska 71
75000 TUZLA

7453 BEGIC, J., Mr - Mining Engr
Rudnici zeljezne rude
78307 LJUBIJA

13194 BENAC, C., Mr - Engng Geologist
"Luka"
Obala jug. mornarice 1
51000 RIJEKA

13195 BIZILJ, P., Mr - Civil Engr
Vlahoviceva 49
61000 LJUBLJANA

2982 BLAGOJEVIC, R., Mr - Engng Geologist
"Geosonda"
17. marta 97-99
11000 BEOGRAD

13196 BOGDANOVIC, H., Mr - Engng Geologist
Institut za ispitivanje materijala
SR Srbije, Bul. vojvode Misica 43
11000 BEOGRAD

7454 BOGOJEVSKI, T., Mr - Civil Engr
Arhitektonsko-gradezen fakultet
Rada Konacara 16
91000 SKOPLJE

13197 BOJANIC, P., Mr - Civil Engr
GP Konstruktor
Ujeviceva 4
58000 SPLIT

2983 BOSNIC, P., Mr - Civil Engr
Zagrebacka 24
71000 SARAJEVO

7455 BOZIC-FAK, D., Mrs - Mining Engr
Titovi rudnici "Kreka-Banovici"
75000 TUZLA

7456 BOZINOVIC, D., Dr - Engng Geologist
* Macvanska 20
\+ 11000 BEOGRAD

2986 BRACINAC, Z., Mr - Engng Geologist
\+ Bul. revolucije 250-11 ulaz
11000 BEOGRAD

2987 BRALIC, J., Mr - Mining Engr
Batajnicki put 2
11080 ZEMUN

7457 BUDIC, S., Mr - Electr. Engr
Rudarsko-geoloski fakultet
Rudarska 71
75000 TUZLA

13198 BUTUROVIC, R., Mr
Ul. Darovalaca krvi 9/111
71000 SARAJEVO

13199 CAGALJ, M., Mr - Civil Engr
Gradjevinski Institut
Janka Rakuse 1
41000 ZAGREB

2993 CALOGOVIC, M., Prof.- Civil Engr
Elektroprojekt
Proleterskih brigata 37
41000 ZAGREB

7461 CAPRIC, D., Mr - Civil Engr
Energoinvest - Higra biro
71000 SARAJEVO - Stup

7463 CAVAR, R., Mr - Mining Engr
RMK Kakanj - pogon Ricica
72240 KAKANJ

2992 CEROVIC, N., Mr - Civil Engr
Energoinvest - Higra biro
71000 SARAJEVO - Stup

2994 COLIC, B., Dr - Engng Geologist
Institut za vodoprivredu "J.Cerni"
Bul. vojvode Misica 43
11000 BEOGRAD

6311 CRNKOVIC, B., Dr - Engng Geologist
Potocnjakova ul. 14
41020 ZAGREB

2980 * CVETKOVIC, M., Dr - Mining Engr
Rudarsko-geoloski fakultet
Djusina 7
11000 BEOGRAD

3122 DEDIC, M., Mr - Civil Engr
Tirsova 14
11080 ZEMUN

13200 DUSI, M., Prof - Mining Engr
Rudarsko-metalurski fakultet
38220 KOSOVSKA MITROVICA

13201 DJORDJEVIC, V., Mr - Engng Geologist
Zavod za geoloska i geofizicka istrazivanjo
Karadjordjeva 48
11000 BEOGRAD

13202 ELEZOVIC, D., Mr - Mining Engr
Tehnicki skolski centar "B".Kidric
38220 KOSOVSKA MITROVICA

13203 FILIPOVIC, M., Mr - Civil Engr
Zavod za projektovanje ZJZ
Nemanjina 6
11000 BEOGRAD

3001 FILOSKI, P., Mr - Civil Engr
Dame Gruev 43
97300 OHRID

13204 FINGERHUT, L., Mr - Mining Engr
Geotehnika
Kupska 2
41000 ZAGREB

3002 FRITZ, F., Dr - Engng Geologist
Institut za geoloska istrazivanja SRH
Sachsova 2
41000 ZAGREB

7465 GAPKOVSKI, N., Mr - Engng Geologist
Institut za geotehnika
Gradezen fakultet
Rada Koncara 16
91000 SKOPJE

3004 GONSIOROVSKI, V., Mr - Civil Engr
V.O. "Velika Morava"
Kneza Milosa 9/V
11000 BEOGRAD

6315 GOPCEVIC, L., Mr - Engng Geologist
Direkcija srednjebosanskih rudnika
72240 KAKANJ

3125 GRUJIC, N., Mr - Geophys.Engr
Energoprojekt
Zeleni venac 18
11000 BEOGRAD

13205 HADZIC, E., Mr
Energoinvest - Higra biro
71000 SARAJEVO - Stup

7468 HALDA, L., Mr - Civil Engr
Zavod za projektovanje ZJZ
Nemanjina 6
11000 BEOGRAD

13206 HEKALO, M., Mr
Ivana Milutinovica 14.F
PLJEVLJA

3005 HERAKOVIC, D., Mr - Civil Engr
GP "Planum"
22. oktobra 15
11080 ZEMUN

7469 HLEBAR, V., Mr - Civil Engr
Save Kovacevica 47
79000 MOSTAR

7470 HODZIC, S., Mr - Mining Engr
Rudarski fakultet
Rudarska 71
75000 TUZLA

6318 IGNTIJEVIC, M., Mr - Geoph.Engr
Jove Ilica 56
11000 BEOGRAD

13207 ILIC, Lj., Mr - Mining Engr
Zavod za naucno istrazivanje i razvoj
R.E.H.K. "Kosovo"
38000 PRISTINA

3007 IVANOV, G., Mr - Engng Geologist
Ul. Mito Hadzivasilev kula 9 stan 35
91000 SKOPLJE

6319 IVANOVIC, K., Mr - Civil Engr
Institut za vodoprivredu "J.Cerni"
Bul. vojvode Misica 43
11000 BEOGRAD

6320 IVANOVIC, S., Mr - Engng Geologist
Bulevar 33 br. 5/60
81000 TITOGRAD

3008 IVOSEVIC, R., Mr - Civil Engr
GP "Planum"
22 oktobra 15
11080 ZEMUN

3009 JAJCANIN, S., Mr - Civil Engr
Rudarsko-geoloski fakultet
Rudarska 71
75000 TUZLA

7474 JAKOVAC, L., Mrs - Mining Engr
Rudarsko-geoloski fakultet
Rudarska 71
75000 TUZLA

7475 JAKOVLJEVIC, T., Mr - Mining Engr
Rudarsko-metalurski kombinat Zenica
Rudnik zeljezne rude
78307 LJUBIJA

13208 * JANEZIC, S., Mr - Civil Engr
Inzenjerski biro Elektroprojekt
Hajdrihova 2
61000 LJUBLJANA

3011 + JANJIC, M., Prof - Engng Geologist
Tolstojeva 5
11000 BEOGRAD

3012 * JASAREVIC, I., Dr - Civil Engr
Janka Rakuse 1
41000 ZAGREB

6322 JOVANOVIC, B., Mr - Engng Geologist
Lipetska 16
11000 BEOGRAD

406 JOVANOVIC, L., Prof - Civil Engr
Institut za vodoprivredu "J.Cerni"
Bul. vojvode Misica 43
11000 BEOGRAD

3014 * JOVANOVIC, R., Prof - Geologist
Zavod za inz.geol. i hidrogeologiju
Gradjevinskog fakulteta
Stjepana Tomica 3
71000 SARAJEVO

6323 + JOVANOVIC, V., Mr - Geophys.Engr
Ivan Begova 5
11000 BEOGRAD

13209 + JURAK, V., Mr
Rudarsko-geolosko-naftni fakultet
Pierottijeva 6
41000 ZAGREB

3123 KATALINIC, I., Mr - Civil Engr
Skopljanska 2
71000 SARAJEVO

3016 KEKIC, A., Dr - Engng Geologist
Ul. 254 br. 6
91000 SKOPLJE

6325 KIKLIC, D., Mrs-Engng Geologist
NH Zdravka Celara 4
11000 BEOGRAD

13210 KOBLISKA-ANTUNOVIC, M., Prof - Mining Engr
Ljube Stojanovic 33
11000 BEOGRAD

3018 KOCAR, F., Mr - Mining Engr
Rudnik mrkog uglja
Sektor za istrazivanje i razvoj
Save Kovacevica 1
72000 ZENICA

3019 KOLAREVIC, R., Mr - Engng Geologist
Kosovska 1
11000 BEOGRAD

405 * KUJUNDZIC, B., Prof - Civil Engr
Institut za vodoprivredu "J.Cerni"
Bul. vojvode Misica 43
11000 BEOGRAD

3036 * LANGOF, Z., Mr - Civil Engr
Zavod za geotehniku i fundiranje
Gradjevinskog fakulteta
Stjepana Tomica 3
71000 SARAJEVO

3025 LANGOVIC, M., Mr - Engng Geologist
Geosonda
27.marta 97-99
11000 BEOGRAD

3037 * LAZAREVIC, D., Prof - Civil Engr
Bul. revolucije 241/34
11000 BEOGRAD

3038 LAZIC, C., Mr - Civil Engr
Institut za vodoprivredu "J.Cerni"
Bul. vojvode Misica 43
11000 BEOGRAD

3041 LESEVIC, Z., Mr - Engng Geologist
Energoprojekt
Zeleni venac 18
11000 BEOGRAD

3039 LISENKO, N., Mr - Technology Engr
Trg francuske republike 6/11
41000 ZAGREB

6326 LOKIN, P., Dr - Engng Geologist
Rudarsko-geoloski fakultet
Djusina 7
11000 BEOGRAD

3042 MAGDALENIC, A., Mr - Engng Geologist
Rudarsko-geolosko-anftni fakultet
Pierottijeva 6
41000 ZAGREB

13211 * MAKSIMOVIC, M., Dr - Civil Engr
Energoprojekt
Zeleni venac 18
11000 BEOGRAD

3044 MANEV, S., Mr - Civil Engr
Ilindenska 17
97300 OHRID

7482 MANDZIC, E., Mr - Mining Engr
*
Rudarsko-geoloski fakultet
Rudarska 71
75000 TUZLA

13212 MANOJLOVIC, M., Mr - Civil Engr
Institut "J.Cerni"
Bul. vojvode Misica 43
11000 BEOGRAD

13213 MARIN, M., Mrs - Mining Engr
Sarhova 32
61000 LJUBLJANA

3045 MARINKOVIC, P., Mr - Engng Geologist
Kosovoprojekt
Zvecanska 1
11000 BEOGRAD

3047 MARKOVIC, O., Mr - Civil Engr
Institut za vodoprivedu "J.Cerni"
Bul. vojvode Misica 43
11000 BEOGRAD

13214 MEHMETI, B., Mr - Mining Engr
Rudnik "Trepca" sa flotacijom
38226 STARI TRG

6327 MIHAJLOVIC, J., Mr - Mining Engr
Mose Pijade 19/38
19210 BOR

3048 MIHAJLOVIC, S., Prof - Civil Engr
Sv. Save 13
11000 BEOGRAD

3049 MIKLAVCIC, E., Mr
Novigradska 7
41000 ZAGREB

7484 MILADINOV, D., Prof - Civil Engr
* Institut za geotehnika
Gradezen fakultet
Rade Koncar 16
91000 SKOPLJE

13216 MILANOVIC, P., Dr - Mining Engr
Takovska 56
11000 BEOGRAD

13215 MILISIC, M., Mr - Civil Engr
Mose Pijade 11/XV
71000 SARAJEVO

7486 MILJANOVIC, T., Mr - Civil Engr
GP Konstruktor
Ujeviceva 4
58000 SPLIT

6330 MILJKOVIC, E., Mr - Civil Engr
Energoinvest Hidrogradjevinski biro
71000 SARAJEVO - Stup

6331 MILKOVIC, D., Mr - Engng Geologist
\+ Zavod za projektovanje ZJZ
Nemanjina 6
11000 BEOGRAD

3053 MILOSEVIC, M., Mr - Civil Engr
II Bulebar 199 - stan 19
11070 NOVI BEOGRAD

3054 MILOVANOVIC, D., Prof - Civil Engr
Bul. oktobarske revolucije 61
11000 BEOGRAD

3055 MITROV, T., Prof - Engng Geologist
Gradjevinski fakultet
Rade Koncar 16
91000 SKOPLJE

13217 MITROVIC, L., Mr - Engng Geologist
Alekse Nenadovica 12-14
11000 BEOGRAD

3058 MOLNAR-SUSIC, J., Mrs - Engng Geologist
\+ Siget 7/11
41000 ZAGREB

6332 MRDAK, L., Mr - Engng Geologist
Sekcija za odrzavanje pruge
Bul. vojvode Misica 6
11000 BEOGRAD

13218 NAJDANOVIC, N., Prof - Civil Engr
* Prote Mateja 61
11000 BEOGRAD

3064 NEDELJKOVSKI, A., Mr - Mining Engr
Cocetova kula 3/15
19210 BOR

3060 NENADIC, N., Mr - Civil Engr
Elektroprivreda BiH
OOUR Zaj. Sluzbe Trebinje
79300 TREBINJE

3062 NIKOLIC, R., Mr - Civil Engr
Institut "Geoexpert"
Koturaska c.47
41000 ZAGREB

3063 NIKOLIC, S., Mr - Civil Engr
"Planum"
22.oktobar 15
11080 ZEMUN

13219 NIKOLIC, Z., Mr - Civil Engr
Energoprojekt
Zeleni venac 18
11000 BEOGRAD

128 NONVEILLER, E., Prof - Civil Engr
* N. Grskovica 28
41000 ZAGREB

13231 NUSRET, S., Mr - Civil Engr
GP Konstruktor
Ujeviceva 4
58000 SPLIT

13220 OBRADOVIC, J., Dr - Engng Geologist
Energoprojekt
Zeleni venac 18
11000 BEOGRAD

13221 OPACIC, D., Mrs - Civil Engr
Zavod za projektovanje ZJZ
Nemanjina 6
11000 BEOGRAD

13222 ORANA, D., Mr - Mining Engr
\+ Tehnicki fakultet
38000 PRISTINA

13223 OSMANAGIC, M., Prof - Mining Engr
Zajedinica za visoko obrazovanje
Vase Miskina 18
71000 SARAJEVO

7492 OSTOJIC, A., Mr - Civil Engr
"Industroprojekt"
Savska 88a
41000 ZAGREB

3069 OSTOJIC, S., Mr - Engng Geologist
\+ Energoprojekt
Zeleni Venac 18
11000 BEOGRAD

3070 PAVLIN, B., Mr - Civil Engr
\+ Elektroprojekt
Proleterskih brigada 37
41000 ZAGREB

3071 PAVLOVIC, M., Mr - Civil Engr
Energoinvest - Higra Biro
71000 SARAJEVO - STUP

7495 PEHLIC, O., Dr - Mining Engr
Institut za rudarska i hemijsko-
tehnoloska
Istrazivanja
75000 TUZLA

6336 PEROVIC, L., Mr
Energoprojekt
Njegoseva 84
11000 BEOGRAD

7499 PETKOVIC, T., Mr - Civil Engr
Dobri Dol 56/VIII
41000 ZAGREB

3024 PETROVIC, D., Mr - Civil Engr
Sarplaninska 3
11000 BEOGRAD

7496 PETROVIC, D., Mrs - Mining Engr
Institut za rudarska i hemijsko-
tehnoloska istrazivanja
75000 TUZLA

3074 PETROVIC, M., Mrs-Engng Geologist
Sumarski fakultet
Kneza Viseslava 1
11000 BEOGRAD

3075 PETROVIC, S., Mr - Civil Engr
Ul. Nikola Parapunov 96
91230 GOSTIVAR

7466 PETROVIC, Z., Mr - Engng Geologist
REIK "Kolubara"
Inv. Grupa "Tamnava"
14230 VRFOCI

13226 PETZEL, M., Mr - Mining Engr
Rudarsko-geolosko-naftni fakultet
Pierottieva 6
41000 ZAGREB

7493 PIRIC, N., Mrs - Mining Engr
Institut za rudarska i hemijsko-
tehnoloska istrazivanja
75000 TUZLA

7494 PJANIC, R., Mr - Mining Engr
Institut za rudarska i hemijsko-
tehnoloska istrazivanja
75000 TUZLA

6337 POPOVIC, B., Mr - Civil Engr
Dzordza Vasingtona 36
11000 BEOGRAD

13224 POPOVIC, M., Mr - Civil Engr
* Zavod za geotehniku i fundiranje
Gradevinskog fakulteta u Sarajevu
Stjepana Tomica 3
71000 SARAJEVO

7691 POPOVIC, R., Mr - Mining Engr
Institut za rudarska i hemijsko-
tehnoloska istrazivanja
75000 TUZLA

13225 POPOVIC, S., Mr - Civil Engr
Zavod za projektovanje ZJZ
Nemanjina 6
11000 BEOGRAD

3078 PROKOPOVIC, S., Mr - Mining Engr
* 11 Armije 32
41040 ZAGREB

7501 PRVULOVIC, J., Mr - Geophy.Engr
Energoprojekt
Zeleni Venac 18
11000 BEOGRAD

3079 RADOJEVIC, J., Mr - Mining Engr
Rudarski institut
Batajnicki put 2
11080 ZEMUN

7502 RADOLOVIC, J., Mr - Civil Engr
Gradevinski fakultet
Kaciceva 26
41000 ZAGREB

3080 RADOSAVLJEVIC, Z., Prof.- Civil Engr
Neznanog junaka 33
11000 BEOGRAD

13227 RADULOVIC, M., Mr - Engng Geologist
Geoloski zavod
81000 TITOGRAD

3081 RADULOVIC, V., Dr - Engng Geologist
+ Dure Salaja 9/a
81000 TITOGRAD

3082 RAJCEVIC, B., Dr - Civil Engr
* Kopitarera Gradina I/IV
11000 BEOGRAD

3083 RAKIC, A., Mr - Physicist
Bul. AVNOJ-a 138
11070 NOVI BEOGRAD

3084 RAKIC, R., Mr - Engng Geologist
Energoprojekt
Zeleni venac 18
11000 BEOGRAD

6338 RASEVIC, M., Mr - Engng Geologist
Visegradska 5
11000 BEOGRAD

6339 REDZEPAGIC, M., Dr - Mining Engr
* Rudarsko-geoloski fakultet
Rudarska 71
75000 TUZLA

3085 ROKIC, D., Mr - Civil Engr
Kernedzijeva 3
11000 BEOGRAD

6340 ROKIC, L., Dr - Engng Geologist
+ Zavod za inzenjersku geologiju i
hidrogeologiju Gradevinskog fakulteta
Stjepana Tomica 3
71000 SARAJEVO

3106 SALOVIC, M., Mr - Mining Engr
Rudarski institut
75000 TUZLA

3085 SAPAC, F., Mr - Mining Engr
Tuskanova 38
41000 ZAGREB

13228 SARAC, Dz., Prof - Civil Engr
* Zavod za geotehniku i fundiranje
Gradevinskog fakulteta
Stjepana Tomica 3
71000 SARAJEVO

13229 SARAC, E., Mr - Civil Engr
Ul. Apikovac 28/a
71000 SARAJEVO

3087 SELIMOVIC, M., Mr - Civil Engr
Petra Drapsina 8/a
79000 MOSTAR

7513 SENDOV, B., Prof - Civil Engr
Arhitektonsko-gradezen fakultet
Rade Koncara 16
91000 SKOPJE

3089 SENICIC, M., Mr - Engng Geologist
Nemanjina 5/11
11000 BEOGRAD

7514 SICAROV, S., Mr - Engng Geologist
Energoinvest, Higra-biro
71000 SARAJEVO - STUP

3107 SIKOVEC, R., Mr - Mining Engr
Zasavski premogovniki
Trg Revolucije 12
61420 TRBOVLJE

13230 SIMIC, M., Mr - Civil Engr
Energoinvest
Hidrogradevinski biro
71000 SARAJEVO - STUP

3091 SLIMAK, S., Dr - Geophys.Engr
Energoprojekt
Zeleni venac 18
11000 BEOGRAD

3092 SMIRNOV, V., Mr - Civil Engr
Solina 4
11000 BEOGRAD

3093 SOBOTKA, D., Mr - Civil Engr
Dobracina 21
11000 BEOGRAD

3095 SOVINC, I., Prof - Civil Engr
* Laboratorij za mehaniko tal
Lepi pot 11
61000 LJUBLJANA

13234 SPANOVIC, S., Mr - Engng Geologist
Geosonda
27. marta 97-99
11000 BEOGRAD

3096 SPEVEC, I., Mr - Civil Engr
Vraniceva 25/b
41000 ZAGREB

3097 SRDIC, R., Mrs. - Engng Geologist
Vuka Karadzica 121
71000 SARAJEVO

6344 STANOJKOVSKI, B., Mr - Mining Engr
Thyssen Schachtbau GmbH
433 Mulheim/Ruhr
Friedrich Ebert Strasse 120
(BRD)

3100 STEFANOVSKI, P., Mr - Civil Engr
Ul. 488 br. 12
91000 SKOPJE

13232 STEVIC, M., Mr - Mining Engr
Rudarsko-geoloski fakultet
Rudarska 71
65000 TUZLA

3126 STOJANOVIC, S., Mr - Engng Geologist
Institut za ispitivanje materijala
SRS
Bulevar Vojvode Misica 43
11000 BEOGRAD

3102 STOJIC, P., Mr - Civil Engr
Hidroelektrane na Trebisnjici
79300 TREBINJE

13233 STOJMIROVIC, B., Mr - Civil Engr
Svetozara Markovica 7/11
11000 BEOGRAD

3104 SUBANOVIC, B., Prof - Civil Engr
Pere Todorovica 10
11000 BEOGRAD

6345 * SUKOVIC, M., Mrs - Engng Geologist
Narodnih heroja 25/V stan 19
11070 NOVI BEOGRAD

7511 SULJKANOVIC, M., Mr - Mining Engr
Rudarsko-geoloski fakultet
Rudarska 71
65000 TUZLA

7518 TATIC, M., Mr - Mining Engr
Rudarsko-geoloski fakultet
Rudarska 71
75000 TUZLA

7520 TINJIC, S., Mr - Mining Engr
Kombinat "Soda-So"
Rudnik Tusanj
75000 TUZLA

6346 TISA, A., Mr - Civil Engr
Seebacherstrasse 102
8052 ZURICH (SCHWEIZ)

7522 TODOROVIC, B., Mr - Engng Geologist
Rudnik "Trepca" sa flotacijom
38226 STARI TRG

3108 + TODOROVSKI, A., Mr - Civil Engr
Hidroelektroprojekt
Nikole Vapcarova 9
91000 SKOPJE

13235 TOMAS, M., Mr - Engng Geologist
Institut za vodoprivredu
"Jaroslav Cerni"
Bulevar Vojvode Misica 43
11000 BEOGRAD

6349 * TORBAROV, V., Mrs - Engng Geologist
Stjepana Tomica 21
71000 SARAJEVO

6348 TORBAROV, K., Dr - Engng Geologist
Zavod za inz. geologiju
Stjepana Tomica 3
71000 SARAJEVO

7692 UZNIK, R., Mr
Strojna tovarna Trbovlje
Vodenjska c.49
61210 TRBOVLJE

3111 VAVPOTIC, K., Mr
Majernova 17
61000 LJUBLJANA

13236 VENTURINI, S., Mr
Lepi pot 23
61000 LJUBLJANA

13237 VESELINOVIC, M., Mr
RZR "Ljubija"
Geoloska sluzba
78300 PRIJEDOR

3113 * VIDIC, F., Mr - Civil Engr
Milcinskega 77
61000 LJUBLJANA

3114 * VINCEK, R., Mr - Civil Engr
Markusevecka c.51
41000 ZAGREB

7527 * VISIC, I., Mr - Civil Engr
Gradevinski institut
OOUR za geotehniku
Janka Rakuse 1
41000 ZAGREB

7528 VITANOV, V., Mr - Civil Engr
Arhitektonsko-gradezen fakultet
Rade Koncara 16
91000 SKOPJE

7529 VRKLJAN, I., Mr - Mining Engr
Rudarsko-geoloski fakultet
Rudarska 71
75000 TUZLA

3116 VUCKOVIC, M., Mr - Civil Engr
Zavod za projektovanje ZJZ
Nemanjina 6
11000 BEOGRAD

3117 VUJEC, S., Prof - Mining Engr
Rudarsko-geolosko-naftni fakultet
Pierrottijeva 6
41000 ZAGREB

13238 VUJOVIC, S., Mr - Civil Engr
Zavod za projektovanje ZJZ
Nemanjina 6
11000 BEOGRAD

3118 VUKCEVIC, V., Mr - Engng Geologist
Preduzece "Jugofund"
Bulevar Lenjina 183
11000 BEOGRAD

13239 ZELEZNIK, S., Mr - Mining Engr
Streliska 22
61000 LJUBLJANA

13240 ZUGIC, O., Mrs
Ul. Fuada Midzica 21
71000 SARAJEVO

SUPPORTING MEMBERS

190 ELEKTROPROJEKT
Proleterskih Brigada 37
ZAGREB

162 INSTITUT ZA VODOPRIVREDU "JAROSLAV CERNI"
Bulevar Vojvode Misica 43
BEOGRAD

ZIMBABWE

Total Number of Corresponding Members - 2

CORRESPONDING MEMBERS

7175 HESLOP, T.G.
5 Russel Drive
Kumalo
BULAWAYO

9015 TAYLOR, H.W.
3, Chinda Heights
SHABANI

STATUTES OF ISRM
Approved by the Council of ISRM
September 26 1978

1 - NAME AND HEADQUARTERS

1.1 - Name

The official name of this Organisation is "International Society for Rock Mechanics", in French "Societe Internationale de Mecanique des Roches", and in German "Internationale Gesellschaft für Felsmechanik". Throughout these statutes, the abbreviation "Society" shall be used.

1.2 - Headquarters

The official headquarters of the Society shall be the seat of the Society's Secretariat, as recommended by the Board and approved by the Council.

2 - OBJECTIVES AND TYPE OF SOCIETY

2.1 - Objects

The objects and purposes of the Society are:

- to encourage and coordinate international cooperation in the science of Rock Mechanics,
- to further international cooperation in Rock Mechanics by organizing international meetings,
- to encourage the international publication of the results of scientific investigations in Rock Mechanics,
- to encourage individuals and local organizations to form National Groups solely or primarily concerned with Rock Mechanics,
- to maintain liaison with other bodies representing fields related to Rock Mechanics, including Geology, Geophysics, Soil Mechanics, Mining Engineering, Petroleum Engineering and Civil Engineering,
- to coordinate meetings concerned with Rock Mechanics and related fields to minimize duplication and conflict, and
- to encourage the teaching of Rock Mechanics.

2.2 - Type of Society

The Society shall be a non-profit Organization, supported by the fees of the Members and unrestricting grants. No Officer or Member shall receive any compensation from Society funds, except the Secretary-General. Support of the Secretariat and all other expenses shall be included in the budget of the Society.

3 - LANGUAGES

3.1 - Official languages

English, French and German shall be the official languages of the Society. All communications from the Secretariat shall be in English and also in French and/or German if appropriate.

3.2 - Other languages

For Congresses and International Symposia of the Society, the language of the host country may also be used for oral presentations, only if translation into an official language of the Society is provided and paid for by the hosting National Group.

3.3 - Authoritative texts

The English text of every statutory document of the Society shall be the sole authoritative version.

4 - MEMBERSHIP

4.1 - Summary

The Society shall be composed of National Groups, Supporting Members and Corresponding Members.

4.2 - Countries

As used in these statutes, the word "country" shall include nations, and dependencies which budget independently for scientific purposes, as well as any multination groups approved by the Board.

4.3 - National Groups

A country shall be represented in the Society by a National Group. This shall be a society or a committee that represents Rock Mechanics, the statutes of which are not inconsistent with those of the Society.

4.4 - Membership details

Membership of the Society shall consist of Ordinary Members through National Groups, Supporting Members, and Corresponding Members.

Ordinary Members are individuals who pay fees to the Society through a National Group.

Supporting Members are societies, associations, companies, and other collective bodies, which are interested in stimulating and supporting scientific progress, and pay fees to the Society either through a National Group or by direct remittance to the Secretariat.

Corresponding Members are individuals who pay fees directly to the Secretariat.

4.5 - Admission

The admission of National Groups, Supporting Members and Corresponding Members shall be decided upon by the Board. The admission of Ordinary Members shall be decided upon by the appropriate National Groups. The Board may also consider the application of Multination Groups. Candidates can appeal to the Council the decision of the Board. The Secretariat shall notify the National Group of any Supporting Member that has submitted a fee directly to the Secretariat.

Admission becomes effective on payment of the first fee.

4.6 - Fees

The annual fees for National Groups, Supporting Members and Corresponding Members shall be based upon a scale of fees revised and approved by the Council from time to time.

The fees for each year are due on January 1. Fees unpaid after April 1 are in arrears.

4.7 Loss of membership

Membership of National Groups, Supporting Members and Corresponding Members may be terminated by the Board. Appeal may be made to the Council.

Termination of Ordinary Membership is reserved to the appropriate National Group.

Termination of membership occurs on the second consecutive April 1 after a National Group or Member fails to pay the annual fee.

5 - ORGANIZATION

5.1 - General

The Society shall accomplish its objectives by means of a Council, a Board and a

Secretariat. These bodies shall be constituted and act as described below.

5.2 - The Board

5.2.1 - The Board shall be composed of the President, the Vice-Presidents, and the Secretary-General. The Secretary-General shall have a voice but not a vote.

5.2.2 - The Board shall have the following duties:

- to decide upon the admission and termination of membership of National Groups, Supporting Members and Corresponding Members,
- to supervise all Society financial matters,
- to act on Society matters when such action becomes necessary between meetings of the Council,
- to advise on all the details of organization of the Congress of the Society,
- to approve the scope, time and location of the International Symposia of the Society,
- to prepare the agenda for Council meetings, and
- to accept grants on behalf of the Society.

5.2.3 The Board decides by a simple majority of those present.

5.2.4 The Board may decide by mail ballot in cases of urgency.

5.2.5 All actions of the Board may be subject to ratification by the Council.

5.3 - The Council

5.3.1 The Council shall consist of the Board, the Past-Presidents of the Society, and one Representative for each National Group.

5.3.2 The Council shall have the following duties:

- to elect the President and Vice-Presidents of the Society,
- to establish the policies of the Society, including technical, administrative and financial matters,
- to decide upon the scope, time, and location of the Congress of the Society, and
- to approve yearly budget statements and closures of yearly accounts.

5.3.3 - The Council decides by a simple majority vote of National-Group Representatives and Proxies, except as stipulated in 10.1 and 11.2 below.

5.3.4 - Each National Group shall have one Representative with one vote. A Representative will have a vote only if all fees for his National Group have been paid. Each National Group may appoint an Alternative to serve on the Council in the absence of the Representative.

5.3.5 - The members of the Board as such and the Past Presidents as such shall have a voice but not a vote, except for the President, who shall vote in the case of a tie ballot (See 6.2 below).

5.3.6 - The Council may decide by mail ballot on matters requiring urgent action.

5.4 - The Secretariat

5.4.1 - The Secretariat shall consist of the Secretary-General appointed by the President with approval by the Board, and administrative and clerical personnel.

5.4.2 - Duties of the Secretariat shall be determined by the President with approval by the Board, and shall include preparation of the annual budget and financial statements.

6 - OFFICERS

6.1 - Officers

The Officers of the Society shall be the President, the Secretary-General, and one Vice-President for each of the following geographic areas: Africa, Asia, Australasia, Europe, North America, and South America.

Vice-Presidents shall be resident in their geographic areas and shall not be resident in the country of the President.

6.2 - The President

The President's authorities and duties shall be:

- to administer the Society,
- to preside at the meetings of the Society,
- to cast the vote necessary to decide a tie ballot,
- to appoint the Secretary-General, and Commissions and Committees with approval by the Board, and
- to officially represent the Society on international matters of Society concern, or designate a Society representative for such matters whenever necessary.

6.3 - The Vice-Presidents

The Vice-Presidents shall be responsible for coordinating activities of common interest in their geographic areas.

6.4 - The First Vice-President

One of the six Vice-Presidents shall be elected by the Board as the First Vice-President.

The First Vice-President shall temporarily assume the duties of the President during any period the elected President is unable to perform these duties. If the President resigns or is deceased, the First Vice-President shall become the President until the next regular election of the Society Officers. In these events a new First Vice-President shall be elected by the Board within ninety days.

6.5 - The Secretary-General

The Secretary-General shall direct the Secretariat. He shall be responsible to the President.

6.6 - Elections

6.6.1 - All Officers, except the Secretary-General, shall be elected at the meeting of the Council held at the time of each Congress of the Society.

6.6.2 - The term of office of elected Officers shall be from the conclusion of the last session of one Congress of the Society until the conclusion of the last session of the following Congress.

6.6.3 - A nominee for office shall be a Member of the Society and he shall be proposed by a National Group.

6.7 - Appointments

6.7.1 - The Secretary-General shall be appointed by the President with approval by the Board.

6.7.2 - If a Vice-President fails to fulfil the above conditions of his appointment, he shall appoint a replacement for the balance of the term.

7 - MEETINGS

7.1 - The Congress of the Society

7.1.1 - The Congress of the Society shall be concerned with general themes of interest to the majority of the membership. A Congress shall be held at intervals of about four years, but the time between two consecutive Congresses shall not exceed five years.

7.1.2 - The scope, time, and location of each Congress shall be decided by the Council at the preceding Congress.

7.1.3 - The Board shall act as the Advisory Committee to each Congress, all the details of organization being submitted for the Board's approval.

7.2 - International Symposia

7.2.1 - The Society shall maintain a program of International Symposia concerned with specialized themes. International Symposia, preferably to be held once a year, shall not convene in the year of a Congress.

7.2.2 - The Board shall approve the scope, time, and location of each International Symposium.

7.3 - Council meetings

7.3.1 - The Council shall meet at the time of the Congress and at other times as the Council decides. In addition, at least thirty per cent of the Council members can request the President to call a meeting of the Council.

7.3.2 - A Council meeting shall have a quorum if more than fifty per cent of Representatives and Proxies are present. No Council member shall hold more than one proxy.

7.4 - Board meetings

7.4.1 - The President or any four Vice-Presidents may call a meeting of the Board at any time and specify the meeting location and agenda.

7.4.2 - Four members of the Board constitute a quorum; the Secretary-General is not counted.

7.5 - Meeting costs

7.5.1 - The direct expenses of conducting a Congress or an International Symposium shall be the responsibility of the National Group organizing the meeting.

7.5.2 - When meetings of the Council, the Board or any Commission of the Society are held at the time and place of a Congress or an International Symposium, the cost of such Council, Board and Commission meetings shall be bourne by the host organization.

7.5.3 - The expenses incurred by Officers in attending Board and Council meetings may be supported by the Society.

8 - COMMISSIONS

8.1 - As provided in 6.2 or as requested by the Council, the President may appoint Commissions for the purpose of studying specific scientific, technological or organizational problems of concern to the Society. Commission members need not necessarily be Members of the Society.

8.2 - The duties and the time period of any Commission of the Society shall be specified by the President with approval by the Board. A Commission may receive financial support to an extent decided by the Board, and based on a budget submitted by the Chairman of the Commission.

9 - PUBLICATIONS

9.1 - The Council shall establish policy concerning publication of the Society.

10 - DISSOLUTION

10.1 - The dissolution or liquidation of the Society can be effected only by a resolution passed by at least a two-thirds majority of all National Groups.

10.2 - After liquidation of the Society is approved, the Council may decide on the distribution of the Society's assets. Such assets, however, may be disposed of only to non-profit organizations the primary interests of which are the advancement of scientific knowledge.

11 - GENERAL

11.1 - Fiscal Year

The fiscal year of the Society shall be January 1 through December 31.

11.2 - Amendments to Statutes

These statutes shall be amended only on the motion of National Groups or Officers of the Society.

Any amendments shall require a two-thirds majority vote of Representatives and Proxies at a Council meeting.

11.3 - Implementation of Statutes

These Statutes shall be implemented by By-Laws approved by the Council.

SOCIETE INTERNATIONALE DE MECANIQUE DES ROCHES

STATUTS

1 - NOM ET SIEGE

1.1 - Nom

Le nom officiel de l'organisme est "Société internationale de mécanique des roches", en anglais "International Society for Rock Mechanics", et en allemand "Internationale Gesellschaft für Felsmechanik". Dans le cours des statuts, le nom abrégé employé est "Société".

1.2 - Siège

Le siège officiel de la Société est le siège du secrétariat de la Société, recommandé par le bureau, et approuvé par l'assemblée générale.

2 - OBJECTIFS ET TYPE DE LA SOCIETE

2.1 - Objectifs

Les objectifs et buts de la Société sont:

- encourager et coordonner la coopération internationale dans la science de la mécanique des roches,
- accroitre la coopération internationale en mécanique des roches en organisant des réunions internationales,
- encourager la publication au niveau international des résultats de la recherche scientifique en mécanique des roches,
- maintenir la liaison avec d'autres organismes consacrés à des domaines voisins de la mécanique des roches, y compris la géologie, la géophysique, la mécanique des sols, le génie minier, le génie pétrolier et le génie civil, et coordonner les réunions consacrées à la mécanique des roches et aux domaines voisins, afin de réduire au minimum les doubles emplois et les conflits.
- encourager l'enseignement de la mécanique des roches.

2.2 - Type de société

La Société est un organisme sans but lucratif, qui a pour ressources les cotisations de ses membres et les dons et legs sans restriction d'emploi. Aucun responsable ni membre ne reçoit d'indemnité sur les fonds de la Société, excepté le secrétaire général. La charge du secrétariat et toutes les autres dépenses font partie du budget de la Société.

3 - LANGUES

3.1 - Langues officielles

L'anglais, le francais et l'allemand sont les langues officielles de la Société. Toutes les communications émanant du secrétariat sont en anglais et aussi, s'il y a lieu, en français et/ou en allemand.

3.2 - Autres langues

Dans les congrès et symposiums internationaux de la Société, la langue du pays hôte peut être employée pour la présentation orale, à condition que la traduction dans une langue officielle de la Société soit fournie à ses frais par le groupe national hôte.

3.3 - Texte de référence

Le texte anglais de tout document statutaire de la Société est celui qui fait autorité.

4 - COMPOSITION ET PARTICIPATION

4.1 - Résumé

La Société est composée de groupes nationaux, de membres bienfaiteurs et de membres correspondants.

4.2 - Pays

Le mot "pays" tel qu'il est employé dans ces statuts désigne les nations et les territoires dotés d'un budget indépendant dans le domaine scientifique, ainsi que des groupes de plusieurs nations approuvés par le bureau.

4.3 - Groupes nationaux

Un pays est représente dans la Société par un groupe national. Celui-ci est une association ou un comité qui représente la mécanique des roches et dont les statuts ne contredisent pas ceux de la Société.

4.4 - Affiliation et catégories de membres

Les membres de la Société sont les membres ordinaires, affiliés par l'intermédiare des groupes nationaux, les membres bienfaiteurs et les membres correspondants.

Les membres ordinaires sont les personnes qui paient leur cotisation à la Société par l'intermédiaire d'un groupe national.

Les membres bienfaiteurs sont des sociétés, associations, compagnies et autres personnes morales décidées à stimuler et aider le progrès scientifique, qui paient une cotisation à la Société, soit par l'intermédiaire d'un groupe national, soit directement au secrétariat.

Les membres correspondants sont des personnes qui paient leurs cotisations directement au secrétariat.

4.5 - Admission

L'admission des groupes nationaux, des membres bienfaiteurs et des membres correspondants est décidée par le bureau. L'admission des membres ordinaires est décidée par leurs groupes nationaux. Le bureau peut aussi prendre en considération la candidature de groupes multinationaux. Les candidats peuvent faire appel de la décision du bureau devant l'assemblée générale. Le secrétariat notifie à chaque groupe national les membres bienfaiteurs de son pays qui paient leur cotisation directement à la Société.

L'admission prend effet au paiement de la première cotisation.

4.6 - Cotisations

Les cotisations annuelles des groupes nationaux, des membres bienfaiteurs et des membres correspondants sont fixées par un tableau approuvé et périodiquement révisé par l'assemblée générale.

Les cotisations de chaque année sont exigibles le ler janvier. Les cotisations non payées après le ler avril sont des arriérés.

4.7 - Radiation

La radiation des groupes nationaux, des membres bienfaiteurs et des membres correspondants peut être décidée par le bureau. Ces membres peuvent faire appel de la décision du bureau devant l'assemblée général.

La radiation d'un membre ordinaire est du domaine de son groupe national.

La radiation d'un groupe national ou d'un membre est automatique le ler avril en cas de non paiement de la cotisation pendant deux années consécutives.

5 - ORGANISATION

5.1 - Généralités

La Société, pour atteindre ses objectifs, met en place un bureau, une assemblée générale et un secrétariat. Ces organes sont constitués et agissent comme il est indiqué ci-dessous.

5.2 - Bureau

5.2.1 - Le Bureau est composé du président, des vice-présidents et du secrétaire général. Le secrétaire général a voix délibérative sans droit de vote.

5.2.2 - Les tâches du bureau sont les suivantes:

- décider l'admission et la radiation des groupes nationaux, des membres bienfaiteurs et des membres correspondants.
- superviser toutes les activités financières de la Société
- prendre toute initiative nécessaire dans l'intervalle des assemblées générales,
- veiller à tous les détails de l'organisation des congrès de la Société,
- approuver les sujets, dates et lieux des symposiums internationaux de la Société,
- préparer l'ordre du jour des assemblées générales et accepter les dons et legs au nom de la Société.

5.2.3 - Le Bureau décide à la majorité simple des membres présents.

5.2.4 - Le bureau peut décider par un vote par correspondance dans les cas d'urgence.

5.3 - Assemblée générale

5.3.1 - L'assemblée générale est composée du bureau, des anciens présidents de la Société, et d'un délégué de chaque groupe national.

5.3.2 - Les tâches de l'assemblée générale sont les suivantes:

- élire le président et les vice-présidents de la Société,
- établir les règles d'action de la Société, y compris dans les domaines technique, administratif et financier.
- décider des sujets, dates et lieux des congrès de la Société, et approuver chaque année le projet de budget et la cloture des comptes.

5.3.3 - L'assemblée générale décide à la majorité simple des délégués et pouvoirs des groupes nationaux, excepté dans les cas fixés ci-dessous en 10.1 et 11.2.

5.3.4 - Chaque groupe national a un représentant, donc une voix. Un délégue ne peut voter que si la cotisation de l'année en cours a été payée. Chaque groupe national peut désigner un suppléant pour siéger à l'assemblée générale en cas d'absence du délégué.

5.3.5 - Les membres du bureau en tant que tels, et les anciens présidents, en tant que tels, ont une voix délibérative sans droit de vote, a l'exception du président qui vote dans les cas de partage égal des voix (voir ci-dessous 6.2).

5.3.6 - L'assemblée générale peut décider par un vote par correspondance sur des sujets qui exigent une action urgente.

5.4 - Secrétariat

5.4.1 - Le Secrétariat est composé du secrétaire général, nommé par le président avec l'approbation du bureau, et du personnel administratif de bureau.

5.4.2 - Les tâches du secrétariat sont fixées par le président, avec l'approbation du bureau et comportent la préparation du budget annuel et la tenue des comptes.

6 - RESPONSABLES

6.1 - Responsables

Les responsables de la Société sont le président, le secrétaire général et un vice-président pour chacune des zones géographiques suivantes: Afrique, Amérique du Nord, Amérique du Sud, Asie, Australasie et Europe.

Les vice-présidents doivent résider dans la zone géographique qu'ils représentent et ne peuvent être du même pays que le président.

6.2 - Président

Les responsabilités et tâches du président sont:

- administrer la Société,
- présider les réunions de la Société,
- décider du vote en cas de partage égal des voix,
- nommer le secrétaire général, les commissions et les comités avec l'approbation du bureau,
- représenter officiellement la Société dans toutes les affaires internationales qui intéressent la Société, ou désigner un représentant de la Société lorsque c'est nécessaire

6.3 - Vice-presidents

Les vice-présidents sont responsables de la coordination des activités d'intérêt commun dans leurs zones géographiques.

6.4 - Premier vice-président

L'un des six vice-présidents est élu par le bureau premier vice-président.

Le premier vice-président assume temporairement les tâches du président pendant toute période où celui-ci est empêché de les remplir. En cas de démission ou de décès du président, le premier vice-président devient président jusqu'à la prochaine élection régulière des responsables de la Société. Dans ce cas, un nouveau vice-président est élu par le bureau dans un délai de quatre vingt dix jours.

6.5 - Secrétaire général

Le secrétaire général dirige le secrétariat. Il est responsable devant le président.

6.6 - Elections

6.6.1 - Tous les responsables, à l'exception du secrétaire général, sont élus a l'assemblée générale qui se tient à l'occasion de chaque congrès de la Société.

6.6.2 - Le mandat des responsables élus va de la fin de la dernière séance d'un congrès de la Société jusqu'à la fin de la dernière séance du congrès suivant.

6.6.3 - Tout candidat doit être membre de la Société et proposé par un groupe national.

6.7 - Nominations

6.7.1 - Le secrétaire général est nommé par le président avec approbation du bureau.

6.7.2 - Si un vice-président cesse de remplir les conditions indiquées ci-dessus, il doit démissionner et le président nomme un remplaçant pour la fin du mandat en cours.

7 - REUNIONS

7.1 - Congrès de la Société

7.7.1 - Les congrès de la Société sont consacrés à des thèmes qui intéressent la majorité des membres. Les congrès se tiennent environ tous les 4 ans, l'intervalle entre deux congrès consécutifs ne pouvant dépasser 5 ans.

7.1.2 - Les sujets, la date et le lieu de chaque congrès sont décidés par l'assemblée générale au précédent congrès.

7.1.3 - Le bureau agit comme comité consultatif pour chaque congrès. Tous les détails de l'organisation sont soumis à son approbation.

7.2 - Symposiums internationaux

7.2.1 - La Société maintient un programme de symposiums internationaux consacrés à des thèmes spécialisés. Les symposiums internationaux sont tenus de préférence une fois par an, mais non l'année où se tient un congrès.

7.2.2 - Le bureau approuvé le sujet, la date et le lieu de chaque symposium international.

7.3 - Assemblée générale

7.3.1 - L'assemblée générale se réunit à chaque congrès et à d'autres dates si elle en décide. En outre, un minimum de trente pour cent des membres peut exiger du président qu'il convoque l'assemblée générale.

7.3.2 - A l'assemblée générale, le quorum est atteint s'il y a plus de cinquante pour cent des délégués et des pouvoirs. Aucun membre de l'assemblée générale ne peut exercer plus d'un pouvoir.

7.4 - Réunions du bureau

7.4.1 - Le président ou bien quatre des vice-présidents peuvent convoquer le bureau à toute époque, en precisant le lieu et l'ordre du jour de la réunion.

7.4.2 - Quatre membres du bureau constituent le quorum, le secrétaire général n'étant pas compté.

7.5 - Coût des réunions

7.5.1 - Les dépenses directes d'organisation d'un congrès ou d'un symposium international sont de la responsabilité du groupe national qui organise la réunion.

7.5.2 - Lorsque des réunions de l'assemblée générale, du bureau ou de toute commission de la Société se tiennent aux mêmes lieux et dates qu'un congrès ou un symposium international, le coût de ces réunions est supporté par le groupe national hôte.

7.5.3 - Les dépenses engagées par les responsables pour assister aux réunions du bureau et de l'assemblée générale peuvent être supportées par la Société.

8 - COMMISSIONS

8.1 - Comme il est prévu en 6.2, ou bien à la demande de l'assemblée générale, le président nomme des commissions pour l'étude de problèmes spécifiques, scientifiques, technologiques ou administratifs, qui intéressent la Société. Les membres de ces commissions ne sont pas nécessairement membres de la Société.

8.2 - Les tâches et la durée de vie des commissions de la Société sont fixées par le président et approuvées par le bureau. Une commission peut recevoir une aide financière dans des limites fixées par le bureau, suivant un budget proposé par le président de la commission.

9 - PUBLICATIONS

9.1 - L'assemblée générale établit les règles concernant les publications de la Société.

10 - DISSOLUTION

10.1 - La dissolution ou liquidation de la Société ne peut être acquise que par une résolution votée par les deux tiers au moins de la totalité des groupes nationaux.

10.2 - Lorsque la liquidation est approuvée, l'assemblée générale décide de l'attribution des avoirs de la Société. Ils ne peuvent être attribués qu'à des organismes sans but lucratif consacrés principalement au progrès de la connaissance scientifique.

11 - GENERALITES

11.1 - Année comptable

L'année comptable de la Société va du 1er janvier au 31 decembre.

11.2 - Modification des statuts

Ces statuts peuvent être modifiés seulement sur la proposition de groupes nationaux ou de responsables de la Société.

Toute modification exige une majorité des deux tiers des délégues et des pouvoirs à l'assemblée générale.

11.3 - Application des statuts

Pour leur applicaticn, ces statuts sont complétés par des règlements approuvés par l'assemblée générale.

INTERNATIONALE GESELLSCHAFT FÜR FELSMECHANIK

STATUTEN

1 - NAME UND SITZ

1.1 - Name

Der offizielle Name dieser Organisation ist "International Society for Rock Mechanics". Französisch lautet der Name "Société Internationale de Mécanique des Roches" und in deutsch "Internationale Gesellschaft für Felsmechanik". In diesen Statuten wird die abgekürzte Form "Gesellschaft" verwendet.

1.2 - Sitz

Der offizielle Sitz der Gesellschaft ist der Sitz des Sekretariats der Gesellschaft, der durch den Vorstand empfohlen und durch den Rat bestätigt wird.

2 - ZIELE UND ART DER GESELLSCHAFT

2.1 - Ziele

Die Ziele und Absichten der Gesellschaft sind:

- Unterstützen und Koordinieren der internationalen Zusammenarbeit auf dem Wissenschaftsgebiet Felsmechanik
- Fördern der internationalen Zusammenarbeit auf dem Gebiet der Felsmechanik durch Veranstaltung internationaler Tagungen
- Fördern der internationalen Publikation von Ergebnissen wissenschaftlicher felsmechanischer Untersuchungen
- Unterstützen einzelner Personen und lokaler Organisationen, Nationale Körperschaften zu bilden, die sich ausschließlich oder primär mit Felsmechanik befassen
- Bewahren der Verbindung mit Körperschaften, die Fachgebiete vertreten, die der Felsmechanik verwandt sind, einschließlich Geologie, Geophysik, Bodenmechanik, Bergbau, Erdöltechnik und Bauingenieurwesen
- Koordinieren von Tagungen, die sich mit Felsmechanik und verwandten Fachgebieten befassen, um Überschneidungen und Konflikte möglichst gering zu halten
- Unterstützen der Lehre in Felsmechanik

2.2 - Art der Gesellschaft

Die Gesellschaft ist eine gemeinnützige Organisation, die durch Mitgliedsbeiträge und vorbehaltlose Förderbeiträge finanziert wird. Abgesehen vom Generalsekretär erhält kein Funktionär oder Mitglied irgendeine Vergütung aus Mitteln der Gesellschaft. Die Finanzierung des Sekretariates und aller anderen Kosten wird im Haushaltsplan der Gesellschaft vorgesehen.

3 - SPRACHEN

3.1 - Offizielle Sprachen

Englisch, Französisch und Deutsch sind die offiziellen Sprachen der Gesellschaft. Alle Mitteilungen des Sekretariats erfolgen in englisch sowie auch in französisch und/oder deutsch, wenn es zweckmäßig erscheint.

3.2 - Andere Sprachen

Auf Kongressen und Internationalen Symposien kann auch für mündliche Darlegungen die Sprache des Gastgeberlandes benutzt werden, wenn die gastgebende Nationale Körperschaft mit eigenen Mitteln in eine offizielle Sprache der Gesellschaft übersetzt.

3.3 - Autoritative Texte

Nur der englische Text eines jeden satzungsgemäßgen Schriftstücks der Gesellschaft ist verbindlich.

4 - MITGLIEDSCHAFT

4.1 - Überblick

Die Gesellschaft setzt sich zusammen aus Nationalen Körperschaften, Unterstützenden Mitgliedern und Korrespondierenden Mitgliedern.

4.2 - Länder

In diesen Statuten umfaßt der Begriff "Land" Nationen und abhängige Gebiete, die ein unabhängiges Budget für wissenschaftliche Zwecke haben, wie auch Multinationale Körperschaften, die vom Vorstand bestätigt worden sind.

4.3 - Nationale Körperschaften

Ein Land ist in der Gesellschaft durch eine Nationale Körperschaft vertreten. Eine Nationale Körperschaft ist eine Gesellschaft oder ein Komitee für Felsmechanik, deren bzw. dessen Statuten den Zielen der Gesellschaft nicht widersprechen.

4.4 - Arten der Mitgliedschaft

Die Mitgliedschaft der Gesellschaft umfaßt Ordentliche Mitglieder über Nationale Körperschaften, Unterstützende Mitglieder und Korrespondierende Mitglieder.

Ordentliche Mitglieder sind Personen, die den Mitgliedsbeitrag der Gesellschaft an eine Nationale Körperschaft zahlen.

Unterstützende Mitglieder sind Gesellschaften, Vereinigungen, Firmen und andere Körperschaften, die interessiert sind, den wissenschaftlichen felsmechanischen Fortschritt zu stimulieren und zu unterstützen. Sie zahlen die Mitgliedsbeiträge an die Gesellschaft entweder über Nationale Körperschaften oder direkt an das Sekretariat.

Korrespondierende Mitglieder sind Personen, die den Mitgliedsbeitrag direkt an das Sekretariat zahlen.

4.5 - Aufnahme

Die Aufnahme Nationaler Körperschaften, Unterstützender Mitglieder und Korrespondierender Mitglieder wird vom Vorstand entschieden. Über die Aufnahme als Ordentliches Mitglied entscheidet die jeweilige Nationale Körperschaft. Der Vorstand entscheidet auch über den Antrag Multinationaler Körperschaften. Kandidaten können gegen die Entscheidung des Vorstandes beim Rat Einspruch erheben. Das Sekretariat benachrichtigt die Nationale Körperschaft des Unterstützenden Mitgliedes, das beantragt hat, den Mitgliedsbeitrag direkt an die Gesellschaft zu zahlen.

Eine Mitgliedschaft wird wirksam, wenn der erste Mitgliedsbeitrag entrichtet worden ist.

4.6 - Mitgliedsbeiträge

Die Jahresmitgliedsbeiträge der Nationalen Körperschaften, Unterstützenden Mitglieder und Korrespondierenden Mitglieder beruhen auf einer vom Rat beschlossenen Gebührenordnung, die von Zeit zu Zeit vom Rat geändert werden kann.

Die jährlichen Mitgliedsbeiträge sind am 1. Januar fällig. Nach dem 1. April nicht bezahlte Mitgliedsbeiträge gelten als Rückstand.

4.7 - Verlust der Mitgliedschaft

Die Mitgliedschaft Nationaler Gruppen, Unterstützender Mitglieder sowie Korrespondierender Mitglieder kann durch den Vorstand beendet werden. Einspruch beim Rat ist möglich.

Die Beendigung einer Ordentlichen Mitgliedschaft bleibt der jeweiligen Nationalen Körperschaft vorbehalten.

Eine Mitgliedschaft ist beendet, wenn eine Nationale Körperschaft oder ein Mitglied an einem 1. April mit dem Jahresbeitrag zwei Jahre im Rückstand ist.

5 - ORGANISATION

5.1 - Allgemeines

Die Gesellschaft verfolgt ihre Ziele mittels eines Rates, eines Vorstandes und eines Sekretariates. Diese Organe sind wie nachstehend zusammengesetzt und arbeiten wie folgt.

5.2 - Vorstand

5.2.1 - Der Vorstand besteht aus dem Präsidenten, den Vizepräsidenten und dem Generalsekretär. Der Generalsekretär hat Stimmrecht, aber kein Wahlrecht.

5.2.2 - Der Vorstand hat die folgenden Aufgaben:

- Entscheiden über Aufnahme und Beendigung der Mitgliedschaft Nationaler Körperschaften, Unterstützender Mitglieder und Korrespondierender Mitglieder
- Überwachen aller finanziellen Angelegenheiten der Gesellschaft
- Bearbeiten der Angelegenheiten der Gesellschaft, die zwischen den Ratssitzungen notwendig werden
- Beraten aller Einzelheiten der Organisation des Kongresses der Gesellschaft
- Bestätigen des Rahmens, der Zeit und des Ortes Internationaler Symposien der Gesellschaft
- Vorbereiten der Tagesordnung von Ratssitzungen
- Annahme von Zuwendungen zu Gunsten der Gesellschaft

5.2.3 - Der Vorstand ist bei einfacher Mehrheit der anwesenden Vorstandsmitglieder beschlußfähig.

5.2.4 - Der Vorstand kann in dringenden Fällen brieflich abstimmen.

5.2.5 - Alle Aktionen des Vorstandes bedürfen der Bestätigung des Rates.

5.3 - Rat

5.3.1 - Der Rat setzt sich zusammen aus dem Vorstand, den Altpräsidenten der Gesellschaft und einem Repräsentanten jeder Nationalen Körperschaft.

5.3.2 - Der Rat hat folgende Aufgaben:

- Wahl des Präsidenten und der Vizepräsidenten der Gesellschaft
- Festlegen der Richtlinien der Gesellschaft, einschließlich technischer, administrativer und finanzieller Angelegenheiten
- Festlegen des Rahmens, der Zeit und des Ortes der Kongresse der Gesellschaft
- Bestätigen des jährlichen Haushaltsplanes und Jahresabschlußberichtes

5.3.3 - Der Rat beschließt mit einfacher Stimmenmehrheit der Vertreter und der bevollmächtigten Stellvertretungen der Nationalen Körperschaften, außer wenn 10.1 oder 11.2 gilt (siehe unten).

5.3.4 - Jede nationale Körperschaft hat einen Vertreter mit einer Stimme. Ein Vertreter ist nur stimmberechtigt, wenn der Mitgliedsbeitrag seiner Nationalen Körperschaft bezahlt ist. Jede Nationale Körperschaft kann einen Stellvertreter ernennen, der sie bei Abwesenheit des Vertreters im Rat vertritt.

5.3.5 - Die Vorstandsmitglieder und Alt-Präsidenten haben in dieser Funktion eine Stimme, aber kein Votum, abgesehen vom Präsidenten, der bei Stimmengleichheit abstimmen soll (siehe 6.2).

5.3.6 - Der Rat kann in dringenden Fällen brieflich abstimmen.

5.4 - Sekretariat

5.4.1 - Das Sekretariat besteht aus dem Generalsekretär, der vom Präsidenten ernannt und vom Vorstand bestätigt wird, sowie Verwaltungs- und Schreibkräften.

5.4.2 - Die Aufgaben des Sekretariats werden vom Präsidenten bestimmt. Sie sind vom Vorstand zu bestätigen. Zu diesen Aufgaben gehören die Vorbereitung des Jahreshaushalts- planes und des Jahresabschlusses.

6 - FUNKTIONÄRE

6.1 - Funktionäre

Die Funktionäre der Gesellschaft sind der Präsident, der Generalsekretär und je ein Vizepräsident für jedes der nachstehend genannten geographischen Gebiete: Afrika, Asien, Austral-Asien, Europa, Nordamerika und Südamerika.

Die Vizepräsidenten sollen in ihren geographischen Gebieten wohnhaft sein. Sie dürfen nicht in dem Land wohnen, aus dem der Präsident kommt.

6.2 - Präsident

Der Präsident hat folgende Rechte und Pflichten:

- Verwalten der Gesellschaft
- Vorsitz bei Tagungen der Gesellschaft
- Stimmenabgabe bei Stimmengleichheit
- Ernennung des Generalsekretärs und der Kommissionen sowie Ausschüssen, die vom Vorstand zu bestätigen sind
- Offizielles Vertreten der Gesellschaft in allen internationalen Angelegenheiten, die die Gesellschaft betreffen, oder, wenn notwendig, Ernennen eines Repräsentanten der Gesellschaf für derartige Angelegenheiten

6.3 - Vizepräsidenten

Die Vizepräsidenten sind in ihren geographischen Gebieten für eine Koordinierung von Aktivitäten von gemeinsamen Interesse verantwortlich.

6.4 - Erster Vizepräsident

Einer der sechs Vizepräsidenten wird vom Vorstand zum Ersten Vizepräsidenten gewählt.

Der Erste Vizepräsident übernimmt zeitweilig die Pflichten des Präsidenten, wenn der Präsident außerstande ist, seinen Pflichten nachzukommen. Legt der Präsident sein Amt nieder oder ist verstorben, wird der Erste Vizepräsident bis zur nächsten regulären Wahl der Funktionäre der Gesellschaft Präsident. In einem solchen Fall wird innerhalb von 90 Tagen vom Vorstand ein neuer Erster Vizepräsident gewählt.

6.5 - Generalsekretär

Der Generalsekretär leitet das Sekretariat. Er ist dem Präsidenten verantwortlich.

6.6 - Wahlen

6.6.1 - Alle Funktionäre, ausgenommen der Generalsekretär, werden auf Ratstagungen gewählt, di während eines jeden Kongresses der Gesellschaft stattfinden.

6.6.2 - Die Amtszeit der gewählten Funktionäre dauert vom Schluß der letzten Sitzung eines Kongresses der Gesellschaft bis zum Ende der letzten Sitzung des folgenden Kongresses.

6.6.3 - Ein Kandidat für eine Funktion muß Mitglied der Gesellschaft sein und von einer Nationalen Körperschaft vorgeschlagen werden.

6.7 - Ernennungen

6.7.1 - Der Generalsekretär wird vom Präsidenten ernannt und vom Vorstand bestätigt.

6.7.2 - Kann ein Vizepräsident die ihm übertragenen Aufgaben nicht erfüllen, ist er angehalten zurückzutreten. Für die verbleibende Amtszeit ernennt der Präsident eine Ersatzperson.

7 - ZUSAMMENKÜNFTE

7.1 - Kongress der Gesellschaft

7.1.1 - Der Kongress der Gesellschaft befaßt sich mit allgemeinen Themen, die für die Mehrheit der Mitglieder von Interesse sind. Die Kongresse sollen etwa alle vier Jahre stattfinden, wobei die Zeitspanne zwischen zwei Kongresses fünf Jahre nicht überschreiten soll.

7.1.2 - Der Rahmen, die Zeit und der Ort eines jeden Kongresses werden vom Rat auf dem letzten Kongress der Gesellschaft festgelegt.

7.1.3 - Der Vorstand fungiert als Beratungsausschuß eines jeden Kongresses. Alle Details der Organisation bedürfen der Bestätigung des Vorstandes.

7.2 - Internationale Symposien

7.2.1 - Die Gesellschaft führt Internationale Symposien durch, die sich mit speziellen Themen befassen. Diese Internationalen Symposien sollen vorzugsweise einmal im Jahr staffinden, jedoch nicht in dem Jahr, in welchem der Kongreß der Gesellschaft abgehalten wird.

7.2.2 - Der Vorstand bestimmt Rahmen, Zeit und Ort eines jeden Internationalen Symposiums.

7.3 - Ratssitzungen

7.3.1 - Der Rat versammelt sich während des Kongresses oder, wenn der Rat es für sinnvoll erachtet, zu einer anderen Zeit. Außerdem kann der Präsident eine Ratssitzung einberufen wenn mindestens 30 % der Ratsmitglieder den Präsidenten um eine Ratssitzung ersuchen.

7.3.2 - Eine Ratsversammlung ist beschlußfähig, wenn mehr als 50 % der Ist-Zahl der Vertreter bzw. bevollmächtigten Stellvertretungen der Nationalen Körperschaften anwesend ist. Kein Vertreter kann mehr als eine bevollmächtigte Stellvertretung übernehmen.

7.4 - Vorstandssitzungen

7.4.1 - Vom Präsidenten oder jedem der vier Vizepräsidenten kann zu jeder Zeit eine Veranstaltung einberufen und der Ort sowie die Tagesordnung festgelegt werden.

7.4.2 - Vier Vorstandsmitglieder sind beschlußfähig, wobei der Generalsekretär nicht zählt.

7.5 - Tagungskosten

7.5.1 - Die unmittelbaren Kosten für die Veranstaltung eines Kongresses oder eines Internationalen Symposiums trägt die Nationale Körperschaft, die die Tagung veranstaltet.

7.5.2 - Findet eine Ratssitzung, Vorstandssitzung oder eine Sitzung einer Kommission der Gesellschaft zu der Zeit und an dem Ort statt, an dem ein Kongress oder ein Internationales Symposium abgehalten wird, sind die Kosten der Ratssitzung, Vorstandssitzung oder Sitzung einer Kommission vom gastgebenden Organisator zu übernehmen.

7.5.3 - Die Ausgaben, die Funktionären auf Vorstands- und Rats- sitzungen entstehen, können von der Gesellschaft getragen werden.

8 - KOMMISSIONEN

8.1 - Wie in 6.2 vermerkt oder wenn vom Rat gefordert, kann der Präsident Kommissionen bilden, um spezifische wissenschaftliche, technische oder organisatorische Probleme untersuchen zu lassen, die die Gesellschaft interessieren. Die Kommissionsmitglieder müssen nicht Mitglieder der Gesellschaft sein.

8.2 - Die Aufgaben und die Dauer der Tätigkeit der Kommissionen der Gesellschaft werden vom Präsidenten festgelegt und vom Vorstand bestätigt. Eine Kommission kann finanziell unterstützt werden. Die Höhe der finanziellen Unterstützung wird aufgrund eines Kostenvoranschlages des Vorsitzenden der Kommission vom Vorstand bestimmt.

9 - PUBLIKATIONEN

9.1 - Der Rat soll Richtlinien über Publikationen der Gesellschaft ausarbeiten.

10 - AUFLÖSUNG

10.1 - Die Auflösung oder Liquidation der Gesellschaft kann nur mit Zweidrittelmehrheit aller Nationaler Körperschaften bewirkt werden.

10.2 - Nach Bestätigung der Liquidation der Gesellschaft entscheidet der Rat über die Verteilung des Gesellschaftsvermögens. Dieses Vermögen können jedoch nur gemeinnützige Organisationen erhalten, die ein primäres Interesse an der Förderung von wissenschaftlichen Erkenntnissen haben.

11 - ALLGEMEINES

11.1 - Finanzjahr

Das Finanzjahr der Gesellschaft währt vom 1. Januar bis zum 31. Dezember.

11.2 - Änderung der Statuten

Diese Statuten können nur auf Antrag Nationaler Körperschaften oder auf Antrag von Funktionären der Gesellschaft geändert werden.

Änderungen erfordern eine Zweidrittel-Stimmenmehrheit der Vertreter und der bevollmächtigte Stellvertretungen einer Ratssitzung.

11.3 - Ergänzungen der Statuten

Diese Statuten werden durch Ausführungsbestimmungen ergänzt, die vom Rat bestätigt sind.

Übersetzung aus dem Englischen

Bochum, Oktober 1978

Beyer/K. W. John

Driltech D80K
It's different.
Driltech's new D80K blast hole drill is truly different. Truck mounted, the D80K features 80,000 pounds of bit loading, 1500 cfm of air rated to 100 psi, maneuvering controls in the operator's cab and is uniquely designed to provide simplified maintenance and servicing.
D80K machinery deck overhead view.
The D80K is designed for one man operation, 8 - 12" holes with fast penetration and easy maintenance. Contact Driltech today.
Driltech Inc.
DRILTECH/ Post Office Box 338/ Alachua, Florida 32615/ (904) 462-4100/ Telex 58-8441

Printed in Dunstable, United Kingdom

91898137R00134